International Symposium on Sun, Earth, and Life (ISSEL 2016)

Journal of Physics: Conference Series Volume 771

Bandung, Indonesia
3 - 4 June 2016

ISBN: 978-1-5108-3386-9
ISSN: 1742-6588

Printed from e-media with permission by:

Curran Associates, Inc.
57 Morehouse Lane
Red Hook, NY 12571

Printed by Curran Associates, Inc. (2017)

For permission requests, please contact the Institute of Physics
at the address below.

Institute of Physics
Dirac House, Temple Back
Bristol BS1 6BE UK

Phone: 44 1 17 929 7481
Fax: 44 1 17 920 0979

techtracking@iop.org

Additional copies of this publication are available from:

Curran Associates, Inc.
57 Morehouse Lane
Red Hook, NY 12571 USA
Phone: 845-758-0400
Fax: 845-758-2633
Email: curran@proceedings.com
Web: www.proceedings.com

Table of contents

Volume 771

International Symposium on Sun, Earth, and Life (ISSEL)
3–4 June 2016, Bandung, Indonesia

Accepted papers received: 8 November 2016
Published online: 26 November 2016

Preface

Papers

International Symposium on Sun, Earth, and Life
June 3-4, 2016, East Hall - ITB

Steering Committee

1. Hendra Gunawan (ITB and AIPI, Indonesia)
2. Thomas Djamaluddin (LAPAN, Indonesia)
3. Edy Tri Baskoro (ITB, Indonesia)
4. Premana W. Premadi (ITB, Indonesia)
5. Budi Dermawan (ITB, Indonesia)
6. Nat Gopalswamy (Goddard Space Flight Center, NASA, USA)
7. J. N. Goswami (Physical Research Laboratory, India)

Scientific Committee

1. Ikbal Arifyanto (ITB, Indonesia)
2. Lucky Puspitarini (ITB, Indonesia)
3. Hendra Gunawan (ITB and AIPI, Indonesia)
4. Premana W. Premadi (ITB, Indonesia)
5. Taufiq Hidayat (ITB, Indonesia)
6. Budi Dermawan (ITB, Indonesia)
7. Mahasena Putra (ITB, Indonesia)
8. Dhani Herdiwijaya (ITB, Indonesia)
9. E. Sungging Mumpuni (LAPAN, Indonesia)
10. Clara Y. Yatini (LAPAN, Indonesia)
11. Nat Gopalswamy (Goddard Space Flight Center, NASA, USA)
12. J. N. Goswami (Physical Research Laboratory, India)
13. Thierry Fouchet (Observatoire de Paris, Paris)

Organizing Committee

1. Dhani Herdiwijaya (ITB, Indonesia) - Chair
2. Lucky Puspitarini (ITB, Indonesia)
3. Anny Sulaswatty (AIPI, Indonesia)
4. Indra Noviandri (ITB, Indonesia)
5. Clara Y. Yatini (LAPAN, Indonesia)
6. Abdul Waris (ITB, Indonesia)
7. Mahasena Putra (ITB, Indonesia)
8. Evan Irawan Akbar (ITB, Indonesia)
9. Ikbal Arifyanto (ITB, Indonesia)
10. Aprillia (ITB, Indonesia)
11. E. Sungging Mumpuni (LAPAN, Indonesia)
12. M. Zamzam Nurzaman (LAPAN, Indonesia)
13. Dede Enan (ITB, Indonesia)

Journal of Physics: Conference Series **771** (2016) 011001 doi:10.1088/1742-6596/771/1/011001

International Symposium on Sun, Earth, and Life
June 3-4, 2016, East Hall - ITB

Keynote Speakers

1. **Prof. Bambang Hidayat** (*Indonesian Academy of Sciences*)
   "A Recollection of Research Merit"
2. **Prof. Nat Gopalswamy** (*Goddard Space Flight Center, NASA, USA*)
   "The Coronal Connection to Space Weather"
3. **Prof. J. M. Pasachoff** (*Williams College, USA*) via Skype
   "Coronal Science and Outreach at Recent and Future Solar Eclipses"
4. **Prof. J. N. Goswami** (*Physical Research Laboratory, India*)
   "Water on the Surface and in the Interior of the Moon"
5. **Prof. Thomas Djamaluddin** (*National Institute of Aeronautics and Space*)
   "New Timau National Observatory: Accelerating Development of Space Science and Technology in Indonesia"
6. **Prof. Taufiq Hidayat** (*Institut Teknologi Bandung*)
   "Nitrile Compounds on Titan as Observed by ALMA"

Organized by

AIPI Akademi Ilmu Pengetahuan Indonesia — Institut Teknologi Bandung 1920 — LAPAN

Sponsored by

TAMARIS HYDRO — BFI FINANCE

Journal of Physics: Conference Series **771** (2016) 011001 doi:10.1088/1742-6596/771/1/011001

Preface

The International Symposium on Sun, Earth, and Life (ISSEL), organized by ITB, Indonesian Academy of Sciences (AIPI), and National Institute of Aeronautics and Space (LAPAN), that was held from, June 3-4, 2016 at East Hall, Institut Teknologi Bandung, Bandung, Indonesia.

I would like to thank our distinguished keynotes for their attendance or contributions: Prof. Nat Gopalswamy from NASA; Prof. Jay Pasachoff from USA; Prof. Goswami from Physical Research Laboratory, India; Prof. Bambang Hidayat from Indonesian Academy of Sciences; Prof. Thomas Djamaluddin from National Institute of Aeronautics and Space; and Prof. Taufiq Hidayat from ITB.

Total Solar Eclipse (TSE) in Indonesia came to a wide variety of issues; people may have arguments that are based on fear or ideology, rather than on what the science says. Thanks to Pak Bambang who has been fought against fear of TSE. Total Solar Eclipse becomes the main purpose of this meeting to clarify those issues. The first Photo Exhibition with 30 selected memorable pictures for TSE and astronomy in Indonesia is proudly organized, in conjungtion with this symposium.

All papers in these Proceedings refer to one from the following topics: solar science, space weather (solar-terrestrial science), atmosphere and aeronomy, earth and planetary Science, astrophysics, astrobiology and astrochemistry, etnoastronomy and archaeoastronomy, related education and public outreach. The program of ISSEL 2016 features 6 invited talks and 78 contributed oral and poster presentations. All full-papers have been reviewed after they were submitted in this event. Selected papers are published in the Institute of Physics (IoP) Conference Series.

On behalf of the organizing committee, we are extremely thankful to all authors for providing their valuable contributions, stimulating discussions, and share their knowledge, ideas, opinions, and meeting with new colleagues. We also express our gratitude to all reviewers for their constructive recommendations and criticism aiding to improve quality of articles.

Finally, I would like to express my deepest appreciation for those who have sponsored this symposium, Tamaris Hidro, and BFI Finance, including all staffs from Faculty of Mathematics and Natural Sciences, LAPAN, AIPI, Bosscha Observatory, LPPM-ITB, Langit Selatan, and students of Astronomy Study Program, who have made unceasing and tremendous efforts with the highest commitment to organize this significant event.

Bandung, June 2016

ISSEL Chairperson
Dhani Herdiwijaya

Journal of Physics: Conference Series **771** (2016) 011001 doi:10.1088/1742-6596/771/1/011001

Keynotes speaker with Rector of ITB, Head of LAPAN, AIPI, and Dean

Keynotes with participants

Excursion to Bosscha Observatory

Journal of Physics: Conference Series **771** (2016) 011001 doi:10.1088/1742-6596/771/1/011001

Bali dance and poster session

Our team

Journal of Physics: Conference Series **771** (2016) 011002 doi:10.1088/1742-6596/771/1/011002

Peer review statement

All papers published in this volume of *Journal of Physics: Conference Series* have been peer reviewed through processes administered by the proceedings Editors. Reviews were conducted by expert referees to the professional and scientific standards expected of a proceedings journal published by IOP Publishing.

An experiment to detect Allais effect around total solar eclipse of 9 March 2016

Putra Mahasena[1,2], Mochamad Irfan[2], Agus Setiawan[2], Maman Sulaeman[2] and Taufiq Hidayat[1,2]
[1]Astronomy Research Division, FMIPA, Institute of Technology Bandung, Indonesia
[2]Bosscha Observatory, FMIPA, Institute of Technology Bandung, Indonesia
E-mail: mahasena@as.itb.ac.id

Abstract. Maurice Allais reported anomalous behavior of pendulums, during solar eclipses in 1954 and 1959. This effect, which is then known as the Allais effect, has been searched for by a number of other researchers. Both positive and negative results had been reported. In this paper we report our pendulum experiment during total eclipse on 9 March 2016, over Indonesia. Apparatus set up and method of measurement are described, and we discuss some points which may contribute to uncertainties of the results.

1. Introduction

In a marathon measurement of several days, before and after the total solar eclipse of 30 June 1954, Maurice Allais had found that during the total solar eclipse, an anomaly in the precession of the plane of oscillation of his paraconical pendulum occured [1]. He observed that, after the eclipse began, the precession of the plane of oscillation reversed its direction and went on to about 13 deg. It then returned to its original direction and precession rate around the eclipse end. Since then many authors had attempted to detect any 'anomaly' that may appears around total solar eclipses; some had succeeded and others had failed. Saxl and Allen [2], for example, observed an anomaly in the period of their torsion pendulum, around the solar eclipse of 7 March 1970. This agrees qualitatively with the observation of Allais [1] in that the greatest change occurs between the onset of the eclipse and its midpoint (see figure 1). However, using an automated Foucault pendulum, on the total solar eclipse of 11 July 2010, Salva [3] found no evidence (within the measurement error) of the Allais effect. Using Saxl-and-Allen type torsion pendulum, Kuusela [4][5] found negative results from measurements during two solar eclipses.

During a solar eclipse in 1995 in India, Mishra and Rao [6] observed a short period temporal variation of $10 - 12$ μgal in their measurement of gravity field with a gravimeter at a region with approximately 80% of eclipse. More positive results about observations with gravimeters during solar eclipses were reported by Chinese team [7] [8]. However, some measurements with gravimeters by others were negative, *e.g.* [9] [10].

Duif [11], and references therein, noted that recent renewed interest in reports about anomalies during solar eclipses might be stimulated by, among other things: (i) realization that our understanding of gravity at galactic scales may be insufficient (giving rise to theories like MOND), (ii) observation of an anomalous acceleration on spacecrafts in the solar system (the Pioneer Anomaly), (iii) anomalous velocity increases of spacecrafts during Earth flybys, and

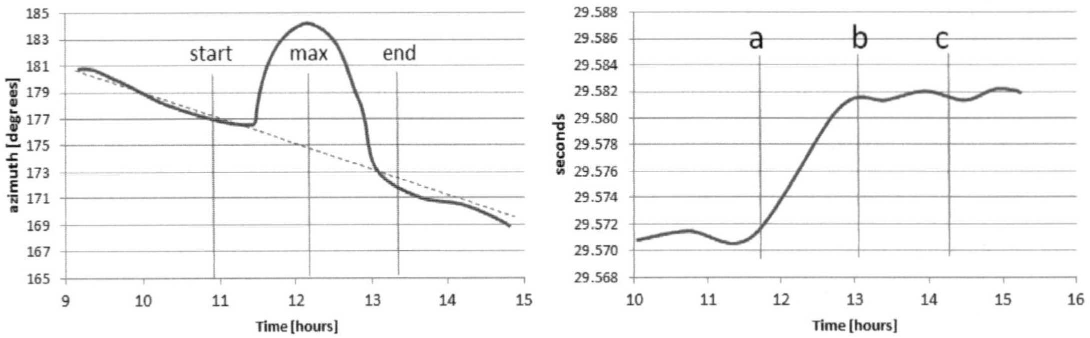

Figure 1. The left panel is adapted from [1], on paraconical pendulum experiment, and the right is adapted from [2], on torsion pendulum experiment. On the left panel, the ordinate is the azimuth of the plane of oscillation of the pendulum, while on the right it is the time used in traversing a constant fixed part of the total vibration path of the oscillating torus on the first swing from rest. Marks of the start, mid point, and end of the eclipse are given in both panels, *i.e.* 'start-max-end' on the left and 'a-b-c' on the right.

Figure 2. Apparatus setup. (a) assembly of accelerating coil, coil power sensor (sensor 1), plane of oscillation sensor (sensor 2), stepper motor, and the controller board; (b) support for the dumping copper ring (it encloses assembly (a)); (c) the complete setup, including a computer communicating with the controller board and the suspended bob.

even (iv) discussions about whether we understand gravity at laboratory scale. In the following we report our attempt to detect gravity anomaly using a Foucault pendulum during a total solar eclipse on 9 March 2016.

2. Apparatus setup

We build a version of automatic pendulum described in Salva *et al.* [12]. The pendulum is designed to be able to oscillate continuously by adding kinetic energy (through the accelerating coil) once every cycle. The electronics, with the help of an Arduino microcontroller, is designed to be able to record the direction of the plane of oscillation as a function of time. Unlike the original design [12], our system lacks the ability to measure the elliptical motion size needed to calculate the precession rate due to Earth rotation alone.

The pendulum consist of a cylindrical bob weighs around 12 kg which is hung from a mandrel-type support using a steel suspension wire around 4.5 m length (see figure 2 (c)). At the bottom of the bob, at the cylinder axis, we put a small strong permanent magnet which is used to accelerate the bob and to detect the orientation of the plane of oscillation. Below the suspended bob is an assembly of accelerating coil, dumping copper ring, Hall sensors, and a stepper motor. In non-moving condition, all the parts need to be centrally aligned carefully. The pendulum is independent from the rest of the driving/measuring system so that we can hang it from different support systems.

An oscillation is started by displacing the bob around 1.2 deg, beyond sensor 1, hold it to a complete stop and then release it. When sensor 1 detect the bob approaching from outside the coil will be turned on for a short time, pulling the bob to the center. The bob will then pass, more or less, above sensor 2. Since both sensor 2 and sensor 1 are carried by a rotatable arm, we can position the arm, using the stepper motor, so that sensor 2 is always "exactly" under the path of the bob, hence record the position. Passing sensor 2, the permanent magnet under the bob will interact with free electrons in the copper ring. This eddy current brake is expected to keep the elliptical motion of the pendulum in an acceptable size.

3. Procedure of measurement

For some possible reasons addressed later, elliptical motion in our system always grew to a point where the driving/measuring system failed, in about one hour after we started the oscillation. We found that the one-hour records of the plane of oscillation direction were very similar if we had started the pendulum from the same azimuthal direction. This latter behavior was consistently observed even when the system was operated at two places with 6 degrees latitude difference, and with different type of supports. Accordingly, we decided to adopt the following procedure of measurement.

(i) Start a measurement from a same azimuthal direction; keep every other condition of the system the same.

(ii) Record the direction of the oscillation plane until the driving mechanism fails.

(iii) Repeat from step (i).

Two example records are shown on the left panels of figure 3. In those plots, $(0, 0)$ corresponds to the start of the oscillation, and the initial azimuthal direction of the oscillation. The pendulum takes some time before reaching a "constant" precession rate in counter clockwise direction. Features on the right end of the curves show the failure of the driving mechanism.

4. Result and discussion

We measured the precession rates of our pendulum at Kalora ($1°12'2''S$, $120°33'44''E$), a village near Poso, in Sulawesi island. Nine records, like that shown in figure 3 (left panels) were obtained, two in 8 March evening and seven in 9 March morning, around the eclipse. For each record we calculated the average slope of the "straight" part of the curve (*i.e.* average precession rate during the measurement). The result is shown in figure 3 (right panel). Although we do not have enough data points, before and after the eclipse, *there is an indication of an increase in average precession rate around the eclipse start.*

Explanations for the anomalous instrument readings during eclipses, based on conventional physics, had been subdivided into 6 kinds by Duif [11], and references therein: (i) seismic disturbances due to increased human activity before and just after an eclipse, (ii) gravitational effects of an increased density air mass spot due to cooling of the upper atmosphere, (iii) tilt due to temperature change of the soil and other atmospheric influences, (iv) tilt due to atmospheric loading, (v) influence of eclipse induced changes of the geomagnetic field, and (vi) instrumental errors. In current experiment our main concern is the instrumental errors.

It is well known that anisotropy of the support system gives rise to elliptical motion. Apparently our support system at Kalora, a tripod made of bamboo, suffers from such anisotropy. We have more rigid support system back in our observatory where we observed less anisotropy during preliminary experiments. The effects of support anisotropy was possibly

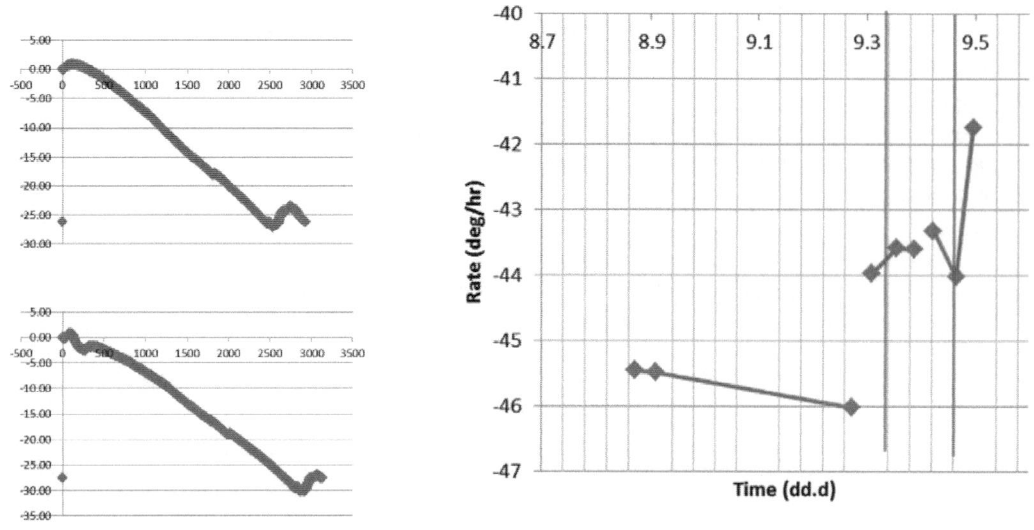

Figure 3. Measurement and result (see text). On the left panels are examples of oscillation plane direction records. The x and y axes are time (in seconds) and azimuthal direction (in degrees), respectively. The right panel shows nine average precession rates around the solar eclipse on 9 March 2016 (UT+8). Two vertical red lines indicate the eclipse start and end.

worsened by slight misalignment of the accelerating coil, and the asymmetry of the coil itself. As our measurement system was unable to compute the size of the elliptical motion, it is impossible to estimate the order of the systematic errors.

Nevertheless, we have have made our first automatic Foucault pendulum and learned a lot about its weaknesses. A plan for improvement is being prepared.

Acknowledgments

We sincerely thank staff at Bosscha Observatory ITB for their technical and administrative support which make this experiment possible, and the conference LOC for their hospitality. Advice from Dimitrie Olenici and Thomas Goodey are greatly appreciated.

References

[1] Allais M 1957 *C. R. Acad. Sci.* **245** 2001
[2] Saxl E J and Allen M 1971 *Phys. Rev. D* **3** 823
[3] Salva H R 2011 *Phys. Rev. D* **83** 067302
[4] Kuusela T 1991 *Phys. Rev. D* **43** 2041
[5] Kuusela T 1992 *Gen. Rel. Gravit.* **24** 543
[6] Mishra D C and Rao M B S 1997 *Curr. Sci.* **72** (11) 783
[7] Wang Q-S, Yang X-S, Wu C-Z, Guo H-G, Liu H-Cand and Hua C-C 2000 *Phys. Rev. D* **62** 041101
[8] Yang X-S and Wang Q-S 2002 *Ap&SS* **282** 245
[9] Ducarme B, Sun H P, d'Oreye N, van Ruymbeke M and Mena Jara J 1999 *JGeod* **73** 53
[10] Mansinha L, Ducarme B, Hinderer J, Meurers B and van Ruymbeke M 2001 *JGSJ* **47** 387
[11] Duif C P 2004 *Preprint* gr-qc/0408023
[12] Salva H R, Benavides R E, Perez J C and Cuscueta D J 2010 *Rev. Sci. Instrum.* **81**, 115102

Influence of Partial Solar Eclipse 2016 on the surface gravity acceleration using photogate sensor on Kater's reversible pendulum

M G Nugraha[1], D Saepuzaman[1], F N Sholihat[1], S Ramayanti[1], A H Setyadin[1], A R Ferahenki[1], A Samsudin[1], J A Utama[1], H Susanti[1], K H Kirana[2]

[1]Departemen Pendidikan Fisika FPMIPA Universitas Pendidikan Indonesia, Jl. Dr. Setiabudhi No. 229, Bandung 40154, Indonesia
[2]Departemen Geofisika FMIPA Universitas Padjadjaran, Jl. Raya Bandung-Sumedang Km. 21, Jatinangor 45363, Indonesia

E-mail: muhamadginanugraha@upi.edu

Abstract. This study was conducted to determine the Earth's surface gravitational acceleration (g) prior to, during, and after a partial solar eclipse. Data was collected in Basic Physics Laboratory Universitas Pendidikan Indonesia, Bandung with coordinates S 6°51'48", E 107°35'40" for three days on March 8 – 10, 2016, in time interval measurement from 6 a.m. to 9 a.m. This research used a standard pendulum, Kater's reversible pendulum, which deviated less than 3° so that the motion can be regarded harmonics oscillation. The period of pendulum oscillation motion is measured by a light sensor (photogate sensor) with accuracy until 10^{-13} seconds. The data analysis shows that there is small difference value of gravity acceleration at the Earth's surface from three days of observation, i.e. in the order of 10^{-3} ms^{-2}. It means, there is a changes in the Earth's surface gravitational acceleration (g) due to the partial solar eclipse but not significant.

1. Introduction

Gravity is a force between two objects of mass that makes the objects attracts each other. Mathematically, Newton reveals that the gravitational force (F) is proportional to the product of the mass (m) of both objects and inversely proportional to the square of the distance (r) between these objects [1].

$$F \approx \frac{m_1 m_2}{r^2} \qquad (1)$$

Gravitation phenomenon of small objects was difficult to observe because it produces very small gravitational force. But in a wider review, the phenomenon of gravity can be observed as the celestial objects and satellites moving on its path, e.g. earth, moon, and sun in the solar system [2, 3].

The research about changes in the gravitational field or gravity acceleration (g) of the Earth's surface during the solar eclipse has been done by some scientists. In 1954, Maurice Allais observed the effects of the solar eclipse on the Earth's gravitational field and he found that the solar eclipse can affect the Earth's gravitational field. Wang et al used Lacoste-Romberg gravimeter to report anomaly of the Earth's gravitational field (7.0 ± 2.7) × 10^{-8} ms^{-2} as a result of a total solar eclipse in 1997 in China [4]. Using the same tools, Yang and Wang reported the decrease of gravitational field during a solar eclipse (contact 1 to contact 4) at 7 ~ 8 μgal [5]. The different results found by Zainuddin, which reported an increase in the gravitational field of the Earth's surface as a result from two event solar eclipse i.e. total solar eclipse 2009 in China with an increase amounted to 18.92%, and a annular solar eclipse 2010 in Maldives with an increase of 7.51% and 8.59% [6]. Based on the research, Wang and Zainuddin found a different conclusion on the effect of solar eclipse to the Earth's surface gravity field.

Research that conducted by Zainuddin uses harmonic oscillations in a simple pendulum which vulnerable of disturbance, such as conic motion of the pendulum and air resistance that could interfere harmonic oscillatory motion of the pendulum. Moreover, the determination the period of oscillation was using a stop watch that is error-prone. In this study, measurement of the Earth's surface gravitational field done using the same concept with Zainuddin research but with different measurement methods, i.e. using Kater's reversible pendulum as oscillate object and a light sensor (photogate sensor) to measure the time period of oscillation.

2. Methods

Total solar eclipse that happen on March 9, 2016 passes through several areas of Indonesia such as Palembang, Bengkulu, Belitung, Palangkaraya, Palu, Ternate and the surrounding areas. Data is collected in Bandung with coordinates 6°51'48"S, 107°35'40"E as one of the areas that receive partial solar eclipse with 85% of the Sun's surface covered by Moon. Data taken during three consecutive days, i.e. on March 8, 9 and 10, 2016 at 06:00 am until 9:00 am which is the time span partial solar eclipse in Bandung.

The gravitational field of the Earth's surface is determined by oscillatory motion of Kater's reversible pendulum that deviated less than 3°, so that the movement considered as a simple harmonic oscillatory motion. In Kater's reversible pendulum experiment, the period measure with two different pendulum position, i.e. at 06:00 - 07:30 am for the initial position with the mass of pendulum object is above and at 07:30 - 09:00 am for the reversed position of pendulum from the initial position. The value of the gravitational field of the Earth's surface using Kater's reversible pendulum is determined by the equation:

$$g = \frac{8\pi^2 \ell}{T_1^2 + T_2^2} \tag{2}$$

Where l is length of Kater's reversible pendulum, T_1 is period of pendulum oscillation for initial position and T_2 is period of pendulum for reversed position.

Period of Pendulum oscillation measure using a light sensor (photogate sensor) with accuracy of 10^{-13} seconds which is connected to a computer so that each oscillation period immediately recorded and stored on the computer. Set of a research instrument shown in figure 1.

3. Result and discussion

The results measurement and analysis shows that the average period of pendulum oscillation for T_1, T_2, and g shown in Table 1. Changes in the Earth's gravitational field surface for each day of observation can be seen in figure 2.

Table 1. The average period of pendulum and the gravitational field of the Earth's surface

Days/Date	Period of pendulum		g (m/s^2)
	Average of T_1 (s)	Average of T_2 (s)	
Tuesday/ 08-03-2016	1,9351292848587	1,9774475097680	10,3226858769144
Wednesday/ 09-03-2016	1,9348869323730	1,9772186279300	10,3251718134668
Thursday/ 10-03-2016	1,9351766705514	1,9772253483534	10,3236234000278

Figure 2 show visible pattern of changes in the gravitational field of the Earth's surface during three days of observation. On the day of solar eclipse (Wednesday, March 9, 2016) changes in the gravity field more fluctuate than the other two days. These results indicate that there are anomalies of the gravitational field of the Earth's surface due to solar eclipse.

When the solar eclipse happened, moon's position is between Earth and the Sun so that the gravity force resultant of the Earth and the objects in it will tend to be heading out of Earth. Thus it predicts that the value of gravitational field of the Earth's surface will decrease. However, the research showed the opposite result, the gravitational field of the Earth's surface that occurred is increase despite the increase was not significant. This is due to the sudden change several environmental parameters during the solar eclipse, such as an increase in the mass distribution of the atmosphere surrounding the research site due to changes in atmospheric temperature when the moon covers the sun (the density of the atmosphere increases), changes in air pressure, disturbance of seismic due to movements mass underground and changes in geomagnetic [7,8]. The changes of environmental parameters during solar eclipse is also indicated on anomalous gravitational field changes the Earth's surface that is more fluctuate as shown in figure 2 (b).

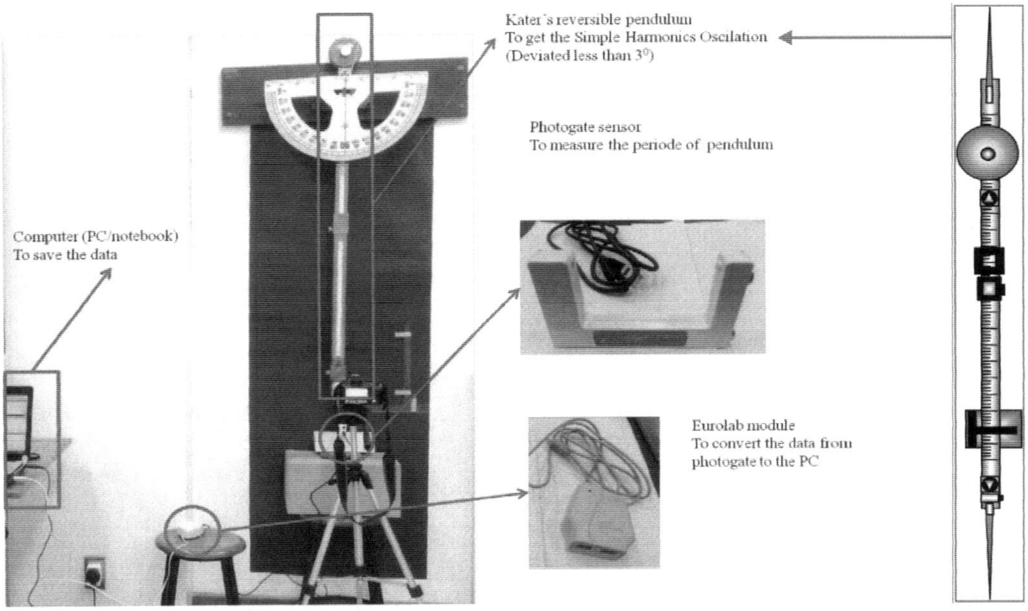

Figure 1. Set of a research instrument

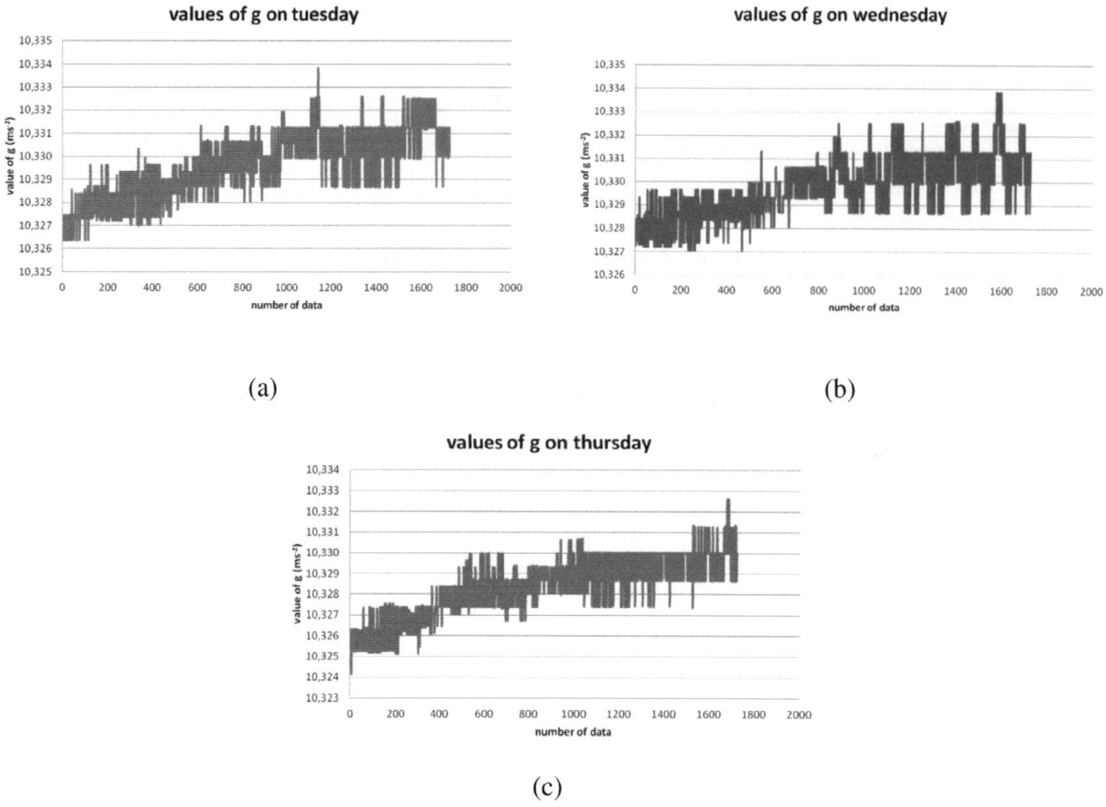

Figure 2. Changes in the gravitational field of the Earth's surface: (a) on Tuesday; (b) on Wednesday; (c) Thursday

4. Conclusion

The research concluded that there is an increase in the gravitational field of the Earth's surface during the solar eclipse although the increasing value is not significant. Nevertheless, it needs further comprehensive studies on the effect of the solar eclipse to the Earth's surface gravitational field, taking into account various parameters that change as a result of the solar eclipse.

Acknowledgments

Thanks to the whole team for their cooperation and constructive discussion. Also, thanks to FPMIPA UPI for support the research fund as well as the Pudak Scientific that facilitates the research instrument.

References

[1] Newton I 1846 *Mathematical Principles of Natural Philosophy* (New York: 45 Liberty Street)
[2] Halliday and Resnick 1977 *Fisika Jilid 1 Edisi ketiga* (Jakarta: Erlangga)
[3] Mikrajuddin A 2016 *Fisika Dasar 1* (Bandung: Penerbit ITB).
[4] Wang Q, Xin-she Y, Chuan-zhen W, Hong-gang G, Hong-chen L and Chang-chai H 2000 *Physical Review D* **62** 1-3.
[5] Yang X and Wang Q 2002 *Astrophysics and Space Science* **282** p 245-253.
[6] Zainuddin M Z, Noorul A A, Mohd S Y and Mohd. H M S 2011 *International Conference on Space Science and Communication* 170-3.
[7] Duif C P 2004 A Review of Conventional Explanation of Anomalous Observation during Solar Eclipses *Preprint* gr-qc/0408023, v5
[8] McKenna P 2009 *New Scientist* (Beijing) p 10

Micro-gravity measurements during the total solar eclipse of 9 March 2016 in Indonesia

Agus Laesanpura[1], Taufiq Hidayat [2], Dady Abdurachman[1], Putra Mahasena[2], Premana W. Premadi [2], Hesti Wulandari[2], Yudi Suharyadi[3], Achmad Sjarmidi[4]

[1]Geophysic Engineering Study Program, ITB
[2]Bosscha Observatory and Astronomy Research Division, FMIPA, ITB
[3]Analysis Research Division, FMIPA, ITB
[4]School of Life Sciences and Technology, ITB

E-mail: laesanpura1@yahoo.com

Abstract. Since 1950s, several authors have reported the so-called anomalous gravity during the total solar eclipses through various experiments. To address this issue, in the moment of the total solar eclipse of 9 March 2016 passing most regions in Indonesia, we undertook microgravity measurements using two precise gravimeters. The measurements were made at two locations: (1) Poso (central Sulawesi), a location close to the centre passage of the total eclipse and (2) Lembang (West Java), the site of Bosscha Observatory, where the partial solar eclipse occurred. The two sites are selected to measure the possible different influence of the eclipse on aligning the three objects. The measurements were recorded three days before and one day after the eclipse at various intervals from 30 seconds to a short time at 5 seconds. The finer recording was performed several hours before and after the total eclipse. This measurement allowed detailed analysis, not only in time but also in frequency range. A detailed analysis is presented in this paper. We consider all possible geophysical as well as atmospheric effects. The residual data show that the shielding effect, usually thought as responsible factor on anomalous gravity, is not significant.

1. Introduction

The anomalous gravity associated with the total solar eclipse (TSE) was reported by several authors (see [1], [2] and references therein) and certainly needs to be confirmed by various measurement. One of the possible explanations is the hypothesis of gravitational shielding [3] occurred during the TSE. This phenomenon may result when there exists an alignment of three massive objects, the Sun, the Moon, and the Earth, particularly during the total solar eclipse [1], [2]. Since the first serendipitous measurement made using the paraconical pendulum–a modification of Foucault pendulum–the so-called Allais effect [4], various methods of gravity measurements have been proposed and performed [1] during the TSE. In fact, variation effects on gravity related to the TSE are subject to many things, besides the solid planetary body. One may also consider the possibility of ocean loading or the phenomena related to atmospheric effects. In general, two methods to observe these effects utilize a kind of Foucault pendulum and gravimeter.

An interesting result of gravitational acceleration measurement using a very precise D-type gravimeter was reported from the 1997 total solar eclipse in Moho (China) [5]. The residual gravity shows a significant change that the authors identified to correspond to the totality of the eclipse, in particular to the first and fourth contacts of the eclipse [5]. However, similar measurement made during other solar eclipses show no evidence of gravity variation, such as in Finland [6]. Meanwhile, the measurement using a sort of Foucault pendulum seems to show another variation [5] of gravity.

In this paper, we report our measurement of vertical gravity to test the possible gravitational shielding effect on the ground during the recent total solar eclipse of 9 March 2016 passing most regions in Indonesia. Two micro-gravimeters were used. One instrument was put in Kalora Village (Poso, Central Sulawesi, at coordinate S 1° 22.862' E 120° 45.221'), that is, a location close to the centre passage of the total eclipse, and the other instrument was put at Lembang, the site of Bosscha Observatory (at coordinate S 6° 49.478' E 107° 36.986'), where the partial solar eclipse can be observed. If the variation of gravity occurs, we expect that the rate could be different at both sites.

2. Observations

We used the CG-5 LaCoste-Romberg micro gravimeter to measure the vertical gravity expressed in μgal. This gravimeter is more sophisticated than D-type gravimeter used in the 1997 observations [5], although both instruments could measure at similar accuracy in μgal.

During the measurement conducted at Kalora (Poso area), the instrument was installed inside a farming house about 300 m away from the main road. The gravimeter was put on a concrete base specially made for this observation, while at Lembang area, the instrument was put inside an isolated room at Bosscha Observatory. During the four days measurement, the temperature variation in the instrument is about 1.02° C, and the ambient room temperature was 29° C and 23° C in Kalora (Poso) and Lembang, respectively. Hence, according to the calibration standard, the effect of temperature variation was negligible during both measurements. Meanwhile, position tilting of the instrument was always remotely monitored and an average tilt of less than 7 arcsec was found. Therefore, the gravimeter worked in stable condition during the whole measurements.

Since Poso is located very close to the passage of the total eclipse, and in fact it is very close to the equator, the data recording in Kalora (Poso) was changed regularly and was made more frequently close to the totality of the eclipse. The CG-5 gravimeter recorded data in flash memory from 2-minute to 5-second intervals during four consecutive days of measurement. When the eclipse was close to the totality at the duration of two hours, the vertical gravity was recorded at 5 second interval each. After the totality on 9 March 2016 at 11:46 am until 10 March 2016, the gravimeter recorded at one minute interval. Subsequently, the measured data were downloaded into a computer for further processing. The data reduction takes into account the effects of tide [8], instrumental drift, data smoothing during the totality of the eclipse, and frequency analysis.

3. Result of observations

The vertical gravity observations during the TSE were performed in two locations, at Kalora (Poso, Central Sulawesi) which is very close to the shore (about 300 m), and at Lembang (West Java) which is in a mountain site at an altitude of about 1300 m. Table 1 shows different local time of the eclipse. During the eclipse, the elevation of the Sun was about 21°.

Table 1. Eclipse of 9 March 2016 in two locations in Indonesia at local time

City	Start time	Start Tot eclipse	Max. eclipse	End tot eclipse	End time	Remark
Kalora/Poso	7:28:10	8:38:21	8:39:41	8:41:01	10:02:00	Total eclipse, near shore
Lembang	6:19	~7:15	~8:59	10:38	11:34	Partial eclipse, in mountain

Figure 1 shows the raw data for the four days of measurement and the tidal effect is clearly seen. Modelling of this effect using a formalism from [6] allows to remove this effect from the data. The box in figure 1 shows the time of the TSE. Comparison of the timing between the eclipse and the tide shows that the tide gave its influence at some later time after the eclipse. So, it needs some time to get a maximum tide, and the influence of the vertical gravity due to the eclipse is therefore low enough.

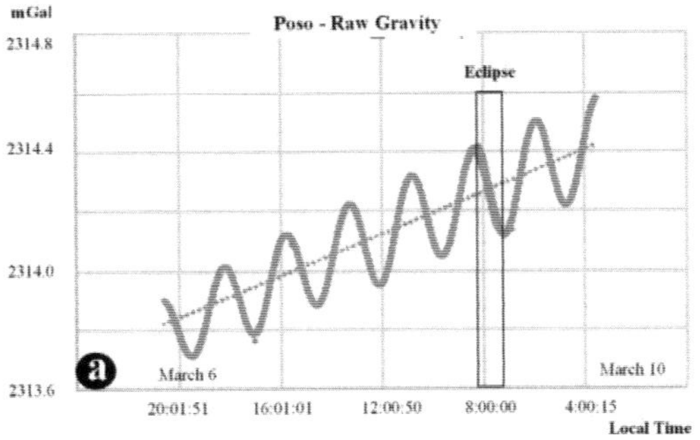

Figure 1. The raw gravity data, at Poso, during four days measurement

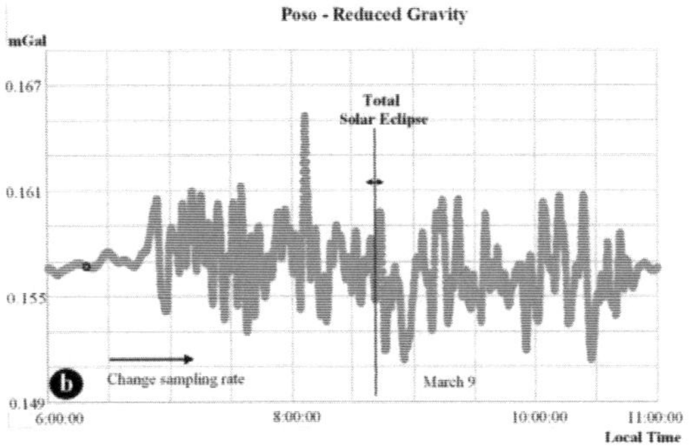

Figure 2. The reduced gravity data, at Poso. The gravity data reduced from tide and drift for five hours.

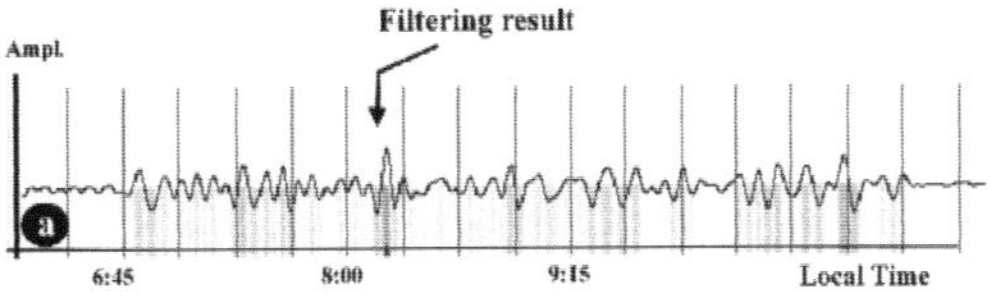

Figure 3. The reduced gravity data during eclipse and frequency respond (2-4 mHz) at Poso

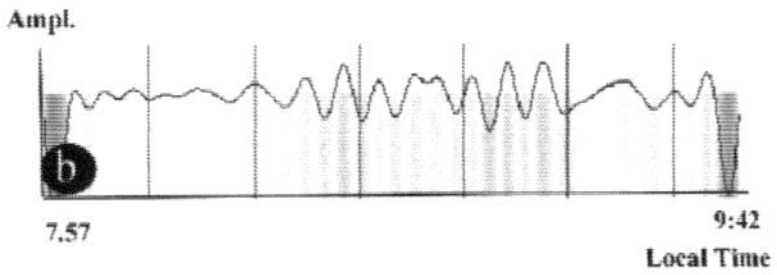

Figure 4. The reduced gravity data during eclipse and frequency respond at Bandung

Subsequently, the remaining question to ask: Is there any microgravity variation observed during the short duration of the eclipse?

As mentioned earlier, anomalous gravity was reported during the total solar eclipse [1], [2]. A precise vertical gravitation measurement during the TSE in Moho region in 1997 was conducted [5], and it is suggested that the shielding effect could probably occur during the TSE to explain the variation of vertical gravity during the totality of the eclipse. However, our measurement, shown in figure 2, does not provide similar results to [5]. The features of gravity signal observed in [5] are rather similar to Mexican hat in the digital signal model. The signals detected in [5] show a significant peak-valley variation of gravity curve near the first and the fourth contact of the eclipse. Our data, shown in figure 2, do not detect this feature.

In our case, the remaining signal appears to be similar to "tremor" in seismology. Alternatively, the signal is similar to accelerograph in seismology for recording micro earthquake, where the unit is in m/s^2 (See the CG-5 Manual or the IDA project using gravimeter, for seismology observation [9]).

Moreover, we can generate the spectrum of our data as displayed in figure 3. The spectrum shows several peaks at certain frequencies obtained during the eclipse. After applying bandpass filtering at (1, 2, 7, 10 mHz), we can get a slightly strong feature for 2-4 mHz, as shown in figure 3. This low frequency signal may correspond to long period seismogram, similar to free oscilation of the Earth (in S or T mode) [10][11]. We infer that it is similar to microtremor for period of 6-7 minute. On the other hand, for data from Lembang observations (in mountain site), the sampling rate taken was at 30 second interval, as shown in figure 4. We can see that the feature is not strong compared to data from Poso measurement. If it is not due to the shielding effect from the eclipse, in the following we propose the possible source of this feature.

4. Discussion

The vertical variation of gravity during the total solar eclipse is important to observe using the very precise gravimeter to investigate the possible existence of gravitational shielding effect [2], [3]. Our measurements suggest that the small variation of vertical gravity is not merely due to the TSE, but we should also take into account the huge ocean movement, very close to measurement site in Poso. In fact, the ocean mass was attracted to move toward the center of the alignment during the eclipse. The emergence of unique anomalous gravity with the shape of *Mexican hat* during the Moho 1997 eclipse and reach 5 μgal of variation certainly raised an important issue. Some authors propose that it could be due to atmospheric effects [1]. However, similar measurement conducted during the TSE in Finland [6] do not detect the same phenomena. Their results are rather similar to our ambient signal detected in our measurement, as shown in figure 1.

We propose that during total solar eclipse, not only the gravity between the three major bodies that can yield the gravity variation, but also wind and ocean that may contribute to the tidal effect. The tide itself will reach a certain gravity value after the eclipse is over. While reaching the peak of the tide, we suppose that the ocean movement hits the surface slowly as "hum" or "little tsunamy". We note that the hum or tsunamy effect on ocean can be detected by seismometer as high as 4-10 mHz [12]. Therefore, we propose that small gravity variation detected in figure 3 is the results of the effect of hum or little Tsunami, instead of due to gravitational shielding.

5. Conclusion

The measurement of vertical gravity during the total solar eclipse in Poso does not support the shielding effect. Instead, a signal of 2-4 mHz detected from our measurement is a result of the ocean movement (hum or little Tsunami) hitting the seashore during tide following the eclipse.

References

[1] Duif C P 2004 A review of conventional explanations of anomalous observations during solar eclipses (*Preprint* gr-qc/0408023)
[2] Van Flandern T 1996 *Astrophys. Space Sci.* **244** p 249
[3] Majorana Q 1920 *Philos.Mag.* **39** pp 488-504
[4] Allais M F C 1957 Mouvement du pendule paraconique et éclipse totale de Soleil du 30 juin 1954 *C.R. Acad. Sci.* **245** p 2001
[5] Wang Q, Yang X,Wu C, Guo H, Liu H, Hual C 2000 *Phys. Rev.* D **62** 041101-1
[6] Makinan J 1990 *Absolute gravity measurements during the July 22, 1990 total solar eclipse in Finland*
[7] Yang X and Wang Q 2002 *Astrophysics and Space Science* **282** pp 245–253
[8] Longman J 1959 *J. Geoph. res.* **64** p 2351
[9] Agnew D C, Berger J, Farell W E, Gilbert J F, Masters G and Miller D 1986 *Eos* **67** p 203
[10] Nishida K 2013 *Ann. Rev. Earth Planet. Sci.* 2013. **41** p 719
[11] Ruymbeke M, Shaoming L, Manshina L, Meureuers B. 2003 *Bulletin d'Information de Marees Terrestres* **138** p 10967
[12] Saito T 2010 *Geophys J. Int.* **182** p 1515

Journal of Physics: Conference Series **771** (2016) 012004 doi:10.1088/1742-6596/771/1/012004

Preliminary results of the solar corona spectroscopic observation of 9th March 2016 Total Solar Eclipse

Emanuel Sungging Mumpuni[1], **Muhamad Zamzam Nurzaman**[1,2], **Nana Suryana**[1]

[1] Center of Space Science, National Institute of Aeronautics and Space (Pussainsa LAPAN), Jl. Dr. Djundjunan 133, Bandung, Indonesia

[2] Dept of Astronomy, FMIPA ITB, Jl. Ganesha 10, Bandung, Indonesia

E-mail: emanuel.sungging@lapan.go.id

Abstract. Spectroscopy observation of solar corona has been carried out from 9th March 2016 Total Solar Eclipse expedition in Maba, East Halmahera. Due to limitation by the weather condition during the observation, the obtained data were not favorable. Because of the low signal to noise ratio, the wavelength calibration was not straightforward. To obtain the optimum results, steps for data reduction were performed as the following: selecting the finest data, extracting the 1D spectrum from 2D spectrum, and calibrating spectrum wavelength (including careful interpolation). In this preliminary result, we discuss the instrumentation, the data under investigation, the extraction process of spectrum, and the polynomial interpolation that has been used for extracting information from our data.

1. Introduction

The total solar eclipse that was observable from Indonesia on March, 9th 2016 has given opportunity to study the corona of the Sun since the solar corona can be observed from ground-based observation at best only during the totality of the solar eclipse.

For the event of 2016 Total Solar Eclipse, we conducted a research on spectroscopy of the corona. Corona spectrum consist of emission lines from excited ions from various sources from plasma that depend on the temperature, and most of these lines are observed as forbidden lines [1]. The spectroscopy of the corona from the eclipse give ample of opportunity to observe specific coronal lines that appear at the moment.

Despite the limitation by weather during the observation, in this preliminary report, we present the data from observation after the totality with good information, the spectrum extraction, and wavelength calibration process for the purpose to have a quick overview on wavelength-calibrated spectrum (Section 2), and finally the overview of the spectrum, also the discussion in Section 3.

2. Data and analysis

We selected Maba, East Halmahera (0° 41' 34" N, 128° 17' 39" E) as the site for the observation, it has the longest duration of totality on the land in Indonesia. Our equipments comprise of telescope with F/10, 9".8, CCD camera 2125 × 1472 pixel square, 6.8μm, and spectrograph with moderate resolution (R $\sim$ 18000) and Littrow design [2]. This telescope design is suitable to observe and study several emission lines from the corona that probably weak according to the recent solar cycle.

However, due to weather condition that limits the data, only small and limited data were obtained from the moment of totality. To extract information from the data, it is important to have calibrated information from similar data but with a better quality during the run of the observation (comparison data).

The data with better quality obtained only prior and after the totality. The comparison data in this work was extracted after the totality, calibrated in the wavelength and features identified. For this part, we expect that we will have strong features, so called *Doublet Na 1 D* as the calibration for the wavelength of the spectrum. The *Doublet Na 1 D* also known as *Sodium doublet* is consist of two lines, the Na D_1 on 5895.92 $\mathring{A}$ and Na D_2 line on 5889.95 $\mathring{A}$. This process is a preliminary result to have spectrum calibrated to the wavelength.

Figure 1 and figure 2 show the particular calibration lamp spectrum for the expected spectrum region and the cross-section of the emission line shown. It can be shown that the data severly tilted to the vertical line and the profile is not flat enough, regardless that strong emission lines appeared. Due to these limitation, the information should be carefully extracted if we want to have proper information from the spectrum. In this work we only have quick overview on calibrated wavelength, so we only focus on the absisca information from the spectrum using IRAF as the software for analysis.

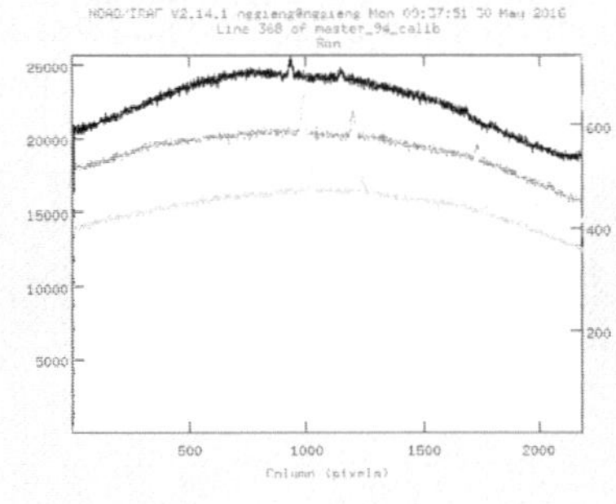

Figure 1. The calibration spectrum data for particular observed wavelength. The bright line on vertical line is the emission spectrum for calibration of the wavelength.

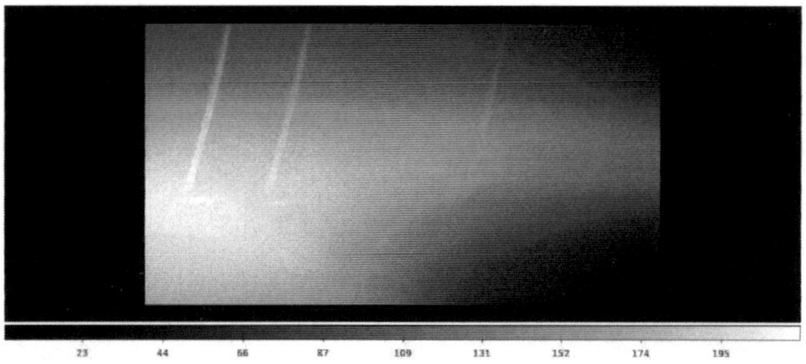

Figure 2. Shows the cross-sectional profile along the line for several lines for the calibration data for selected spectrum region as shown from left figure. Different colours represent the different lines for the image data with unit in pixels.

After standard calibration process (dark substraction and flat fielding), we followed standard spectroscopy extraction and identify the important emission line, as shown in figure 3. The emission lines were identified with related wavelength from standard *He-Ne-Ar* atlas.

We then found the relation between pixels to wavelength by fitting with spline polynomial, as shown in figure 4. We also tried to see the stability of the relation by fitting the second order relation as shown in figure 5. It can be shown that the data are quiet well fit between the two relation, so we have good confidence that the wavelength is well related to particular pixels. Next, we applied the identification wavelength to spectrum.

3. Result and discussion

Figure 6 shows the result of calibrated spectrum under investigation, the expected features appear clearly as strong absorption lines (marked by red rectangle)

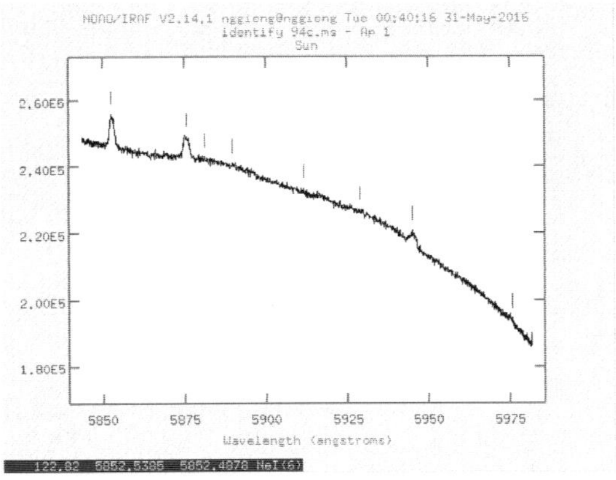

Figure 3. Identification of several emission line identified shown with vertical blue line above the spectrum.

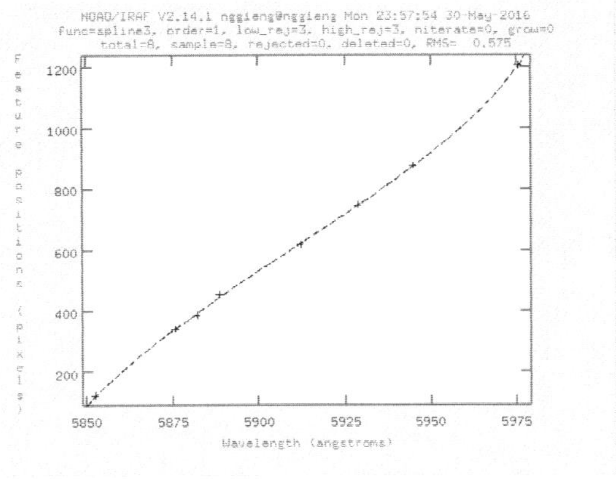

Figure 4. The fitting of the relation between wavelength (absisca) and pixels value (ordinate).

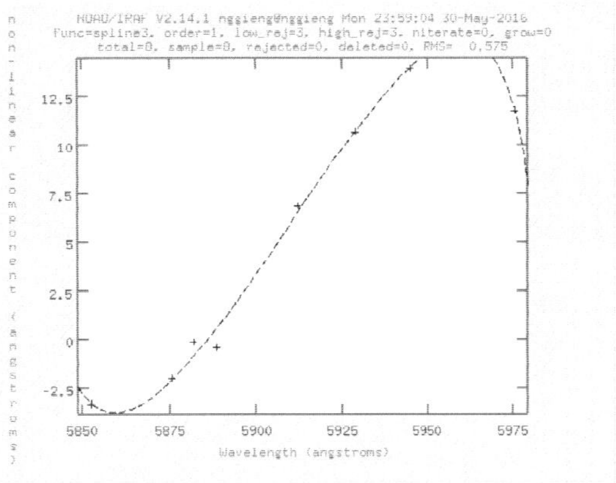

Figure 5. The non-linear part of the relation between wavelength and pixels.

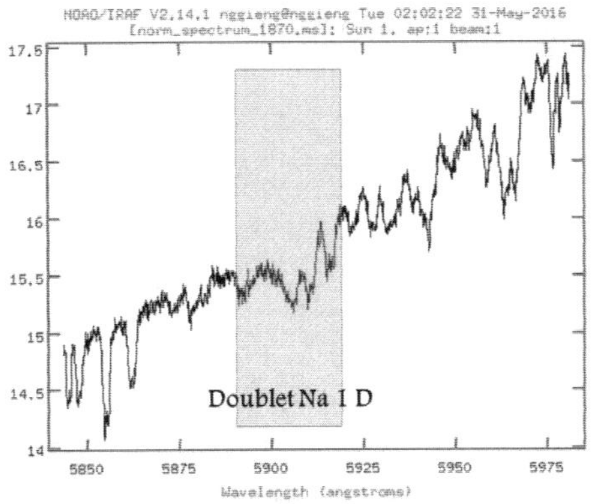

Figure 6. *Doublet Na 1 D* is marked in the red rectangle.

However to have the proper spectrum information, the result still needs more corrections. As seen in figure 1, it is necessary to apply geometrical correction to the spectrum. We can choose between calibrating flux or normalization, however we have to be certain about the proper coordinate on ordinate direction, so the geometrical correction still need to be done by long slit transformation method.

Another aspect to be considered is the spectrum from the corona is located outside the selected domain and in poor quality data, so the identification of the expected lines on different wavelenght region might be done by polynomial fitting. The expected coronal line is not on exactly on sodium doublet region. Some method can be consider, such as Legendre, Chebyshev, or Spline method to fit. However this selection of the method will rely on the quality of the extracted data.

4. Conclusion

We have shown that our instrument was capable to obtain the spectrum with good quality even in the condition that could severly limit to obtain the best data. As we have demonstrated, we still could extract the expected *Doublet Na 1 D* lines. Since the spectrum coverage is very narow, if the expected spectrum line is located on different region that is on very poor condition, the result from this extraction can be used to fit the expected line as the additional information to give the proper information.

Acknowledgments

This research was supported by Pusat Sains Antariksa grant for eclipse expedition 2016.

References

[1] Mumpuni E S 2015 Korona Matahari - Laboratorium Astrofisika dari Gerhana Matahari Total *Media Dirgantara* **10 3** 7-10.

[2] Thizy O 2007 Lhires III High Resolution Spectrograph *Proceedings for the 26th Annual Conference of the Society for Astronomical Sciences* 31-36.

Journal of Physics: Conference Series **771** (2016) 012005 doi:10.1088/1742-6596/771/1/012005

Ludendorff coronal flattening index of the total solar eclipse on March 9, 2016

Tiar Dani[1], Rhorom Priyatikanto[1], and Abdul Rachman[1]

[1] Space Science Center, National Institute of Aeronautics and Space, LAPAN Jl. Dr. Djundjunan 133 Bandung 40173, Indonesia

E-mail: `tiar.dani@lapan.go.id`

Abstract. Ludendorff coronal flattening index of the Total Solar Eclipse (TSE) on March 9, 2016, was calculated at various distances in solar radius. As a result, we obtained the coronal flattening index ($\epsilon = a + b$) at a distance of 2 solar radii is 0.16. The 24th solar cycle phase based on the 2016 TSE event obtained -0.64 which showed the corona is pre-minimum type. Resulted coronal flattening index value gives a predicted maximum amplitude of the monthly sunspot number (W_{max}) for the 25th solar cycle to be 70 ± 65. Therefore, the solar activity for 25th solar cycle predicted to be lower than the current solar cycle, which has a maximum sunspot number value of 146 in February 2014

1. Introduction

Space-based observations provide opportunity to study the physics of solar corona [1], but ground-based observations conducted during total solar eclipses give another chances to understand the lower part of corona [2] since the global and local magnetic field of the sun inner part influenced the solar coronal shapes. Consequently, the structure will change according to the solar cycle. Corona at minimum solar activity has asymmetrical brightness distribution characteristics around the solar disc with two very bright streamers around the equator. While corona at maximum solar activity has characteristics of uniform brightness with the streamer around the sun [3]. Photometry of the K and F corona observed using the white-light filter during TSE events [4]. K-corona is the photosphere light scattered by the free electron gas in the corona while F-corona is the photosphere light scattered by dust particles of small size in the ecliptic plane [5].

One way to quantify the global structure of the corona is using Ludendorff coronal flattening index (ϵ) [6]. Several studies calculated coronal flattening index for TSE event have been conducted [7, 4, 3, 8, 6, 9]. From the compilation of measurements, variation of coronal flattening index over solar cycle can be modeled and the characteristics of upcoming solar cycle (e.g. time of maximum and the amplitude) can be predicted [8]. In this study, we obtained coronal flattening index by analyzed the solar image isophotes during totality on the 2016 TSE. Henceforth, the sunspot number maximum amplitude for the next solar cycle is predicted using the coronal flattening index. In addition, the type of corona determined based on the current solar cycle phase.

2. Data and method

We conducted the 2016 TSE observation at Ngata Baru ($0^o55'27$" South, $119^o57'27$" East, $320m$ asl.), Central Sulawesi, Indonesia. At this point, the duration of totality is 2m 14s while the general sky condition was clear. Four images were acquired using 12 Mpx camera with various field of views and exposure times. All these images were captured using ISO 800 setting with exposure from $1/60s, 1/100s, 1/160s, and 1/200s$. Taken from 07:39:17 until 07:40:41 local time.

No special data reduction was applied, but the images were rotated to get the proper orientation and matched with reference images (SDO/AIA 193 from http://sdo.gsfc.nasa.gov and SOHO/LASCO C2 from http://sohodata/nascom.nasa.gov). Observed features such as prominence and streamer became visual guides for the matching process. Additional data of monthly sunspot number from WDC-SILSO from Royal Observatory of Belgium (http://www.sidc.be/silso) and multi-year forecast sunspot number from SWPC NOAA (http://www.swpc.noaa.gov) was used to determine the phase of current solar cycle.

Ludendorff flattening index was determined from the isophotal contours (isophote in brief) with various equatorial distances from the limb. The isophote was constructed for each grayscale images that was properly oriented to the solar equator. Image with different exposures enable us to construct several sets of isophote that give the value of flattening indexes at various distances. The flattening index itself is defined as

$$\epsilon = \frac{d_0 + d_1 + d_2}{D_0 + D_1 + D_2} - 1 \tag{1}$$

where d_0 is the diameter of isophote at the equator, d_1 and d_2 is the diameter at an angle of $\pm 22.5^o$ axis parallel to the equator. While D_0, D_1, and D_2 is the diameter isophote in the polar axis direction. The value of ϵ increase linearly from the limb to a certain distance, $R_{max} = 1.5 - 2.5\ R_{sun}$ then decrease up to a distance $R_{min} = 3.5 - 5.5\ R_{sun}$. Ludendorff (and the citing authors) use ϵ at $r = 2\ R_{sun}$ as the photometric flattening index. To obtain this number, extrapolation is often used since the low intensity of the outer part of the solar corona.

The value of ϵ is also a function of the solar cycle phase (Φ) as introduced by [10] with following equation:

$$\Phi = \frac{T_{ecl} - T_{min}}{|T_{max} - T_{min}|} \tag{2}$$

where T_{ecl} is the time of the eclipse (in years), T_{max} and T_{min} is the time of maximum and minimum solar cycle near T_{ecl}, respectively. T_{max} identified from WDC-SILSO data and T_{min} identified from SWPC NOAA. Phase of solar activity ϕ changes from -1 to 1 and can be interpreted as the phase of solar cycle from maximum to maximum. [8] obtained a linear fitting equation for the maximum monthly sunspot number (W_{max}) with flattening index (ϵ) when approaching the minimum solar cycle using data from 60 events of TSE. This prediction equation has large 1σ-uncertainty (± 65) due mainly to the large sensitivity of the flattening index value to many factors, such as number of innermost isophotes selected for linear approximation, possible error in the orientation of coronal images, etc.([8])

$$W_{max} = -2.8 + 466.1\epsilon \tag{3}$$

By using equation 3, the maximum amplitude of the monthly sunspot numbers for the next solar cycle can be calculated. We also compared our results of the coronal flattening index, TSE phase, and W_{max} with other studies.

3. Results and discussion

The north-south direction of the four TSE 2016 observation images were rotated by 66° clockwise (figure 1: Left panel) to get the proper orientation and matched with reference images (see Section 2)

Figure 1. Left panel: TSE image is superimposed on the relevant X-ray image from SDO/AIA193 (inner) and streamer structure from SOHO/LASCO to confirm the exact rotation. Middle panel: Example of the white-light corona during 2016 TSE from our observation with f/5.6 aperture and 1/60s exposure. Right panel: isophotes images with marking line for measurement of flattening index .

Twenty isophotes in various solar radius obtained from these four images. Then each of the isophotes calculates its coronal flattening index using equation 1 (figure 1: Right panel). Coronal flattening index for various solar radius from 20 isophote are plotted and shown in figure 2. The isophotes construction is limited to the distance of $1.7R_{sun}$ due to low signal to noise ratio outside this distance. To overcome this limitation, Ludendorff coronal flattening index at $r = 2R_{sun}$ is extrapolated using the linear equation $\epsilon = 0.0471R_{sun} + 0.061$ obtained from linear regression as displayed in figure 2. Ludendorff coronal flattening index for 2016 TSE is 0.16.

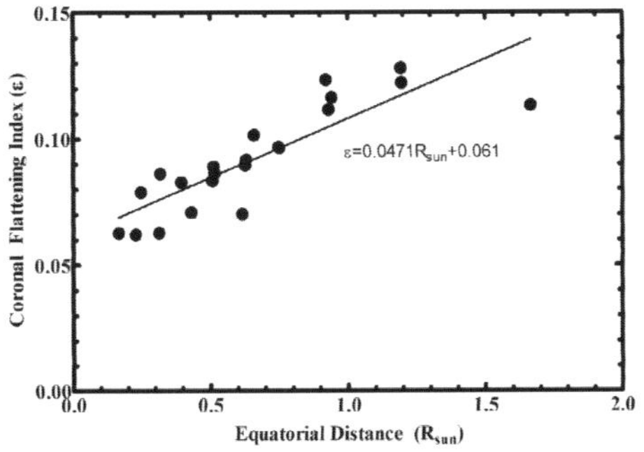

Figure 2. Coronal Flattening Index of the solar corona during 2016 TSE.

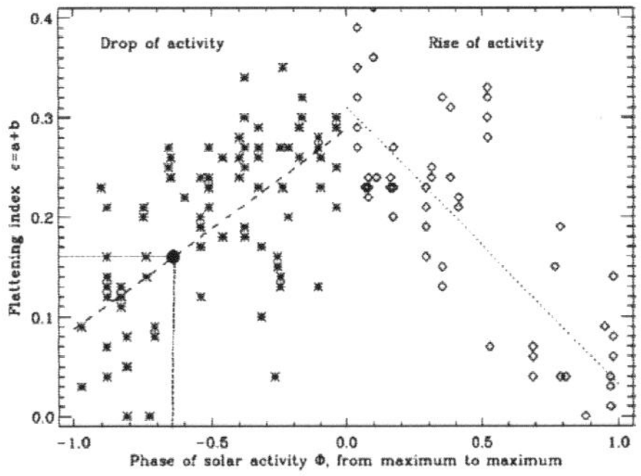

Figure 3. Dependence of the flattening index on the phase of solar activity [8]. Result of TSE 2016 shown in black dot mark.

The 2016 TSE corona phase for 24th solar cycle then calculated using equation 2. The peak time (in year) of the 24th solar cycle (T_{max}) based on data from WDC-SILSO Royal Observatory of Belgium, Brussels, the maximum monthly sunspot number for the 24th solar cycle is 146.1 which occurred in February 2014 ($T_{max} = 2014.123$). While the 24th solar cycle minimum based on multi-year predictions of monthly sunspot numbers from NOAA SWPC occurred in December 2019 with the monthly sunspot number values is 4.1, so $T_{min} = 2019.923$. And for the eclipse time (T_{ecl}) occurred on March 2016 is 2016.206. 24th solar cycle phase then calculated using equation 2, obtained $\Phi = -0.64$. Keep in mind that the solar cycle phase Φ value is determined by the prediction of T_{min} which might be still get corrected. The value of ϵ and ϕ for TSE in 2016 were added to the plot of coronal flattening index and the solar cycle phase which compile based on the 60 TSE events from 1851 to 2010 as presented in the figure 3.

Figure 3 shows that 2016 TSE occurs when solar activity went to its minimum value hence the observed corona is a pre-minimum type corona. Also the 2016 TSE corona shapes are nearly symmetric based on the flattening index value despite a fairly large streamer near the solar south pole on 2016 TSE which affects the solar poles diameter but in this case, does not really influence the final value of the coronal flattening index.

Furthermore, the prediction for maximum amplitude of the sunspot numbers for 25th solar cycle obtained by using equation 3. So the predicted of maximum sunspot number for the 25th solar cycle (W_{max}) is 70 ± 65. The flattening index factors used to predict the maximum amplitude sunspot numbers for the next cycle can be regarded as an indirect characteristic of the solar poles magnetic field ([8]). Our result means that the 25th solar cycle activity is predicted to be lower than the current cycle, which has peaked in February 2014 with a sunspot number value of 146. This can be compared with results from other studies. For example, [11] stated that the next solar cycle will reach maximum on October 2023 with amplitude of 109 which is inside 1σ-uncertainty of our calculated amplitude. [12] also gave slightly higher predicted maximum amplitude of sunspot number in cycle 25 which is 132. This maximum value will be reached on April 2023. [13] used long-term record of the sunspot groups area to study the solar cycle behavior. The implication of that study is the predicted maximum amplitude of 25th solar cycle which is as low as 50 ± 10.

4. Conclusion

The coronal flattening index is 0.16 and the 24th solar cycle phase is -0.64 based on 2016 TSE observations. This results suggest that the corona is a pre-minimum type. This gives the predicted maximum of the monthly sunspot number for the 25th solar cycle to be 70 ± 65. Therefore, the solar activity for 25th solar cycle is predicted to be lower than current solar cycle.

Acknowledgments

The authors wish to express their thanks to WDC SILSO, Royal Observatory of Belgium, Brussels, US SWPC NOAA, NASA/SDO and the AIA, EVE, and HMI science teams, NASA and ESA SOHO for providing the solar data.

References

[1] Brueckner G E, Howard R A, Korendyke M J K C M, Michels D J, Moses J D, Socker D G, Dere K P, Lamy P L, Llebaria A, Bout M V, Schwenn R, Simnett G M, Bedford D K and Eyles C J 1995 *Physics, Solar* **162** 356–402
[2] Pasachoff J M, Rusin V, Drucmuller M and Saniga M 2007 *Journal, The Astrophysical* **665** 824–829
[3] Sadovenko I V and Pishkalo M I 2007 *Proc. of Contributed Papers WDS07* III pp 32–35
[4] Stoeva P, Stoev A, Kuzin S, Shopov Y, Kiskinova N, N N S and Pertsov A 2007 *Atmospheric and Solar Terestrial Physics* **70** 414–419
[5] Sungging E M 2015 *Media Dirgantara* **10** 7–10
[6] Stoeva P V, Stoev A D and Kuzin S V 2011 *Sun and Geopshere* **6** 36–38
[7] Gulyaev R A 1997 *Astronomical and Transactions, Astrophysical* **13** 137–144
[8] Pishkalo M I 2011 *Physics, Solar* 347–363
[9] Stoeva P S A and S K 2012 *Sun and Geosphere* **7** 81–84
[10] Ludendorff H 1928 vol 16 (Sitzzungsber. Preuss. Akad. Wiss. and Phys.-Math. Klasse and Berlin) pp 185–192
[11] Li K J, Feng W and Li F Y 2015 *Atmospheric and Solar-Terrestrial Physics* **135** 72–76
[12] Rigozo N R, Echer M P S, Evangelista H, Nordemann D J R and Echer E 2011 *Atmospheric and Solar-Terrestrial Physics* **73** 1294–1299
[13] Javaraiah J 2015 *Astronomy, New* **34** 54–64

Imaging and spectroscopic observations of the 9 March 2016 Total Solar Eclipse in Palangkaraya

Abdul Majid Al Kholish[1,2], Imanul Jihad[1,2], Irham Taufik Andika[1,2], Evaria Puspitaningrum[1,2], Fathin Q. Ainy[1,2], Sahlan Ramadhan[1,2], M. Ikbal Arifyanto[1,2] & Hakim L. Malasan[1,2]

[1]Astronomy Study Program, Faculty of Mathematics and Natural Sciences, Institut Teknologi Bandung, Bandung, Indonesia
[2]National Astronomy Olympiad Coaching Team (TPOA), Bandung, Indonesia

E-mail: majid.alkholish@gmail.com

Abstract. The March 9th 2016 total solar eclipse observation was carried out at Universitas Palangkaraya, Central Kalimantan. Time-resolved imaging of the Sun has been conducted before, after, and during totality of eclipse while optical spectroscopic observation has been carried out only at the totality. The imaging observation in white light was done to take high resolution images of solar corona. The images were taken with a DSLR camera that is attached to a refractor telescope (d=66 mm, f/5.9). Despite cloudy weather during the eclipse moments, we managed to obtain the images with lower signal-to-noise ratio, including identifiable diamond ring, prominence and coronal structure. The images were processed using standard reduction procedure to increase the signal-to-noise ratio and to enhance the corona. Then, the coronal structure is determined and compared with ultraviolet data from SOHO to analyze the correlation between visual and ultraviolet corona. The spectroscopic observation was conducted using a slit-less spectrograph and a DSLR camera to obtain solar flash spectra. The flash spectra taken during the eclipse show emissions of H 4861 Å, He I 5876 Å, and H 6563 Å. The Fe XIV 5303 Å and Fe X 6374 Å lines are hardly detected due to low signal-to-noise ratio. Spectral reduction and analysis are conducted to derive the emission lines intensity relative to continuum intensity. We use the measured parameters to determine the temperature of solar chromosphere.

1. Introduction

We conducted imaging and spectroscopic observation before, after and during totality. We intended to observe the coronal structure up to two solar radii that is not observed by Solar Heliospheric Observatory (SOHO). Using visual image combined with other wavelength such as extreme ultraviolet (EUV) and X-ray, we can determine more detailed coronal structure.

The spectroscopic observation can be used to determine the coronal electron temperature. E-corona in visual wavelength contains Fe XIV line (5303 Å - ionization potential 355 eV) and Fe X line (6374 Å - ionization potential 235 eV), where intensity ratio of Fe XIV and Fe X is sensitive to temperature of solar corona.

The observation was conducted at Universitas Palangkaraya in Palangkaraya, Central Kalimantan. The weather was rainy at the early morning and thick cloud covered the Sun at the time of totality. However, we managed to get some images of the corona and the flash spectrum with low signal-to-noise ratio.

Our goal is to determine the temperature of the solar chromosphere and to complete the ultraviolet data with our visual result to get the correlation between solar corona and the solar activity. However, due to poor quality of images which are contaminated with high noise from the clouds, we only focus on the spectroscopic analysis in determining the temperature of the solar chromosphere.

2. Method

On March 9th 2016, we observed Total Solar Eclipse in Palangkaraya (2°12'36" S, 113°55'12" E). Total Solar Eclipse in that location was occurred at 00:28:56 until 00:31:26 (UT) or lasted 2 minutes and 30 seconds (http://xjubier.free.fr/en/site_pages/SolarEclipseCalc_Diagram.html). In imaging observation we used instrument configuration: (1) William-Optics telescope with diameter 66 mm, f/5.9, (2) *Neutral Density 5* filter, (3) DSLR Canon EOS 500D. On the other hand, we used instrument configuration: (1) Star analyzer SA-100 (100 grooves/mm, R~100) transmission grating, (2) DSLR Canon EOS 1100D in spectroscopy observation. Those observations used EOS Utility 2 as remote camera.

Imaging observation was conducted by taking a number of images of the Sun from the first contact until the fourth contact. At the time of the second contact to the third contact we conducted video recording observation in order to obtain the dynamics of the corona and get the image more than the usual image acquisition. Spectroscopic observation was conducted by taking a number of images of the Sun on a few seconds before the second contact and few seconds after the third contact so we got flash spectrum and Corona spectrum as the result.

3. Result

3.1. The corona image

Instead of 2 minutes and 30 seconds of totality, we just recorded the totality in 30 seconds before 3rd contact with video mode. After that, we stacked the video in a picture and enhanced it. The image result is shown in figure 1.

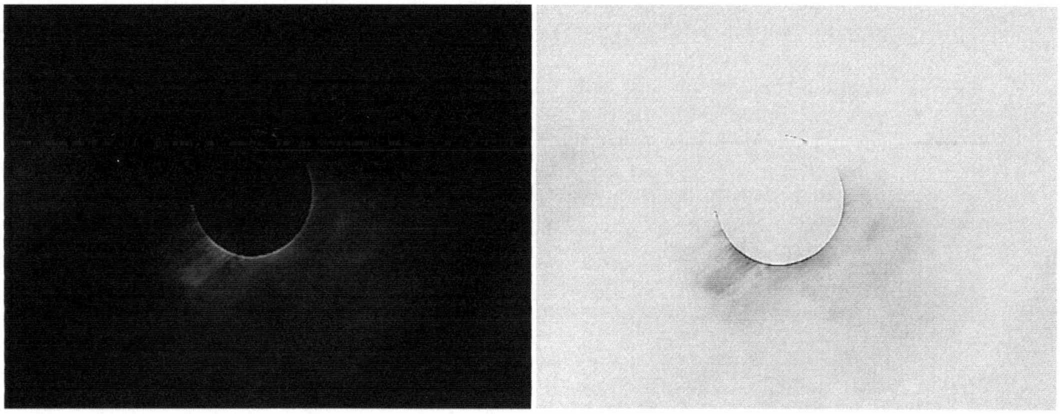

Figure 1. *Left*: image of original stacked and enhanced 30 s totality video . *Right:* inverted image.

3.2. The flash spectrum

The data were taken during third contact phase of total solar eclipse. Due to cloudy weather, we only managed to obtain the spectrum with low signal-to-noise ratio. Figure 2 shows the flash spectrum shows strong emission lines from Hβ (4861 Å), He I (5876 Å), and Hα (6563 Å). These lines are emitted from solar chromosphere with large height range. The Fe XIV (5303 Å) and Fe X (6374 Å) lines, which are produced by solar corona, are hardly detected due to low signal-to-noise ratio.

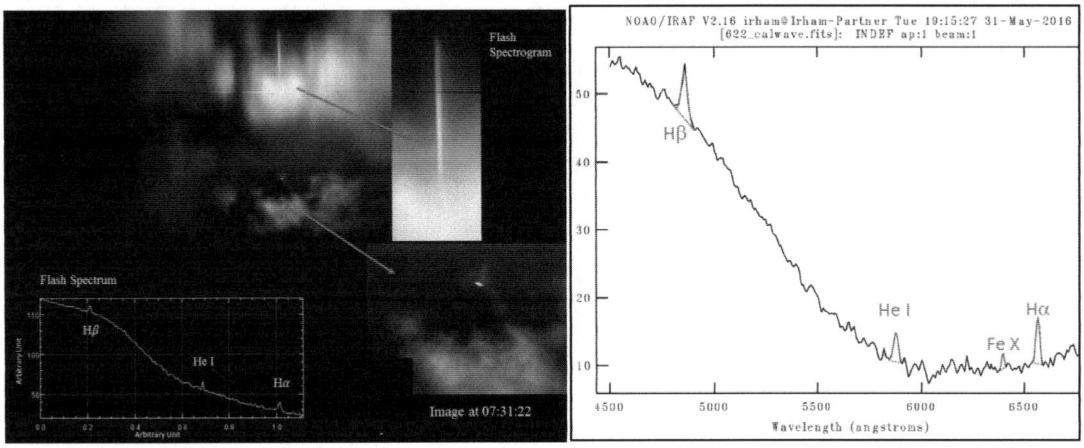

Figure 2. *Left:* flash spectrogram and spectrum that were taken at 07:31:22 LT. *Right:* flash spectrum and detected lines that were taken at 07:31:24 LT.

The local continuum underneath each line was fitted by interpolating the continuum windows around the line to measure I_c. We fitted each line using single Gaussian function to measure I_λ. We measured the relative intensity of each emission line using equation $I = (I_\lambda - I_c)/I_c$ where I_λ is the integrated intensity underneath the line profiles and I_c is the corresponding continuum intensity of continuum background. We detected 3-4 emission lines only in 3 images. The relative intensities of the observed lines are tabulated in table 1.

Table 1. Relative intensity of the observed lines in the flash spectrum.

Line	Relative Intensity	Error	Time (local)
Hβ	4.73	0.03	
He I	10.26	0.12	7:31:20
Hα	17.05	0.17	
Fe X	2.90	0.11	
Hβ	1.17	0.01	
He I	1.65	0.03	7:31:21
Hα	11.47	0.06	
Hβ	0.09	0.01	
He I	0.52	0.02	7:31:22
Hα	9.06	0.05	

4. Analysis

4.1. Imaging

The coronal structure in our image cannot be seen clearly because the solar disk was covered by clouds. Two images in Extreme Ultraviolet (EUV) 193 Å and 304 Å that were taken from SDO AIA are shown in figure 3. Image of 193 Å represents the lower solar corona and the image of 304 Å represents the solar chromosphere. We detected a prominence in 193 Å and 304 Å. However, coronal structure in 193 Å and chromosphere activities in 304 Å could not be detected in our visual image. So, our visual observation cannot be combined with these two EUV images and the correlation that shows solar activity cannot be determined.

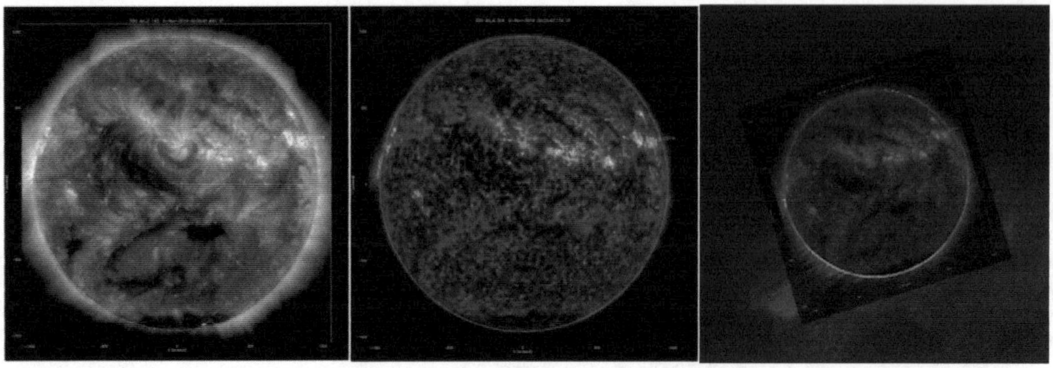

Figure 3. Left: Image in Extreme Ultraviolet (EUV) 193 Å from SDO AIA. *Center*: Image in Extreme Ultraviolet (EUV) 304 Å from SDO AIA. *Right*: Combined observed image with EUV images cannot be used to find a correlation of solar activity.

4.2. Spectroscopy

We only observed Fe X in our spectrum, while Fe XIV is not detected due to low signal-to-noise ratio. Therefore, we cannot determine the ratio of Fe XIV and Fe X to determine the temperature of the corona by using [1]. However, the detected Fe X still gives us information that the minimum temperature of the corona is 1.2×10^6 K because minimum temperature that is needed to produce Fe X is above that value[2].

Acknowledgements

We are grateful to Prof. John. H. Seiradakis (Aristotle Univ. of Thessaloniki, Greece), Mr. Robert B. Slobins (Yahoo! Solar Eclipse, USA) for encouragements and valuable advices. We thank FMIPA ITB, Universitas Palangkaraya, Astana Ilmu, Perdana Consulting and other members of Olympiad coaching team for valuable support.

References

[1] Brickhouse N, Esser R, Habbal S R 1995 Demonstrating the limitation of line ratio temperature diagnostic using [Fe X] and [Fe XIV] spectral line intensity observations *International Solar Wind 8 Conference* 69

[2] Nikolskaia K I, Utorbin V G 1984 The coronal disturbence (W 90 deg, N 25 deg) in the eclipse spectra of July 31, 1981 *Solar Physics* **91** pp 141-157

[3] Voulgaris A, Athanasiadis T, Seiradakis J H and Pasachoff J M 2010 A comparison of the red and green coronal line at the 29 March 2006 and the 1 August 2008 total solar eclipse: consideration of the temperature of the solar corona *Solar Physics* **264** pp 45-55

Prominence measurement of total solar eclipse: March 9th 2016, Ternate, Indonesia

Luthfi Naufal[1], Fargiza Abdan Malikul Mulki[1], Siti Fatima[1], Widyanita[1], Saffanah Zahirah[1], Christoforus Dimas Satrya[2] and Dhani Herdiwijaya[1]

[1] Astronomy Department, InstitutTeknologi Bandung, Bandung 40132, Indonesia
[2] Physics Department, InstitutTeknologi Bandung, Bandung 40132, Indonesia

Email: luthfi.naufal@gmail.com

Abstract. Prominence is a hot gas that consists of electrically charged particles, which is propelled outward Solar surface. It can happen when the activity of sunspot rising up. Ground-based observation of prominence is rather difficult due to the brightness of the Sun, which it needs appropriate method. During total solar eclipse (TSE), prominence is easier to observe. In Ternate, Maluku islands (0°4632.0"N 127°2305.7"E), we observed solar prominence in visual and infrared spectra performed by using regular-DSLR camera for visual wavelength and modified-DSLR (IR-cut filter removal) for infrared wavelength, respectively. Next, we calculated the length of prominence from TSE images, and found that the real minimum length of prominence in red is 1.05 times longer than green and blue ones. And, the real maximum length of prominence is 1.08 times longer than blue and 1.04 times longer than green. The intensity of red is brighter 1.37 times to green and 1.23 times to blue. The causes of the results due to the different temperatures of the prominences and the radiation from H-alpha emission. We succeeded to observe directly the prominence with simple cameras and measured its length and brightness.

1. Introduction

One of the most exotic features that we can observe during Total Solar Eclipse (TSE) is solar prominence. Solar prominence is a most common part of the solar atmosphere, which can be seen in large region of corona. The term of prominence is refer to long-lived cool and dense plasma curtain in the hot and rarefied outer solar atmosphere [1]. Prominence has one hundred times temperature lower but denser than that of corona. It means that it is thermally isolated from their surrounding regions [2]. Generally, there are two ways to classify prominence: based on its location in solar surface and based on its dynamics [3]. Classification by the location is divided into two types: quiscent and plage (active region) prominences. Whereas, classification by dynamics are filament and active prominences. Solar prominence activity could be seen when TSE is happened. Fortunately, we could detected two major parts of prominence from images of TSE which occurred in March 9th 2016. Main path of the TSE is through 12 cities in Indonesia and the longest duration of TSE is about 3m 7s. We observed The TSE in Ternate, Indonesia. Ternate was one of the best place where we could observed the TSE clearly and the period of the TSE was about 2m 45s. Prominence can be observed in wide wavelength range, but its image clarity depends on microparticle temperature and surrounding atmospheric regions. Visual limb-prominence can be observed even by amateur astronomers using a coronagraph, or at total solar eclipse. Whereas, filaments are rather different where they will be easily found and observed through a Hα filter at regular time (not only TSE). In H-alpha band, we can easily distinguish prominence from whole solar disk. In fact, Hα filter blocks all wavelength from the Sun except the wavelength, which is emitted by the excited hydrogen atom at particular energy on 6563 Å wavelength. This hydrogen emission is responsible for the special characteristic of color appearance of prominence and chromosphere.The duration of TSE is very short comparing to the existence period of the prominences that is for days to weeks. Thus, in this work, we tried to measure the size and brightness of the prominence from the images captured by simple cameras.

1.1. The characteristic of solar prominence

Prominence is propelled as one effect of magnetic dynamic and thermal activity in solar atmosphere. Prominences are frequently found inside coronal cavities, which are tunnel-like elliptical dark regions [4] at the solar limb, with up to 40 % density depletion [5], around the prominences. Regarding to the prominence scale, we consider two selections: large-scale structure and fine-structure prominence. In case of quiscent prominence, large-scale structure is unchanged while the fine-structure is changed rapidly [6]. There are three types of quiescent prominences based on the location of figuration prominence [3]. Those are type A which is associated with active region in solar surface, type B which is related to the region between two active regions, and type C which spans over polar crown neutral line and near an expanding active regions. The dynamics of prominence can be understood as a contribution of magnetic-force resistance to gravitational attraction of interior and solar surface. They typically have a length of 6×10^4 m to 6×10^5 m, a height spanning from 1.5×10^4 m to 1×10^5 m above chromosphere and a width ranging from 5×10^3 m to 1.5×10^4 m. The temperature range is about 4300-8500 K [7, 8].

2. Method

Prominence was observed by using three DSLR cameras. For obtaining infrared (IR) spectrum, we used modified Canon EOS 550D camera. Modification was made by removing IR-cut filter, which blocked infrared spectrum, from front of camera then the camera could observed the wavelength from visible to infra-red. Furthermore, the filter that was hampered visible light was used to obtain infrared spectra. In order to obtain visible light, we used Canon EOS 5D Mark II and Canon EOS 700 D cameras.

In collecting image data using Canon 550 D (IR) and 700 D (visible), the cameras were used in video mode 8-bit standard compression. For Canon 5D mark II, 14-bit raw-video was used. The prominence images were taken by using frames of videos. Moreover, the frames were separated into Red (R), Green (G) and Blue (B) channels for next further analysis: calculating the brightness and dimensions of prominences.

3. Results and discussion

Figure 1 shows results from frames of videos in multiple channel. Figure 1a shows the red channel, it can be seen that the prominence is brighter and longer than in other channels: green (figure 1b) and blue (figure 1c). In red channel, there are some saturation of images, and the two prominences are indistinguishable because of its bright intensity and it is different from green and blue spectra in which two prominences are observed clearly. Images from IR (figure 1d) shows that the prominence and corona are not emitting IR wavelength, only, we observed some radiation in range red to near-infrared from the Sun.

Figure 2 is a plot (intensity profile) of the brightness vs pixel position of prominences, representing intensity profile from photosphere to the near edge of prominence. Only in red channel (figure 2b), there are saturation in pixel 1 and in pixel 4 to pixel 16. Because we just concern in measuring the dimensions of prominences, only pixels in edge of prominences are taken into account, and no saturation in those pixels. Figure 2a shows how to calculate the intensity profile from images to enumerate intensity for each channel in each pixels in order to estimate the real length of prominences.

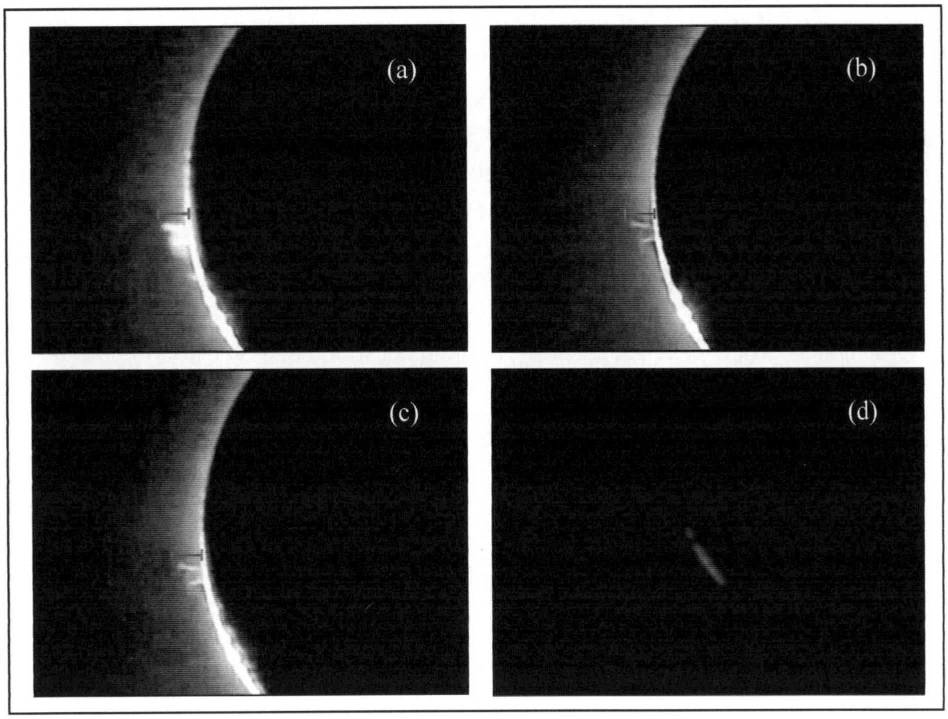

Figure 1. Images of solar prominence in (a) red (b) green (c) blue (d) infrared spectra.

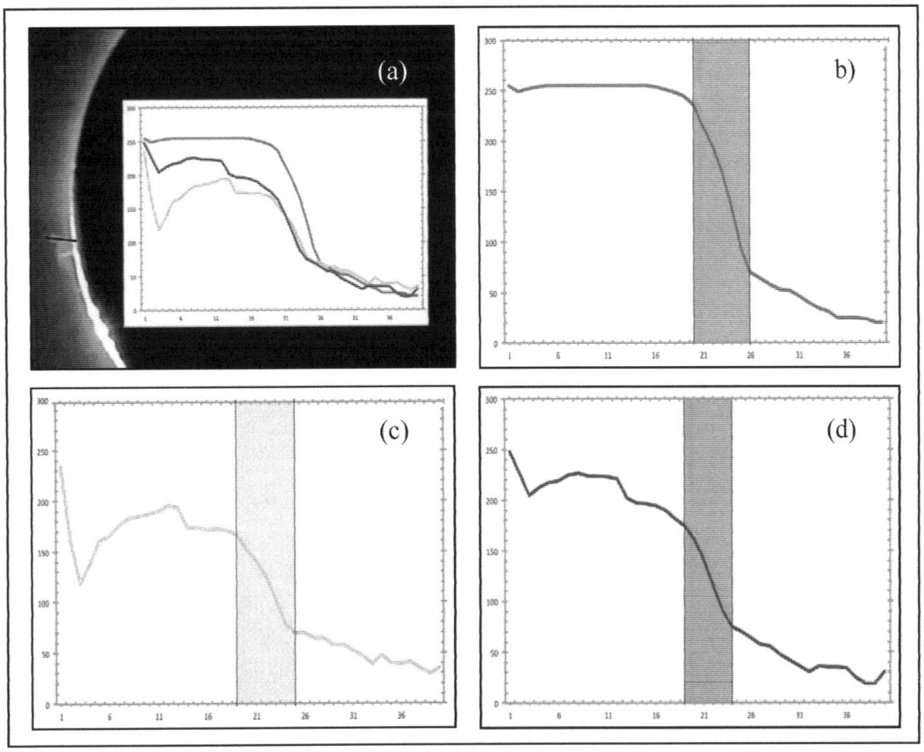

Figure 2. The intensity profile of prominence (a) in red (b) green (c) and blue (d).

Table 1. Pixel length and its real length of prominence.

Spectra	Minimum Length Pixel (px)	Real Minimun Length (km)	Maximum Length Pixel (px)	Real Maximum Length (km)
Red (6200A-7500 Å)	20	28958.66819	26	37646.26897
Green (4950A-5700 Å)	19	27510.73475	25	36198.33549
Blue (4500A-4950 Å)	19	27510.73475	24	34750.40202
Infrared (> 7200 Å)	0	0	0	0

The pixel length of prominence is calculated to the pixel where the intensity drops drastically. Due to unclear edge of prominences, the pixel length of prominence is calculated in ranges (minimum length and maximum length). Tabulation of pixel length and its transformation to real length of prominence is showed in table 1. From intensity profile in figure 2 and results in table 1, it is observed clearly that the red wavelength is brighter and longer than that of other wavelength and it confirmed that the main-radiation sources of prominence is H-alpha [9].

4. Conclusion

Based from our observation and calculation (figure 2 and table 1), we found that the intensity in red wavelength is brighter than that of blue and green, 1.37 times to green and 1.23 times to blue. The real mininimum length of prominence in red is 1.05 times longer than green and blue ones. And, the real maximum length of prominence is 1.08 times longer than blue and 1.04 times longer than green. These results are due to the different temperatures of the prominences and the radiation from H-alpha emission.

Acknowledgment

We thank to Astronomy Department, Faculty of Mathematics and Natural Sciences (FMIPA), ITB, for funding support. We also thank to all friends for beneficial discussion.

References

[1] Xia C and Keppens R 2016 Formation and plasma circulation of solar prominences *APJ* **823** 1
[2] S Parenti 2014 Solar Prominences: Observations *Living Rev. Solar Phys.* **11** 1
[3] P I Papics 2007 *Solar Prominences, Solar Flares and Coronal Mass Ejections* http://papics.eu/SolarPhenomena.pdf
[4] Gibson, et al. 2010 Three-Dimensonal Morphology of A Coronal Prominence Cavity *APJ* **724** 1133-1146
[5] Fuller, et al. 2008 Observing the Unobservable Modelling Coronal Cavity Densities *APJ* **678** 515-530
[6] P Heinzel, I Dorotovic and R J Rutten 2007 The Physics of Chromospheric Plasmas *ASP Conference Series* **368** 27-48
[7] V Bommier , E L Degl'Innocenti, J L Leroy and S Sahal-Brechot 1994 Complete determination of the magnetic field vector and of the electron density in 14 prominences from linear polarizaton measurements in the HeI D3 and Hα lines *Sol. Phys.* **154** 231-260
[8] J L Leroy, V Bommier and S Sahal-Brechot 1983 The magnetic field in the prominences of the polar crown *Sol. Phys.* **83** 135-142
[9] A I Poland and R H Munro 1976 Interpretation of broad-band polarimetry of solar coronal transients - Importance of H-alpha emission *Astrophysical Journal* **209** 927-934

White light corona during total solar eclipse on March 9, 2016

Irfan Imaduddin[1], Evan I Akbar[1, 2] and Gerhana P Putri[2]

[1] Astronomy Study Program, Institut Teknologi Bandung, Ganesha 10, Bandung, 40132 Indonesia
[2] Bosscha Observatory, Institut Teknologi Bandung, Lembang, 40391 Indonesia

E-mail: irfanimaduddin@gmail.com

Abstract. We observed the white-light corona during the total solar eclipse of 2016 March 9 from *Corong Beach*, East Borneo. The solar corona is nearly circular with exception on the southern part of the Sun. Coronal structures are clearly seen. Based on the data, we obtained the Ludendorff flattening index, Nikolsky geometric flattening index, and phases of solar activity (Φ and P) are 0.129, 1.32, -0.716, and 0.573 respectively. Relation between Ludendorff and Nikolsky index, sunspot number, and phases of solar activity were discussed. We also predicted the amplitude of solar cycle 25 to be 196 ± 52 (based on 13-month smoothed monthly data) and 130 ± 42 (based on monthly sunspot number data).

1. Introduction

The solar corona is the outermost layer of the Sun's atmosphere. The bright white corona is visible due to the scattered light of the photosphere by free electrons in the solar corona. Since the brightness is only one-millionth of the photosphere, white light corona (particularly the innermost part, up to about 1.5 solar radii) can only be observed during total solar eclipse or using K-coronagraph. Thus, the TSE has given an opportunity to investigate the corona.

It is widely accepted that the shape and structure of solar corona is changing by time, depends on solar activity or magnetic field variability. On the global scale, magnetic fields are shaping the corona to become asymmetric during minimum solar activity and nearly circular during maximum [1], while on the small scale they are generating basic structures (streamers, helmet streamers, plumes, etc.) and fine structures (voids, loops, radial threads, etc.) of the corona [2].

There is also a relation between shape and brightness of corona and sunspot number. The corona is bright and uniform when the sunspot number is maximum. It is getting dimmer and more elongated towards equator when the sunspot number is minimum [3].

One of the quantitative analyses of coronal shape and structure is proposed by Ludendorff. He was measuring ratio of isophotes diameter on equatorial and polar directions [4]. Another one is proposed by Nikolsky in 1956, he was measuring geometrical characteristics of polar rays and streamers [5].

In this paper we will discuss the structures of white light corona during total solar eclipse on the March 9, 2016 and show the measurement results of the Ludendorff and Nikolsky flattening indexes. We will also study the correlation between Ludendorff flattening index, Nikolsky geometric flattening index, sunspot number and phases of solar activity using the past 60 total solar eclipses in 1851-2010 from data compilation by Pishkalo [6, 7] and monthly sunspot number from World Data Center-SILSO (http://www.sidc.be/SILSO/) [8]. Lastly, we will predict the amplitude of solar cycle 25.

2. Observation and data reduction

The observation of total solar eclipse was carried out on March 9, 2016 at *Corong Beach*, Penajam Paser Utara, East Borneo, Indonesia (1°24′12.4″ S, 116°39′14″ E), 1.5 m above sea level. The totality was last for 1 minutes 51 seconds. The altitude of the Sun was 32,7° in the East. Nineteen images of white light corona were obtained using William Optic Zenithstar 66SD (66 mm, f/5.9) and Nikon D5100. Six best frames were chosen, combined, and processed using Adobe Photoshop CC 2015 and Adobe Ligthroom to bring out the details of coronal structures.

3. Results and discussions

3.1. Coronal structure

The composite image of white light corona shows an almost circular corona with an exception around south pole of the Sun. This suggests that the eclipse occurred more or less at the maximum of solar cycle. All structures of coronal formation are clearly seen (figure 1). At the north and south poles, polar plumes are well developed. The helmet streamers are visible on the high latitude of western side of the Sun, extended up to three solar radius. The helmet streamers are believed to be related to the chromospheric activity [9]. Small prominence and void are visible on the eastern side of the Sun.

3.2. Ludendorff flattening index (ε)

The Ludendorff flattening index ε is an ellipticity parameter of white light corona. To calculate the index, we produced isophotes of white light corona as displayed in figure 1. Since we did not take sky flat field image during the observation, the image is suffered from non-uniform sky background. We measured the index by,

$$\varepsilon = \frac{E_1+E_2+E_3}{P_1+P_2+P_3} - 1 \qquad (1)$$

where E_1, E_2, E_3 and P_1, P_2, P_3 are defined as isophotes diameter on the equator and poles that separated at angles of 22.5°.

The rise of ε is properly approximated with a linear function,

$$\varepsilon = a + b(r_e - 1) \qquad (2)$$

where r_e is the mean radius of the Sun. The value of ε equals to $a + b$ at $r_e = 2$. Therefore, we fitted the flattening index values based on the isophotes and obtained the value of ε is 0.129 at $r_e = 2$ (figure 1).

There is a linear relation between Ludendorff flattening index and monthly sunspot number as shown by Pishkalo using data of 60 total solar eclipses on 1851 to 2012 [6]. We are re-calculated Ludendorff flattening index – daily (W_d) and monthly (W_m) sunspot number relation by adding the new value of ε (see figure 2a and 2b). Daily (W_d) and monthly (W_m) sunspot number datas were taken from WDC-SILSO. The results are quite similar, $a + b = 0.2357 - 0.0008\ W_d$ for daily and $a + b = 0.2327 - 0.0007\ W_m$ for monthly sunspot number.

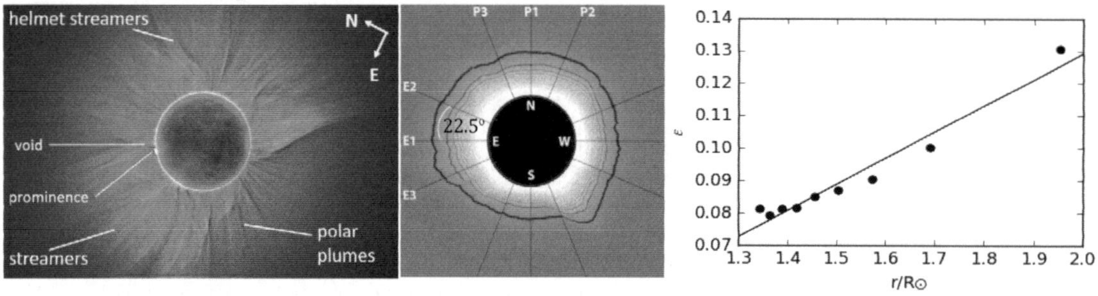

Figure 1. Left: Composite image of white light corona during total solar eclipse on March 9, 2016, with coronal structures marks. Center: Isophotes of the white light corona. Right: Plots of Ludendorff flattening index ε as a function of solar radius measured from isophotes.

We also study the relation of Ludendorff flattening index and solar cycle. Ludendorff introduced the phase of solar activity Φ and P as follows,

$$\Phi = \frac{T_{ecl} - T_{min}}{|T_{max} - T_{min}|} \tag{3}$$

$$P = \frac{T_{ecl} - T_{min1}}{T_{min2} - T_{min1}} \tag{4}$$

where T_{ecl} is the time of eclipse; T_{max} and T_{min} are the times of solar cycle maximum and minimum adjacent to the eclipse; and T_{min1} and T_{min2} are the time of solar cycle minimum at the beginning and and the end of the cycle respectively [4,7]. We used data from WDC-SILSO to determine the time of solar cycle minimum and maximum, yielding $\Phi = -0.716$ and $P = 0.573$. These results suggest that the eclipse occurred after the maximum of solar cycle. Using the datas from Pishkalo and WDC-SILSO, we found the relation between Ludendorff flattening index and phases of solar activity. The results are shown in figure 2c and 2d, giving approximation of harmonic functions $a + b = 0.2385 - 0.0193\Phi - 0.1973\Phi^2$ and $a + b = 0.2681 - 0.6334P + 0.6707P^2$.

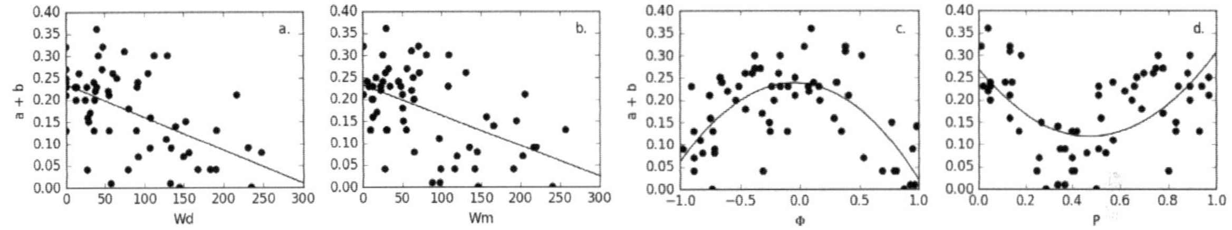

Figure 2. Relation between Ludendorff flattening index and daily sunspot number (2a), monthly sunspot number (2b), phases of solar cycle (2c and 2d).

3.3. *Nikolsky geometric flattening index (H)*

The Nikolsky geometric flattening index is calculated based on the extent and position of streamer and polar rays. It is expressed as

$$H = \frac{\Pi_N + \Pi_S}{2\pi} + \frac{\pi}{\frac{1}{n}\Sigma(\varphi_n + \beta_n) + \frac{\pi}{2}} \tag{5}$$

where Π_N and Π_S are the extents of the northern and southern polar ray systems (plumes) along the limb, n is the number of streamers, φ is the heliographic latitude of the bases of streamers; β is the angle of streamers to the equator. From the composite image on figure 1, we obtained H is equal to 1.32.

We used the above value along with 77 data of TSEs in 1860-2010 [7] and sunspot number data from WDC-SILSO to find the relation in figure 3a and 3b. In general, the maximum value of H is expected during minimum sunspot number according to the relation $H = 1.6095 - 0.0013\, W_d$ for daily sunspot number and $H = 1.6514 - 0.0018\, W_m$ for monthly sunspot number. We calculated phases of solar activity Φ and P using equation (3) and (4), yielding best fit of harmonic function $H = 1.6219 - 0.1794\Phi - 0.4192\Phi^2$ and $H = 2.0138 - 3.3562P + 3.46893P^2$ as shown in figure 3c and 3d.

3.4. *Relation between ε and H*

As it it done by Pishkalo [7], we investigated the relation between Ludendorff and Nikolsky flattening index. As shown in figure 3e, a linear relation is, $H = 0.6661\ \varepsilon + 1.3853$. This figure is useful to calculate Ludendorff flattening index in case the white light corona image is heavily suffered from non-uniform background or foreground (e.g. partially cloudy weather during total solar eclipse).

3.5. *Prediction of solar cycle 25*

The relation between Ludendorff flattening index, sunspot number and phases of solar activity as shown before, suggest that the flattening index can be used to forecast the amplitude of the next solar cycle. To do this, we plotted Ludendorff flattening index from 10 total solar eclipses which occurred near the maximum solar cycle during 1851-2012. Then used two datas of sunspot number from WDC-SILSO, i.e. monthly and 13-month smoothed monthly sunspot number to construct figure 3f and 3g. The best approximation of maximum sunspot number based on monthly and 13-month smoothed monthly sunspot number data are $Wmax = 159.22(a + b) + 109.43$ and $Wmax = 264.1(a + b) + 162.61$ respectively. The results are 196 ± 52 for 13-month smoothed monthly data and 130 ± 42 for monthly sunspot number. In general, these values are higher compared to the published prediction of other authors [10, 11, 12, 13].

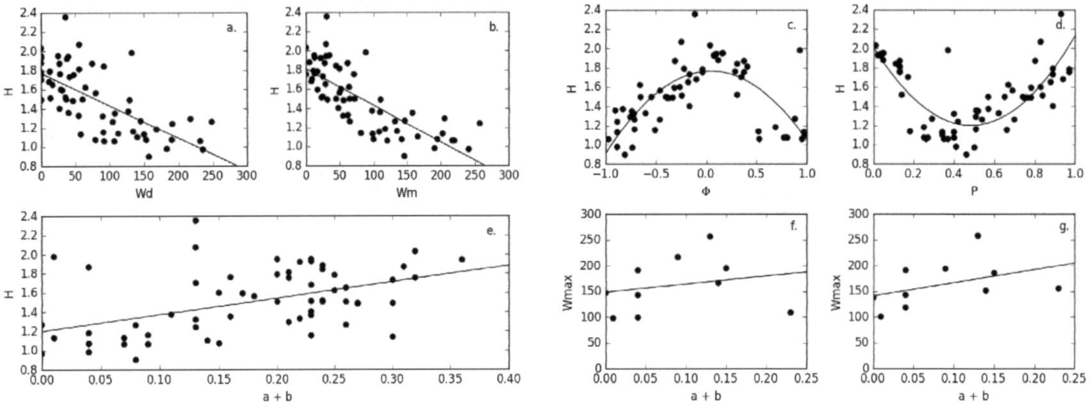

Figure 3. Relation between Nikolsky flattening index and daily sunspot number (3a), monthly sunspot number (3b), phases of solar cycle (3c and 3d), relation between Ludendorff and Nikolsky flattening index (3e), plots of Ludendorff flattening index and maximum 13-month smoothed monthly sunspot number (3f), and plots of Ludendorff flattening index and maximum monthly sunspot number (3g).

Acknowledgments

We thank Muhammad Rayhan for his technical assistance in image processing.

References

[1] Loucif M L and Koutchmy S 1989 *J. Astron. Astrophys. Suppl. Ser.* **77** 45–66
[2] Rusin V, Druckmüller M, Minaroviech M and Saniga M 2008 *J. Astroph and Space Science* **313** 345-49
[3] Hansky A 1897 *Bull. de l'Academie des Sciences* (St. Petersbourg) **6** 251
[4] Ludendorff H 1928 *Uber die Abhangigkeit der Form der Sonnenkorona ven der Sonnenfleckenhaufigkeit* (Berlin: Sitzzungsber. Preuss. Akad. Wiss.) pp 185–92
[5] Nikolsky G M 1956 *J. Astron. Zh.* **33** 84–6
[6] Pishkalo M I 2011 *J. Solar Phys.* **270** 347–62
[7] Pishkalo M I 2012 *J. Kinematics Phys. Celestial Bodies* **8** 183-88
[8] SILSO World Data Center 2016 *on-line International Sunspot Number Catalogue (Royal Observatory of Belgium)* http://www.sidc.be/SILSO/
[9] Stoev A, Shopov Y, Muglova P, Kiskinova N, Varbunova Y, Michaylova Z I and Zelkov R 2002 *Proc. Int. Conf. Varna (Bulgaria)* ed. D N Mishev and K J H Phillips p 57
[10] Pishkalo M I 2008 *J. Kinematics Phys. Celestial Bodies* **24** 242-47
[11] Hamid R H and Galal A A 2013 *J. Advanced Research* **4** 275-78
[12] Li K J, Feng W and Li F Y 2015 *J. Atmos. Solar-Terrestrial Phys.* **135** 72-6
[13] Rigozo N R, Souza Echer M P, Evangelista H, Nordemann D J R and Echer E 2011 *J. Atmos. Solar-Terrestrial Phys.* **73** 1294-99

Analysis on atmospheric pressure, temperature, and wind speed profiles during total solar eclipse 9 March 2016 using time series clustering

Lala Septem Riza[1], Yaya Wihardi[1], Enjang Ali Nurdin[1], Nanang Dwi Ardi[2], Cahyo Puji Asmoro[2], Agus Fany Chandra Wijaya[2], Judhistira Aria Utama[2], and Asep Bayu Dani Nandiyanto[3]

[1] Department of Computer Science Education, Universitas Pendidikan Indonesia
[2] Department of Physics Education, Universitas Pendidikan Indonesia
[3] Department of Chemistry Education, Universitas Pendidikan Indonesia

E-mail: `lala.s.riza@upi.edu`

Abstract. Air temperature, pressure, and wind speed measurements on the surface taken during the Total Solar Eclipse (TSE) of March 9, 2016, are made. They were taken in Terentang Beach, Bangka Island, Indonesia. In this paper, we propose to analyze them by using time series clustering. The following steps are conducted: data collecting, splitting, smoothing, distance calculation, and clustering. The final results show cluster memberships of the three parameters on 3 time frames: one day before, the TSE day, and one day after. After doing some simulations, it can be seen that the profiles of temperature and pressure on the TSE day are on the same cluster while the wind-speed profile on the TSE day is the same as on the one day after.

1. Introduction

Analysis on temperature, pressure, and wind speed profiles during the solar eclipse has been done by some researchers, such as in [1–3]. In other words, it is an important task to do in astrophysics since solar eclipse, especially for TSE, provides rare opportunity for studying on the eclipse effects focused on the upper and lower atmosphere.

Basically, similar to the previous studies, this research is also aimed to analyze the profiles during TSE of March 9, 2016 in Bangka Island, Indonesia. However, this task is not done based on common approaches in the astrophysics field, but we will discover and analyze it from another perspective, which is based on machine-learning (ML) approaches for time series analysis. ML can be defined as the field of scientific study focusing on algorithms that are able to learn from data [4]. In this research, we use an essential technique in ML: clustering. Furthermore, to accomplish the experiment, we utilize the programming language *R* [5].

The remainder of this paper is structured as follows. Section 2 briefly introduces to *R* ecosystems and time series clustering. In Section 3 and 4, we describe the experimental design, its implementation on *R*, and results. Finally, Section 5 concludes the paper.

2. R ecosystem and time series clustering

R ecosystem provides a programming language and scientific-analysis environment that consists of more than 8000 packages for statistics, visualization, machine learning, etc. Furthermore, interested readers can find some useful books introducing *R*, such as [6]. In *R*, there are two following repositories saving *R* packages: the Comprehensive *R* Archive Network (CRAN, http://cran.r-project.org/) and Bioconductor (http://www.bioconductor.org/). For example, in CRAN we can find following packages included in the Task View "Machine Learning:" *frbs* [7, 8] and *RoughSets* [9].

Time series clustering is an unsupervised learning aimed to separate a set of time series data according to their distances or similarities. In this case, we are using an *R* package, called *TSclust* [10]. It contains four sets of dissimilarity measures: model-free (e.g., dynamic time warping (DTW)), model-based approaches (e.g., Maharaj distance), complexity-based approaches (e.g., compression-based dissimilarity measures), and prediction-based approaches. Moreover, interesting articles that discuss time series clustering can be found in the literature, such as in [11].

3. Experimental design

This research follows the experimental design illustrated in figure 1. Basically, it contains 5 steps as follows:

(i) Data acquisition: In this step, we take the data consisting of temperature (in $^{\circ}C$), pressure (in *milibar*), and wind speed (in *inchHg*) at Terentang Beach (i.e., $2^{\circ}26'38.77''$ South, $106^{\circ}19'36.01''$ East and on the height of 2.5 m above sea level). The data, which contain 163893 points, are available from at 6:10:41 WIT (Western Indonesia Time, (UTC+8)) on March 8, 2016 until at 9:52:01 WIT on March 10, 2016.

(ii) Data splitting: This step is aimed to divide the data into 3 time frames corresponding to 3 days (i.e., Before, TSE, and After) with the time range: 6 : 10 : 41 and 9 : 52 : 02 WIT. This range already covers the TSE's time. So, we will obtain 6 curves for all parameters.

(iii) Data filtering/smoothing: To make the profiles smooth, we use the linear filtering technique on time series data.

(iv) Dissimilarity calculation: We calculate distances among 3 time frames for each parameter. To ensure the results, we conduct 7 dissimilarity methods: Euclidean (EUCL), DTW, temporal correlation coefficient (CORT), integrated periodograms (INT.PER), invertible ARIMA (AR.PIC), invertible and stationary ARMA (AR.MAH), and complexity invariant (CID).

(v) Data clustering: In this research, we are using Hierarchical Clustering, which is a method of cluster analysis to build a hierarchy of clusters.

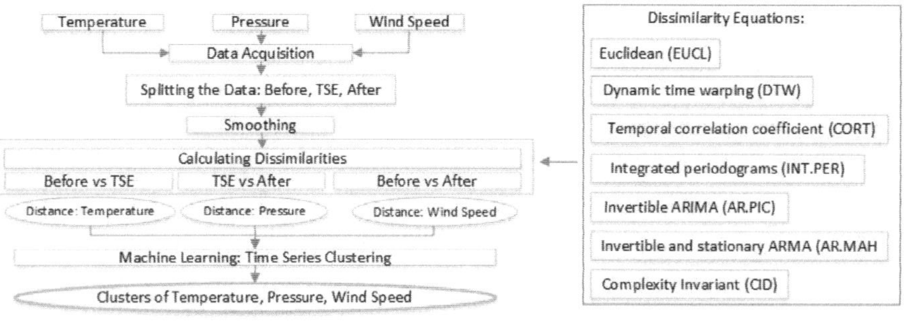

Figure 1. Experimental design using time series clustering.

4. Experiments and results

In this experiment, the *R* package *TSclust* is used to analyze our data. So, the following codes are run on the *R* environment. Furthermore, by considering 5 steps as illustrated in Section 3 we conduct the following experiments:

(i) Data acquisition: We assume that the data containing temperature, pressure, and wind speed are saved in *datasetTSE.csv*, and will be loaded into the *R* object *data* by the below code: `R> data <- read.csv("datasetGMt.csv", header=FALSE)`

(ii) Splitting into 3 time frames: It can be done by the following code:
```
R> dayBefore <- data[start[1]  :  end[1], ]
```
It can be seen that we define starting and ending points, which are at $6:10:41$ and $9:52:02$ for each day.

(iii) Filtering the data: After obtaining the profiles of temperature, pressure, and wind speed for each day, we do smoothing by typing the below code:
```
R> dayBefore_FilterT <- filter(dayBefore[, 2], filter = rep(1/500, 500))
```
It should be noted that the above code is only used for filtering the temperature profile of the first time frame (one day before). For other parameters, we can do the similar code. We can also plot the profile, for example for the temperature curve (see figure 2):
```
R> plot(dataT, type="l")
```

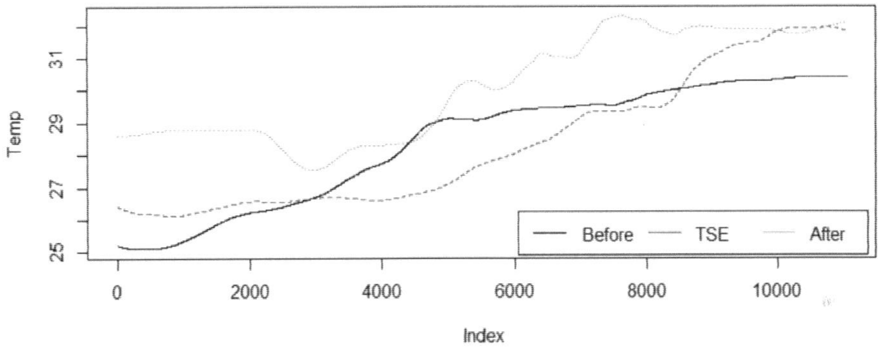

Figure 2. The profiles of temperature after splitting and smoothing.

(iv) Calculating Dissimilarities: The distances among the time frames for each parameter will be determined by the following code:
```
R> T_comb <- cbind(T_Before, T_TSE, T_After)
R> T.eucl <- diss(T_comb, "EUCL")
```
It should be noted that because of the limited space, we just show the code for calculating dissimilarity of temperature by using Euclidean equation. The other parameters can be done by similar codes by following the ways on [10]. The result of this step can be seen in table 1.

(v) Clustering on each Parameter: After obtaining the distances, we can now determine the membership of clusters by typing the following code:
```
R> T.eucl.hclus <- cutree(hclust(T.eucl), k = 2)
```
For the other methods of dissimilarities can be done by executing similar codes. The final result can be summarized as illustrated in table 2. It can be seen that we can state that the temperature and pressure profiles on the TSE day are quite similar to the ones on one day before. On the other hand, the wind speed profile on the TSE day is closed to the one on one day after.

Table 1. The distances of temperature, pressure, and wind speed among 3 time frames using the Euclidean equation.

Distances of Temp. (T)			Distances of Press. (P)			Distances of Wind Speed (WS)		
	T_Before	T_TSE		P_Before	P_TSE		WS_Before	WS_TSE
T_TSE	**111.392**		P_TSE	**19.067**		WS_TSE	121.576	
T_After	208.011	208.323	P_After	215.670	219.899	WS_After	153.632	**93.600**

Table 2. The cluster membership of temperature, pressure, and wind speed for 3 time frames.

	Cl. Membership of T			Cl. Membership of P			Cl. Membership of WS		
	Before	TSE	After	Before	TSE	After	Before	TSE	After
EUCL_hclust	1	1	2	1	1	2	1	2	2
CORT_hclust	1	1	2	1	1	2	1	2	2
DWT_hclust	1	1	2	1	1	2	1	2	2
INT.PER_hclust	1	1	2	1	1	2	1	2	2
AR.PIC_hclust	1	1	2	1	2	2	1	1	2
AR.MAH_hclust	1	1	2	1	2	2	1	1	2
CID_hclust	1	1	2	1	1	2	1	2	2

5. Conclusion and future work

The research can be summarized as follows: (i) a strategy to analysis temperature, pressure, and wind speed by using time series clustering has been presented; (ii) according to the experiments, the profiles of temperature and pressure on the TSE day are the same behavior, represented by the same cluster while the wind-speed profile on the TSE day is the same characteristic as on the one day after. As future work, we plan to extend the analysis on other areas of Indonesia influenced by partial solar eclipse and total solar eclipse, so that we will have a comprehensive understanding on the profile changes.

References

[1] Anderson R, Keefer D and Myers O 1972 *Journal of the Atmospheric Sciences* **29** 583–587

[2] FernÁndez W, Castro V and Hidalgo H 1993 *Earth, Moon, and Planets* **63** 133–147

[3] Wang K Y and Liu C H 2010 *Geophysical Research Letters* **37**

[4] Kovahi R 1998 *Machine Learning* **30** 271–274

[5] Ihaka R and Gentleman R 1996 *Journal of computational and graphical statistics* **5** 299–314

[6] Team R D C 2008 *An Introduction to R* (Vienna, Austria: R Foundation for Statistical Computing)

[7] Riza L S, Bergmeir C, Herrera F and Benítez J M 2015 *Journal of Statistical Software* **65** 1–30

[8] Riza L S, Bergmeir C, Herrera F and Benítez J M 2014 *Fuzzy Systems (FUZZ-IEEE), 2014 IEEE International Conference on* (IEEE) pp 2149–2155

[9] Riza L S, Janusz A, Bergmeir C, Cornelis C, Herrera F, Ślezak D, Benítez J M *et al.* 2014 *Information Sciences* **287** 68–89

[10] Montero P and Vilar J 2014 *Journal of Statistical Software* **62** 1–43

[11] Warren Liao T 2005 *Pattern Recognition* **38** 1857–1874

The influences of solar radiation changes on the meteorological variables during the total solar eclipse of 9th March 2016 in Central Bangka, Indonesia

Ryantika Gandini[1], NanangDwi Ardi[1], M. Iid Mujtahiddin[2]

[1]Physics Department, Universitas Pendidikan Indonesia,
Jl. Dr. Setiabudhi No 229 Bandung 40135
[2]BMKG StasiunGeofisika Bandung, Jl. Cemara No. 66 Bandung

E-mail: ryantikagandini@gmail.com, nanang_dwiardi@upi.edu

Abstract. Observations of the meteorological variables have been conducted in Terentang coastal area, Central Bangka which is located in totality path of Total Solar Eclipse on March 9, 2016. These measurements were made before, during, and after the solar eclipse using a portable automatic weather station with 1 Hz data recording to investigate the influence of total solar eclipse on the incoming solar radiation and other weather variables. Due to the cloudiness at the first phase of the eclipse, the reduction of the radiation was not proportional to the percentage of the solar obscuration. Along with the disappearance of thick clouds, solar radiation changes reach the lowest value from 123 W/m2 to zero point at 5 minutes before the maximum phase. In contrast to the radiation decrement, decreasing air temperature was not comparable to the solar obscuration because it was not only determined by radiation, but also surrounding environment. High relative humidity, on average 89%, was discovered as an effect of the low atmospheric pressure which was affected by the insolation decline. The effect of total solar eclipse can be seen more clearly with spectrum analysis using Fourier transformation to identify periodogram patterns of each meteorology variable. This transformation produced a spectrum's peak totality which is higher than before and after the event on the insolation. The spectrum represents that total solar eclipse has considerable effect to the incoming solar radiation and others which is indicated by the change of the amplitude.

1. Introduction

The abrupt change of the incoming solar radiation (insolation) during solar eclipse has inspired meteorologist to study the response of the atmosphere [1], the study includes meteorology variables such as solar radiation, air temperature, atmospheric pressure, relative humidity, wind speed and wind direction. The dramatic reduction of solar radiation reach the lowest value at maximum phase or contact, analogues with solar obscuration that has been found by [2], similar to the results obtained by [1]. Although the results of most meteorological studies provide similar pattern of temperature changes, the solar eclipse is unique event because the precise drop may differ depending on several factor (e.g. timing, synoptic situation, surrounding environment, percentage of solar occultation etc) [3].

This research aimed to study of the effect of total solar eclipse (TSE) on March 9, 2016 by monitoring changes in the weather parameters before, during and after TSE by observing the changes and performing spectrum analysis using Fourier transformation. Besides, the observation aims to correct a reduction of insolation model [4] which does not take the atmosphere and other factors into account [5]. We aim to obtain insolation reduction model which considering the correction of the atmosphere and clouds. Calculation of percentage numbers on this was study done by comparing it with normal day data.

2. Measurements sites, weather background and instrumentation

Bangka is one of the islands in Indonesia which is traversed by the totality path of 9th March's solar eclipse [6]. This research was conducted in Terentang Coastal area, Central Bangka (-2.44S, 106.32E). It is located far from huge building and surrounded by Natuna Sea in the north and east.

In general, on the day of the eclipse, the sky was covered with thick clouds. Thick clouds covered the sky 26 minutes before the eclipse until 12 minutes after the first contact. The thick clouds gradually disappeared nearly in the totality phase. The observations of the meteorological variables were conducted using portable automatic weather station with 1 Hz data recording. The occurrence times of total solar eclipse in Terentang Beach are shown in table 1.

Table 1. Occurence times of TSE in Terentang Beach [7].

Event	Time (UTC)	Time (Local Time)
Start of partial eclipse	23:20:58	06:20:58
Start of total eclipse	00:22:11	07:22:11
Maximum eclipse	00:23:07	07:23:07
End of total eclipse	00:24:02	07:24:02

3. Results and analyses

Our data are changes in meteorological parameters. Total solar eclipse in Terentang Beach shows significant effects on solar radiation or insolation which control the weather [8][9]. Generally there are changes (fluctuations) in the increasing insolation pattern during the eclipse (marked in red line in figure 1(a)). In the initial phase of the eclipse, the solar insolation decreased by 61.36% compared with the previous day insolation. In 2nd and 3rd phase, it decreased by 100%, while in 4th phase it decreased by 58%.

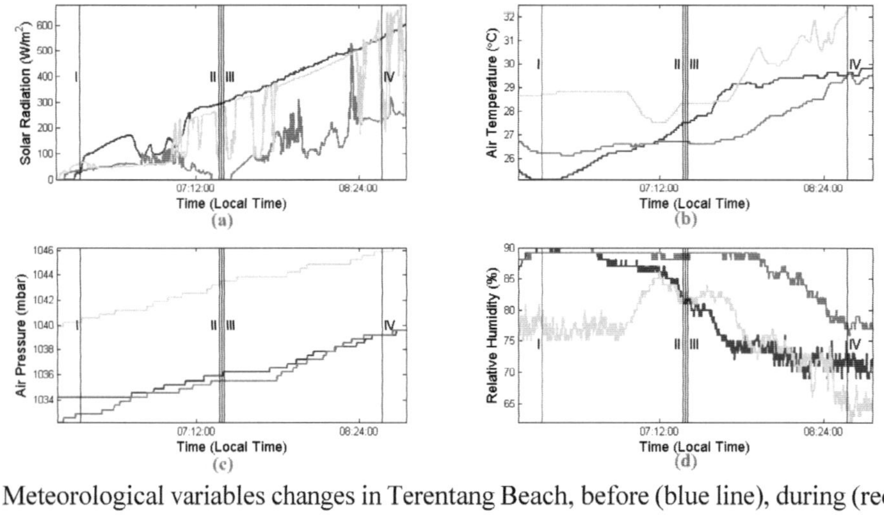

Figure 1. Meteorological variables changes in Terentang Beach, before (blue line), during (red line) and after (green line) TSE March 9, 2016. (a) Solar radiation/insolation, (b) Air temperature, (c) Air pressure, (d) Humidity.

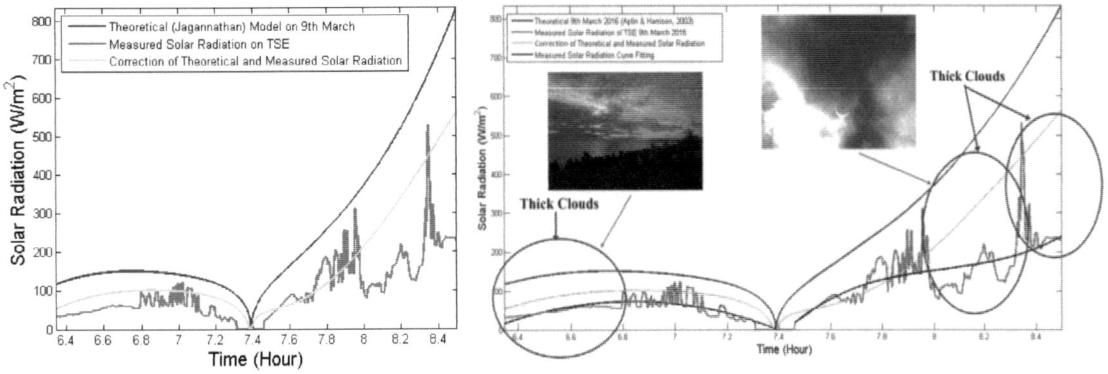

Figure 2. The results of corrected model (left). The existence of thick clouds formed during TSE indicated by a green circle (right).

Corrected model which has been carried out generates the insolation reduction model which has similar but lower value than Jagannathan model [5] (figure 2 (left)). Fluctuations of insolation for 20 minutes at the initial phase of eclipse occur due to thick clouds in the first phase of the eclipse. The presence of thick clouds during the measurement also is demonstrated by the drastic reduction in insolation compared to the theory and models (indicated by a green circle in figure 2 (right)).

Based on the corrected model, due to thick clouds during the TSE, the insolation reduces around 50.98%-66.45%. This value indicates that the presence of the clouds during the solar eclipse was not able to remove all radiation. Thus direct observation without any protection, despite the thick clouds, can harm the eyes.

The reduction of solar insolation has caused the decrease of energy received by the Earth so that the air temperature dropped by 0.1^0C 11m16s after the totality. The pattern of temperature decrease is influenced by the circumstances on the site. It is situated on the beach surrounded by oceans which causes differences in the rate of temperature decline due to the heat capacity of the sand and water.

In addition, it is also due to the eclipse time that was in the early morning when the air temperature was still cold and the nearby vegetation which was capable of storing heat. Thus the air temperature did not decrease significantly at that time. While the air pressure (figure 1(c)) decreased by 0.334mb at the totality time. The pattern of air pressure was lower than the normal air pressure, or in other words the air pressure was more inert than in normal day.

Decrease in air temperature accompanied by decrease in air pressure resulted significant changes to humidity, which continued to decline toward noon. The high value of humidity (89%) began to decline at 7:55:35 AM (UTC+7) (figure 1(d)). The high value of the air humidity signifies the rich content of the water vapour due to the absence of the solar energy that reached the Earth. So that at the totality, air humidity has greater value than in a normal day.

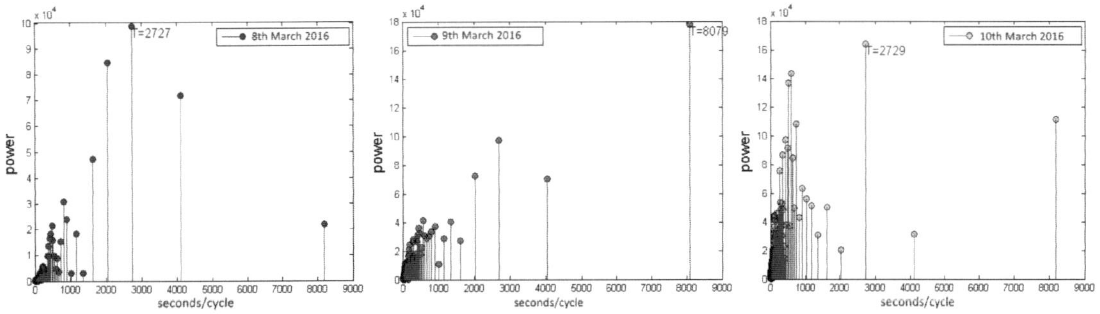

Figure 3. Periodogram of solar radiation. Period on the graphs are in seconds.

Solar insolation reduction has changed the periodogram pattern of insolation, air temperature, pressure and relative humidity. Through Fourier transformation, we found changes in weather parameters during TSE. Total solar eclipse changed the period of maximum insolation into every 134 minutes 39 seconds. This indicates a major influence of TSE on the pattern of maximum insolation changes, around 66% compared to the normal day. Period changes of other meteorological variables are shown in table 2 which differ only 1.25% from normal day.

Table 2. Period of Spectrum Pattern of Each variables

Meteorological Variables	Period		
	8th March	9th March	10th March
Solar Radiation	45m 27s	134m 39s	45m 48s
Temperature	68m10s	67m19s	68m13s
Pressure	45m 27s	44m 52s	45m 48s
Relative Humidity	68m 10s	67m19s	68m13s

Changes of period of maximum activity that occurs during the TSE indicate TSE influence on weather parameters besides the solar radiation reduction, which is characterized by a reduction in the period of air temperature, air pressure and relative humidity.

4. Conclusions

Corrected models have shown that the presence of clouds during TSE reduce solar radiation by 50.98% -66.45%. This shows that the presence of thick clouds will not be able to protect the eyes from solar radiation. We conclude that TSE significantly influence the weather. Changes due to TSE are clearly seen after conducting Fourier transformation to all data which show a drastic change in the period periodogram insolation by 66%, while other weather parameters only showed a change of 1.25%.

References

[1] Founda D et al. 2007 Atmospheric Chemistry and Physics 7 5543–53
[2] Ahrens D, Iziomon M G, Jaeger L, Matzarakis A, and Mayer H 2000 *Meteorologische Zeitschrift* **10(3)** 215-223
[3] Gerasopoulus E *et al.* 2008 *Atmospheric, Chemistry and Physics* **5** 5205-20
[4] Aplin K L and Harrison R G 2003 Meteorological effects of the eclipse of 11 August 1999 in cloudy and clear conditions *Proceedings: Mathematical, Physical and Engineering Sciences* 353-371
[5] Jagannathan P, Chacko O, and Venkiteshwaran S P 1956 *Indian J. Meteorology and Geophys* 93-98
[6] BMKG 2016 http://www.bmkg.go.id/BMKG_Pusat/Geofisika/Tanda_Waktu /GERHANA_MATAHARI_TOTAL_9_MARET_2016.bmkg
[7] Xjubier 2016 http://www.xjubier.free.fr (accessed on 2016)
[8] Tjasyono B and Harijono S W 2008 *Meteorologi Indonesia vol 2* (Jakarta: Badan Meteorologi dan Geofisika) p 1
[9] Tjasyono B 2004 *Klimatologi* 2 ed (Bandung: Penerbit ITB) p 11

A meteorological study of the sea and land breezes in Bangka Indonesia during the total solar eclipse on March 9, 2016

Nanang Dwi Ardi[1], Yuyu Rachmat Tayubi[1], Ryantika Gandini[1], Cahyo Puji Asmoro[1], Dini Nurfiani[2], Agus Fany Chandra Wijaya[1], Taufik Ramlan Ramalis[1]

[1]Earth and Space Laboratory, Physics Department, Universitas Pendidikan Indonesia, Jl. Dr. Setiabudhi No 229 Bandung 40154
[2]Earth Observatory of Singapore (EOS), Nanyang Technological University
50 Nanyang Avenue, Block N2-01a-15,
Singapore 639798

E-mail: nanang_dwiardi@upi.edu

Abstract. Of all mesoscale phenomena, sea and land breezes have been the most studied, both observationally and theoretically. During the total solar eclipse (TSE), some parameters vary anomalously in response to the occurrence of TSE on 9th March 2016 in Bangka, Indonesia. As it occurred during morning hours, it is considered to be much more significant because of its effect in sea and land breezes pattern, especially in term of lag time. We use Portlog Rainwise Automatic Weather Station which is able to measure data in 1 Hz accuracy. Comparison of hourly observations during first contact until final contact result a slow cooling temperature reduction of 0.1°C until totality phase. A significant weakening in wind speed of about 2.5 ms^{-1} was observed clearly from the instrument. Unfortunately, the reduction is not coincided with the change of wind direction until the third contact. A near constancy in temperature and the wind's behaviour during TSE result delay time in sea and land breezes pattern after the totality. The lag time occurred in three days measurement data, especially sea breeze onset on the next morning day. The lag time in sea breeze was about one hour retardation. The effect of eclipse on the onset timing of sea and land breeze may be attributed to the proportion of heat accumulation due to the obscuration.

1. Introduction

One of the natural phenomena that have a dominant influence on parameter changes in coastal areas and small islands is the meteorological phenomenon. The influential phenomenon can be monitored as early as possible through a variety of field measurements [1].

Sea and land breeze is one result of the impact of changes in meteorological parameters. The sea breeze system begins at some distance offshore from the coast and expands further offshore [2]. The sea breeze is a recognized occurrence, which disturbs seaside areas primarily during the warm season. It is driven by the temperature change between land and sea, which characteristically peaks during the afternoon hours. Effects of the sea breeze include local changes in temperature, humidity, wind speed, wind direction, cloud cover, and sometimes precipitation.

Changes in solar irradiance, temperature, wind and other variables caused by a total solar eclipse are not quite the same as those occurring during the transition from day to night, because during the diurnal changes the variation of the solar elevation angle is large and gradual. During an eclipse, the Sun is hidden by the moon in a few minutes; therefore its elevation angle does not change much [3]. The upsurge of wind speed was attributed to a cold front in the penumbra and deceleration of wind speed was attributed to stabilization of surface layer following the temperature drop and suppression of turbulent process [4]. Moreover, retrospective analysis in deduced wind and temperature change was introduced by Clayton more than a century ago [5].

2. Methods

Central Bangka Regency is regency in the province of Bangka Belitung, Indonesia. This area was chosen due to the best view in term of Total Solar Eclipse (TSE) path on March 9, 2016. The measurement site is selected are shown below which is based on obscuration predicted model by Jubier, 2016 [6] (figure 1), where the umbral path is also shown. By applying *Portable Weather Logger* (*Portlog™) Rainwise Automatic Weather Station* (PRAWS) in 1 Hz accuracy for 163893 data (figure 2), solar radiation, air temperature, air pressure, and relative humidity have been successfully measured in three days (figure 5) [7]. This instrument was prepared on seashore of Terentang Beach with longitude 106°19'34.31"E and latitude 2°26'37.38"S, roughly 10 meters from coastline.

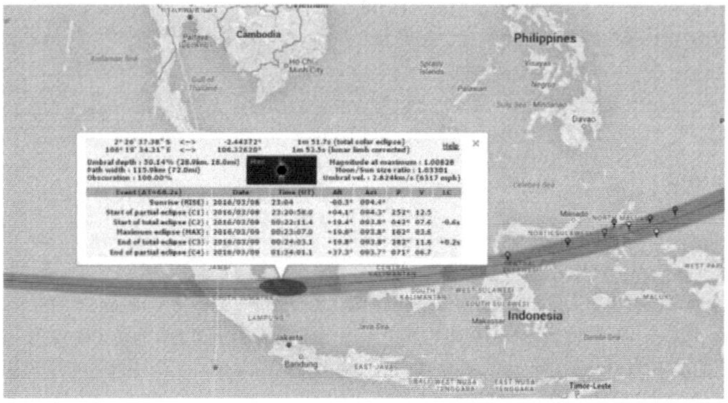

Figure 1. Measurement site of weather parameters in Central Bangka (http://xjubier.free.fr/)

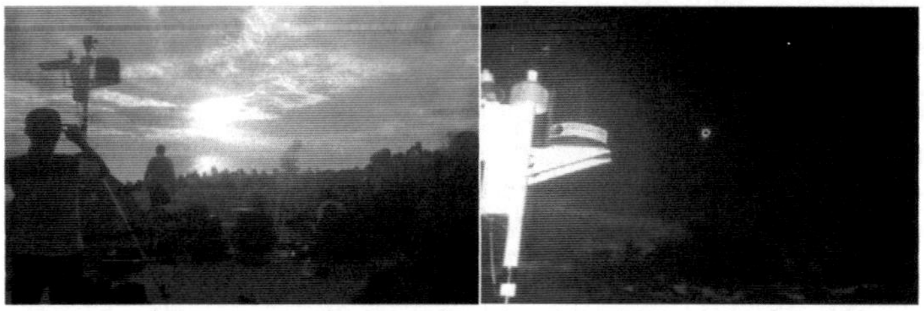

Figure 2. Portlog™ Rainwise Automatic Weather Station (PRAWS) in 1 Hz accuracy; in between thousand visitors (left side); during totality of TSE (right side).

3. Result and discussion

Figures 3 and 4 show the change of wind behaviour in three days measurements. The behaviour was indicated by its speed and direction. For the speed, wind data shows a deceleration from 1.9 m.s^{-1} at 2:07:57 Local Time (LT) and becomes zero at 7:27:00 LT in March 9, 2016. The wind drop value during transition phase of sea and land breeze affected on lag time, especially in speed term. For example, after 4 minutes totality, wind speed returned to the actual pattern of the previous day. Unfortunately, we did not have an exact time on average onset of sea and land breeze in the site. Not only in morning time, the wind at noon was also disturbed in its speed. Then, it induced that the sea and land breeze have experienced retardation more than one hour. Extension of time occurs in land breeze when swapping to the sea breeze and ends when the land gets warm at the end of totality [8]. Unfortunately, this time

Journal of Physics: Conference Series **771** (2016) 012011	doi:10.1088/1742-6596/771/1/012011

extension was not accompanied by a significant drop in temperature (only 0.1 °C) during the total phase (figure 5).

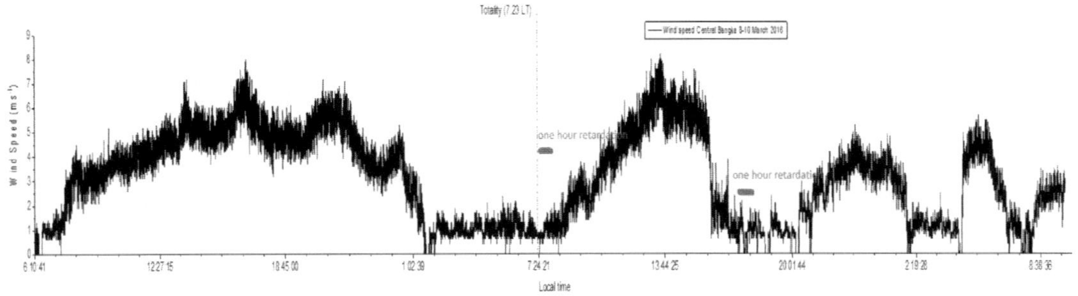

Figure 3. Day-to-day variations in wind speed from 8 until 10 March 2016.

The disturbance in wind condition was clearly seen on three days variation data (figure 4) which is presented in wind rose diagram mode. The diagram depicts accumulation in speed and direction of the wind. At the measurement site, we remarked that prevailing direction of the wind was either from the northwest or from the southwest and the mean of wind speed was 3.1 $m.s^{-1}$ with the maximum at 8.2 $m.s^{-1}$. The condition would be valid on one day before and one day after totality (figure 4). However, this condition did not apply to the total solar eclipse phase which overall showed wind direction to the southwest and the peak dropped to 4 $m.s^{-1}$. The reversal of wind direction during the totality is a response induced by the cooling of the surrounding environment. Then, we found that the onset of sea breeze over the site was delayed on the next day about one hour. It seems that the time delay is the shift period between sea breeze dropping to zero power and land breeze reaching modest wind speed.

In general, the lag time of sea breeze is affected by slow cooling during obscuration. Fortunately, this common assumption is indicated by several weather parameters data which have been simultaneously measured by the wind sensor (figure 5). Nevertheless, the delayed on sea and land breeze might be caused by several environmental factors like cloudiness, day-to-day variation of weather parameters and solar obscuration (indication was clearly figured on stability during first contact until the end of obscuration) (figure 5). The three parameters (air temperature, air pressure, relative humidity) show a static stability which is caused wind remaining stable on the fixed orientation during totality. For example, the temperature and the pressure are a near constancy during the embezzlement. Similarly, humidity has a stable value at 90% before the end of 4th phase. In contrast to the three previous parameters, solar radiation turns to zero value during the totality phase. This behaviour was not in accordance with the daily variation of irradiance in the mornings which it tends to be linear pattern. This condition during totality might decrease in fluctuation generated turbulence in the air condition compared to normally diurnal variation. The sign of reduction turbulence was emphasized by wind reduction in speed and orientation.

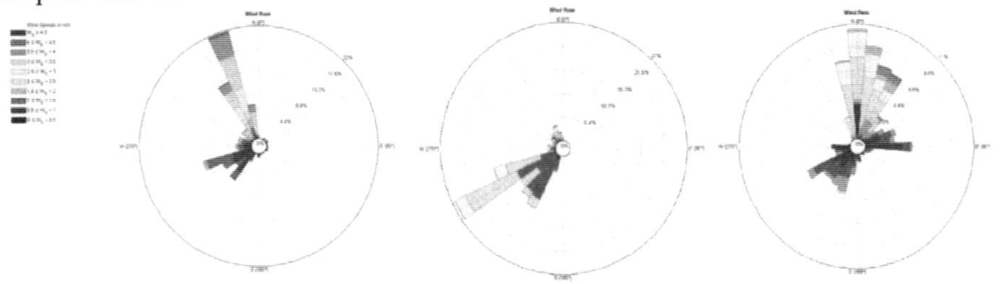

Figure 4. Wind direction data variations in three days measurement, at 8th March (*left side*), 9th March of totality (*center*), and 10th March (*right side*).

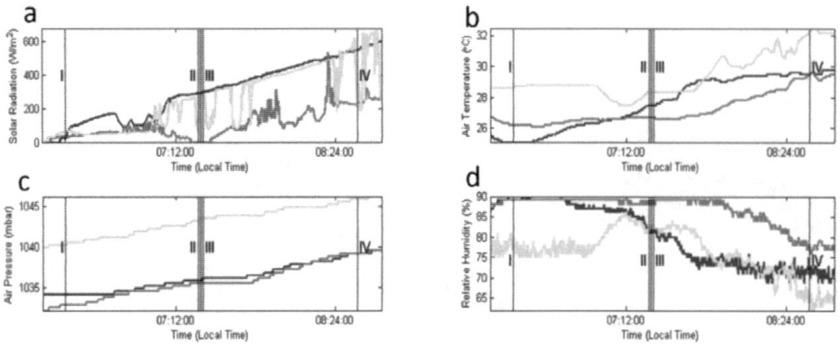

Figure 5. Temporal variation of (a) solar radiation, (b) air temperature, (c) air pressure, and (d) relative humidity during 1st contact (I) until last contact (IV) at 8th March (*blue line*), 9th March (*red line*), and 10th March (*green line*).

4. Conclusions

As in case of solar irradiance, decrease in temperature as the eclipse developed toward totality was inferred. As a result of the drop, the wind speed decreased and also the orientation of local wind has changed to the southwest. Delay of one to two-hour onset sea breeze and land breeze as a result of cooling on the eclipse day was detected on the day after the eclipse. This delay was made possible because of the static stability perturbation by total solar eclipse for daily conditions at the measurement site. Lastly, from measurements at Central Bangka, we discovered a substantial eclipse induced influence on all the surface layer parameters.

References

[1] Supriharyono 2000 *Pelestarian dan pengelolaan sumberdaya alam di wilayah pesisir tropis* (Jakarta: PT Gramedia Pustaka Utama)

[2] Laughlan G P 1997 *The User's Guide to the Australian Coast* (New Holland: Frenchs Forest) p 213

[3] Fernandez, Castro W, Wright J, Hidalgo H and Sfienz A 1993 *J. Earth, Moon, and Planets* **63** 119-132

[4] Aplin K L and Harrison R G 2003 *Proc. Royal Society* (London) p 353-371

[5] Clayton H 1901 *The eclipse cyclone and the diurnal cyclones Ann. Astron. Observ* (The United State: Harvard College) pp 5–22

[6] http://xjubier.free.fr/en/site_pages/solar_eclipses/TSE_2016_GoogleMapFull.html (accessed on February 2016)

[7] http://www.rainwise.com/products/attachments/6735/20150710010448.pdf (accessed on February 2016)

[8] Murthy B S, Latha R, Sreeja P and Dharmaraj T 2012 *J. of Atmos. and Solar-Terrestrial Phys.* **89** 33–39

Zenith sky brightness and celestial objects visibility during total solar eclipse on March 9, 2016 at Terentang Beach Bangka Island

A F C Wijaya[1], C P Asmoro[1], A A Rochman[1], T R Ramalis[1], J A Utama[1], N D Ardi[1], Amsor[1], M G Nugraha[1], D Saepuzaman[1], A Sutiadi[1] and D Nurfiani[2]

[1]Department of Physics Education, Faculty of Mathematics and Natural Sciences Education, Universitas Pendidikan Indonesia, Jl. Dr. Setiabudhi 229 Bandung 40154 Jawa Barat Indonesia
[2]Earth Observatory of Singapore, Nanyang Technological University, Singapore

Abstract. This paper endeavor to describe *sky brightness measurements* which was carried out by a team of total solar eclipse observers (TOGEMA) on 9th March 2016. The observations took place at Tarentang Beach, Bangka Island and it utilized the SQM-LU instrument (Sky Quality Meter–USB Connector) with 1 second time interval data. During total phase that lasted about 1 minute 52 seconds, the instrument recorded the brightness of the sky of 12.88 mag/["]^2 as the dimmest value. This value is approximately 500 times brighter than the dimmest night sky conditions at the same location, obtained on the previous observation. It was found that the brightest sky that could be measured by SQM–LU during Total Solar Eclipse (TSE) 2016 was 5.91 mag/["]^2. The activity with digital camera also captured the appearance of Venus and Mercury. The appearance of Venus (–3.71 mag) confirmed naked eye limited magnitude theory. This may explain the inability of observers to perceive Mercury (0.46 mag) using naked eye during the total phase of solar eclipse.

1. Introduction

Indonesia was the only country with a land traversed the path of totality in the event of total solar eclipse (TSE) on March 9, 2016. According to the US Space Agency, NASA, the TSE on March 9, 2016 was the 10th nature phenomenon which crosses Indonesian territory, where the 1st event was recorded back in 1901. Looking at the great potential that the phenomenon would contribute, Indonesian Government, through its Ministry of Tourism and Creative Industry, had designed an event dedicated to facilitate TSE observation all across Indonesian archipelago (shown in figure 1)

It is known that solar eclipse occurs when the Moon position is close to one point of intersection of the orbital plane of the Moon with the ecliptic. Meanwhile, TSE occurs when the angular size of the Moon observed from the Earth is equal to or greater than the angular size of the angle of the sun. The angular size of the Moon relative to the Sun will determine the duration of the total phase, and this event can be observed from a point on Earth's surface. Theoretically, the calculation for the longest duration of the total phase and the widest Moon umbra projected on the surface of the Earth respectively can reach 7.5 minutes and 270 km. During TSE on 9th Marth 2016the total phase lasted the longest in the western Pacific Ocean region for ~ 4 minutes with a Moon umbra width of ~155 km.

Such phenomenon also provides great research possibilities. First, TSE National Committee under the command of the National Institute of Aeronautics and Space (LAPAN) was established in 2015 to coordinate research activities of various research groups both inside and outside the country as well as becoming the center scientific information to the public and the media. At UPI (Universitas Pendidikan Indonesia), The Laboratory of Earth and Space the Department of Physics Educationalso formed a Solar Eclipse Observation Team (TOGEMA), aimed to conduct observations and research regarding this event. Therefore, the paper reported one of the observations and measurements made by TOGEMA related to the brightness of the sky during the phases of the solar eclipse.

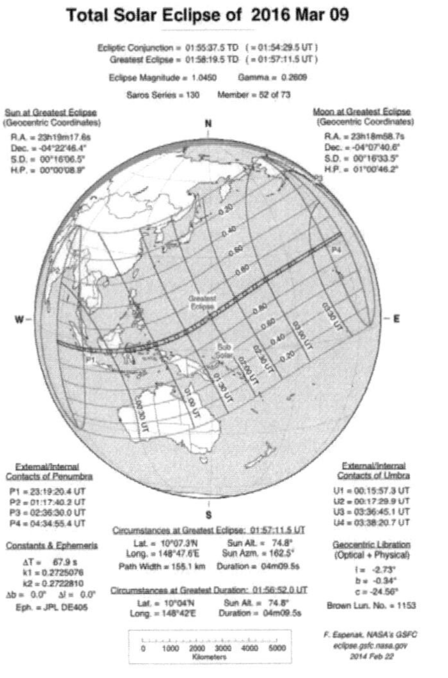

Figure 1. Total solar eclipse visibility chart and eclipse circumstances on March 9th, 2016.

2. Observations and measurements

TOGEMAs' TSE observations were held on the Terentang beach (2° 26 '39 "south latitude, 106° 19' 36" east longitude, 2.5 meters above sea level), Bangka Island, Bangka-Belitung. The location was selected since Bangka has the closest location to Java traversed path of eclipse totality. Moreover, the local government had also prepared adequate public facilities for tourists and researchers who came to the site.

It was noted that several contacts were made during the eclipse. The 1st Contact (C1), which was when the Moon and Sun disc-outer contacted, to mark the start of the eclipse, occurred at 06:20:58 local time. At the time of the height of the Sun on the horizon ~ 4°. This phase was not successfully investigated and enshrined as there were clouds covering the horizon in the direction of sight to the sun. After the first contact, the Sun disk was slowly covered by the Moon disk, where it eventually closed. This was called the 2nd contact (C2), which marks the beginning of the total phase, when the disk of the Moon and Sun in contact for the first time. This was recorded at 07:22:11 local time. At that time, the Sun elevated ~ 19° on the horizon, and it was not obscured by clouds. . The duration of the total phase at the site of the observation lasted about 1 minute 52 seconds and ended simultaneously with the 3rd contact (C3), that was when the Moon and Sun contacted for the second time at 07:24:03 local time. The end of the eclipse was marked by the 4th contact that was when the Moon and Sun in contact-beyond the second time, noted at 08:34:01 local time. TOGEMA team had managed to observe and capture the phases of TSE with pictures and videos.

During the phases of the eclipse, the team had successfully performed measurement of the change in sky brightness in the direction of the zenith, by means of instrument SQM-LU (Sky Quality Meter-USB Connector), connected to a laptop with 1 second time interval data. The SQM-LU stated the sky brightness in units of magnitude per arc second squared (mag / ["]^2), which is the unit commonly used for the extended light sources. The electronic guide book, which can be accessed on the manufacturer page (http://www.unihedron.com), mentioned that this tool has been calibrated with an accuracy of 10% (± 0.1 mag / ["]^2) and darkest sky conditions that can be measured is 21.80 mag / ["]^2.

3. Results and discussion

The zenith sky brightness variations during the phases of the eclipse are shown in figure 2. Measuring the brightness of the sky in the direction of the zenith had actually been carried out since the night before. This measurement was to obtain information on the condition of the dimmest night sky on the observation site, to be compared with the current situation of the total TSE phases. The dimmest night on Terentang beach occurred just after midnight, which was at 03:07:34 local time, with the results of measurements of 20.66 mag / ["]^2. Towards the rising Sun on March 9, 2016 at 06:01:10 local time, the condition of the horizon at sunrise azimuth and zenith direction gradually becomes clearer. SQM-LU could still perform before experiencing saturation due to the brightening of the environmental conditions, were 5.91 mag / ["]^2. The similar sky brightness recouped ~ 9 minutes into the 2nd contact and ~ 7 minutes after the 3rd contact. From the measurement results, it was concluded that the condition of the sky toward the zenith which was able to be measured by SQM-LU at 2016 GMT during the observation was 5.91 mag / ["]^2. This equated to 700 cd/m^2, which was similar to typical Saturn surface brightness with a magnitude of -0.4 and an angular diameter of 20.5" [2].

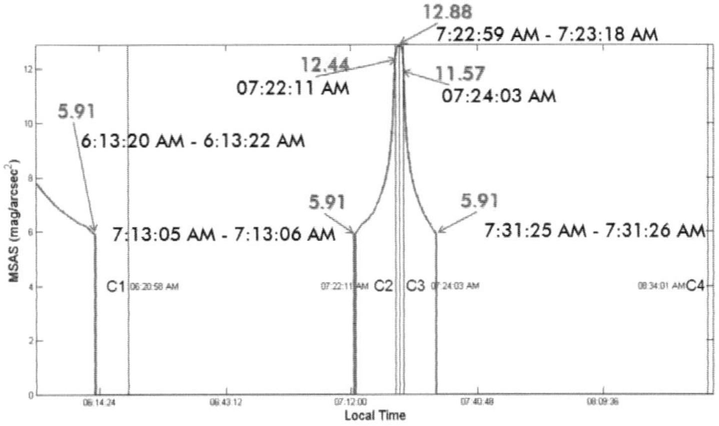

Figure 2. Data plotting of SQM–LU measurement during TSE March 9, 2016.

Meanwhile, the dimmest condition of the sky in the direction of the zenith recorded by SQM LU during the total phase lasts was 12.88 mag / ["]^2, which lasted only for 19 seconds (07:22:59 - 07:23:18 local time). In magnitude, the greater the number means the dimmer object or condition of the sky. The results of these measurements showed that the dimmest conditions during the phase of total eclipse was still lighter (~ 500x) than the condition of the darkest of the night sky that has been measured on the previous night. It is because of the contribution of the distribution of light (light scattering) originating from outside the Moon's umbra [3]. For other locations on the Earth where partial solar eclipse (PSE) occurred, the conditions at the peak of the eclipse would relatively be lighter since the Sun light was still at sight.

During the TSE observations, one of the team members was able to observe the presence of Venus with unaided eye. The position of the Sun in the sky is in the direction of Aquarius, but none of the brightest stars of the zodiac members were visible. The Landscape digital image of the Eclipse also identified the appearance of Venus and Mercury. Reports of success Venus observation (-3.71 mag) corresponded to the magnitude of the threshold value of the unaided eye (-0.8 mag) based on the brightness of the sky (12.88 mag / ["]^2) during the total phase. The magnitude were obtained with a correlation derived by K. Fisher, one of the SQM, which is available on the manufacturer's webpage (http://www.unihedron.com/projects/darksky/NELM2BCalc.html):

$$V_{naked-eye} = 7.93 - 5 \times \log\left[10^{\left(4.316-\frac{m_{sky}}{5}\right)} + 1\right] \quad (1)$$

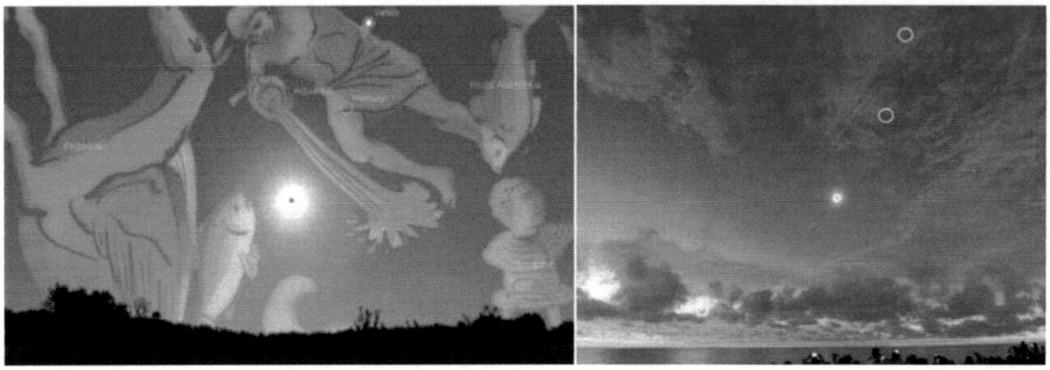

Figure 3. *(Left panel)* Sky simulation at the moment of total phase at Terentang beach using Stellarium software. *(Right panel)* Identification of planet Venus and Mercury (yellow circle) in digital image obtained using pocket digital camera NIKON Coolpix P520 f/3.1, 1/8 second, ISO 450 during total phase.

With a smaller magnitude than the unaided eye magnitude threshold, Venus was easy to be observed during the total phase. On the other hand, it was difficult to observe Mercury at the time, since it had a magnitude which was fainter than magnitude threshold, namely 0.46. To ensure that the observed was Venus, the team members compared the digital image obtained on the resulting simulation software Stellarium (www.stelarium.org) as shown in figure 3.

4. Conclusion

The results of the measurements of sky brightness at the zenith direction during TSE 2016 phases showed that brightness values changed rapidly, particularly towards the peak of the eclipse (the 2nd contact to the peak of the eclipse), and sometime after the peak of the eclipse (the 3rd contact). Not all sky brightness during phases of the eclipse were successfully measured due to the limited value sensor. The maximum brightness is 5.91 mag / ["]^2. Conditions of the darkest skies in the direction of the zenith during the total phase was 12.88 mag / ["]^2, which was equivalent to the magnitude of the threshold by unaided eyes - 0.8 mag. With the threshold value of this magnitude, the appearance of Venus during the total phase lasts had been successfully verified. Meanwhile, the appearance of Mercury which could only be observed through the digital imaging, was described, since this object had a magnitude fainter than unaided eye magnitude threshold.

Acknowledgments

2016 TSE expedition activities, socialization, and public outreach conducted TOGEMA funded by Indonesian University of Education (UPI) through the Office of the Vice Rector for Academic and Student Affairs and Office of the Vice Rector for Research, Partnerships, and Operating Funds. Our thanks also go to the Government of Central Bangka Regency on technical assistance and facilities provided during the expedition.

References

[1] Pasachoff J M 2009 Scientific observations at total solar eclipses *Research in Astron. Astrophys.* 9 6 613–34

[2] Clark R N 1990 *Visual Astronomy of the Deep Sky* (Cambridge: Cambridge University Press) p 11

[3] Sharp W E, Silverman S M and Lloyd J W F 1971 Summary of sky brightness measurements during eclipses of the sun *Appl. Opt.* 10 1207–10

Journal of Physics: Conference Series **771** (2016) 012013 doi:10.1088/1742-6596/771/1/012013

The sky brightness measurement during the 2016 solar eclipse in Ternate

Yudhiakto Pramudya and Muchlas Arkanuddin

Program Studi Magister Pendidikan Fisika Universitas Ahmad Dahlan, Jl. Pramuka 42, Yogyakarta 55161, Indonesia

Pusat Studi Astronomi Universitas Ahmad Dahlan, Jl. Gondosuli 1, Yogyakarta 55166, Indonesia

E-mail: yudhiakto.pramudya@pfis.uad.ac.id

Abstract. Obscuration of the Sun by the Moon during total solar eclipse generates the changing of the sky brightness. Sky Quality Meter (SQM) was employed to measure the sky brightness during the 2016 total solar eclipse. The sky was still bright at the first contact time. It is shown by the SQM value of zero. Approaching the second contact time, the SQM value is rising up started at the value of 5.92 mpsas. The curvature profile of the SQM measurementvalue is similar to the curvature profile of the SQM measurement at the dawn and dusk. However, the flatness part of the curvature is much shorter than night time value of SQM. The maximum of SQM measurement value is lower than the SQM measurement value during the night in Ternate. It is 12.47 mpsas and happened at the maximum phase of the eclipse. It was confirmed by the fact that at the time of totality, the sky close to the horizon was still bright. There is a discrepancy between the predicted and actual second and third contact and maximum eclipse time. By assigning the maximum of SQM measurement value as the reference value of maximum eclipse time, the actual second and third time can be calculated. The shape of curvature between the actual second and third contact time is symmetry.

1. Introduction

The total solar eclipse path of March 9th, 2016 was stretches across Indonesia. The total solar eclipse occured since early morning in the western part of Indonesia until mid-day in the eastern part of Indonesia. Hence, the sun altitude vary across Indonesia. The sky brightness depends on the sun altitude. During the night, when the sun altitude is negative, the sky will be dark. The sun disk is covered by the moon disk during the totality. Hence, the sun rays is blocked. The sky begins to darken to be a night-like sky.

Number of research activites has been done to observe the solar eclipse phenomenon. The sky brightness during the total solar eclipse has been measured on number of occasion such as 1954 Sweden Eclipse [1], 1970 African Eclipse [2] and 1973 American Eclipse [3]. The radio telescope was employed on the 1954 Sweden Eclipse. The photometer was used on the 1970 and 1973 eclipse. The sky brightness measured on the zenith direction that was measured during the 1970 Eclipse showed decreasing value up to four order of magnitude compared to normal day sky. The sun altitude was 46° from the horizon. Similar results was also obtained from the measurement of 1973 Eclipse with the sun altitude was 37° from the horizon.

Another factor contributing to the sky brightness is the number of particle in the atmosphere. The particles will give the effect of scattering and absorption. The model has also been calculated to investigate the scattering contribution to the sky radiance during the total solar eclipse [4]. Hence, it is important to measure the sky brightness to understand characteristics of the atmosphere at the time of eclipse. This research is intended to measure the sky brightness changes during totality.

2. Instrumentation

The sky brightness measurement has been done using Unihedron Sky Quality Meter (SQM). It is inexpensive and compact size sky brightness photometer. It provides the value of sky brightness. It only measures the visible light and blocks the infrared light.The value is expressed in the unit of magnitude per square arc second (mpsas) on a logarithmic scale. The mpsas means the brightness is spread over an area of squared arc second. The higher value of SQM measurement means the darker sky. Since, it is related to the apparent magnitude of the star, higher value of apparent magnitude dimmer the star.

The SQM has two types. They are SQM and SQM-L. The SQM-L covers the 20° angular region of sky [5]. The SQM was installed to the zenith direction. It is connected to the computer via USB cable since it is SQM-LU. The computer is AXIOO with Windows 8 operating system, Intel(R) Celeron(R) CPU N2940 1.83 GHz processor and 2.00 GB RAM. For the rest of the article, the SQM-L is written as SQM only. On this research, the SQM took the data every 1 second.

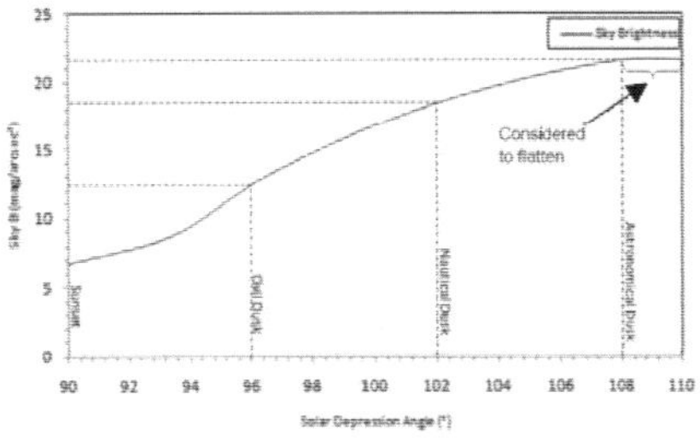

Figure 1. The sky brightness as a function solar depression angle [6]

The SQM is usually employed to measure the night sky brigthness quality. Since the sky brightness depends on sun altitude, the SQM is able to measure the changes of sky brightness in function of time. As the time changes, the sun altitude or the solar depression angle changes as shown in figure 1 [6]. The change can be used to perform analysis of twilight time.

To provide additional data about the sky condition, two cameras were employed during the totality. The first camera is digital camera of CANON EOS 600D. The second one is the ASUS Zenfone 5 camera smartphone 8 Megapixels.

3. Observations

The observation has been done in the terrace of Masjid Al Munawar, Ternate, Indonesia. The latitude is 0.7898° North and longitude is 127.3907°. It is located at 2 meters above sea level. The location position data is obtained using google maps. The observation location is in the totality path. However, it is about 42 kilometers north of the center line of totality path.

There were 2 times of measurement. The first one was measuring the sky brightness before the dawn. The second one was running the SQM during the eclipse. The first measurement can be used to compare the sky brightness during the totality to the night sky brightness. The time of measurement has been synchronized with the offiicial time of BMKG (Indonesian Agency of Meteorological, Climatological, and Geophysical). The Local Time (LT) is Universal Time (UT) + 9 hours. The first, second, third and fourth contact was occured at 08:36:04.8 LT, 09:51:41.9

LT, 09:54:22:0 LT and 11:20:53.0 LT respectively according to the NASA prediction data. The sun altitude also is obtained from NASA calculation. The sun altitude of maximum eclipse was 47.7° at 09:53:01.6 LT. It is similar to the 1970 Elipse sun altitude measurement. The predicted duration of totality was 2 minutes and 40 seconds.

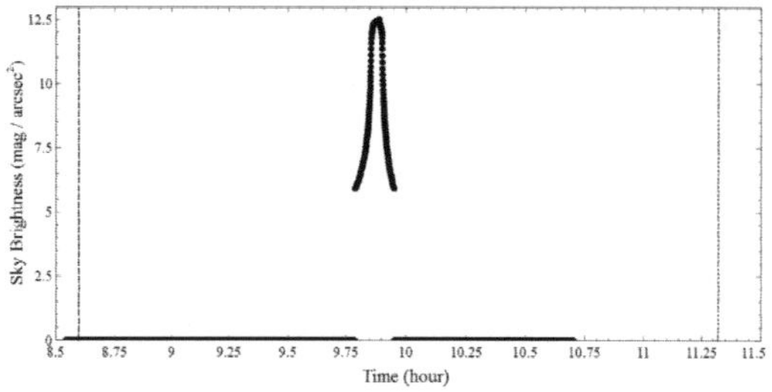

Figure 2. The SQM measurement during the solar eclipse. The vertical dashed line on the left side marked the first contact and the vertical dot line on the right side marked the fourth contact of the solar eclipse.

4. Result and analysis

The SQM measurement during the solar eclipse is shown in figure 2. The predicted first contact time is shown by the vertical dased line on the left side. The sky was still bright, hence the SQM reading is zero. The vertical dot line on the right side give the information the predicted time of the fourth contact. However, the observation has been stopped before the fourth contact. By extrapolation, the SQM reading is zero at that time. It means, the sky is bright like the normal day sky.

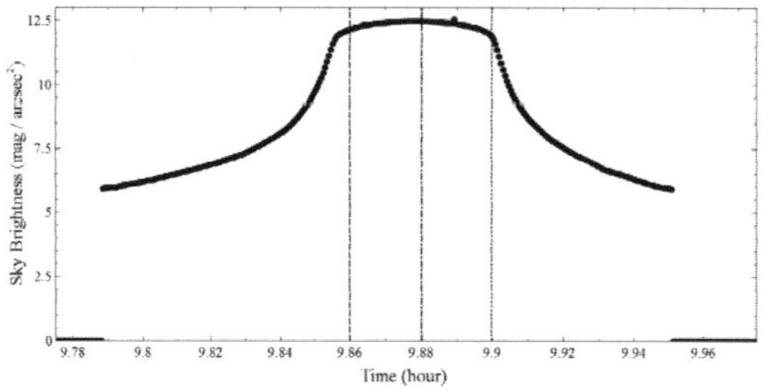

Figure 3. The SQM measurement during the solar eclipse. The vertical dashed line on the left side marked the predicted second contact time, the vertical dashed-dot line marked the predicted maximum eclipse time and the vertical dot line on the right side marked the predicted fourth contact time of the solar eclipse.

To investigate the second and the third contact and maximum eclipse, figure 2 is zoomed in as shown in figure 3. The vertical dashed line on the left is the predicted second contact time. The vertical dased-dot line in the middle is the predicted maximum eclipse time. The vertical dot line in the right side is the predicted third contact time. The predicted maximum eclipse time was occured after the SQM measurement reach the maximum value. The maximum value is 12.47 mpsas at 9:52:41 LT. Ignoring the spike that occurs in the right due to electronical signal transmission. There is a discrepancy between the predicted and actual time. The position of the predicted second and third contact time are asymmetry.

Hence, the actual time is used as reference to calculate the actual second and third contact time. The predicted duration is also used to generate the results. Hence, the calculated second and third contact time are 9:51:21 LT and 9:54:1 LT, respectively. Those times correspond to the SQM value of 11.75 mpsas and 11.76 mpsas, respectively. Based on this calculation, by keeping fix duration of totality, the SQM value of the second and third contact are similar.

The maximum SQM value is compared to the SQM measurement before dawn on the same day. Based on the profile as shown in figure 1, the maximum and flat SQM value is 19.6 mpsas. Hence, the difference between the eclipse sky brightness and normal night sky brightness is 7.13 mpsas. The sky brightness during totality was not completely dark. The sky close to the horizon was still bright as shown in figure 4. It was similar to the condition at the dawn or dusk.

Figure 4. The sky brightness condition during the totality as captured by digital camera. The sun at the totality phase can be seen on the camera display

The SQM measurement curve is asymmetry. This is due to the apperance of cloud. The cloudy condition occured before and after the totality. At the time of the totality, the sky was clear. The SQM mesaurement value jumps from zero to 5.92 mpsas due to sensitivy threshold of instrument. However, the curvature of SQM measurement value is similar to the profile as shown in figure 1. The maximum of SQM measurement value and and its curvature profile can be used for analysing particle scattering and absorption in the atmosphere that influence the sky brightness.

5. Conclusion

The sky brightness during the total solar eclipse has been done in Ternate using SQM-LU. There is discrepancy between the predicted and actual contact time. The maximum of SQM measurement value is lower than the normal night sky brightness. These phenomenon can be explained by off-center line of totality path and particle scattering and absorption in the atmosphere. The curvature profile is similar to the curvature profile of dawn and dusk.

6. Acknowledgment

This research was supported by the research funding from Lembaga Penelitian dan Pengembangan Universitas Ahmad Dahlan and assisted by Universitas Muhammadiyah Maluku Utara.

References

[1] Coates R J, Gibson J E and Hagen J P 1958 *Astrophysicsal Journal* **128** 406
[2] Dandekar B and Turtle J *Applied Optics* **10** 1220
[3] Shaw G *Applied Optics* **14** 388
[4] Shaw G *Applied Optics* **17** 272
[5] Birriel J and Adkins J *JAAVSO* **38** 221
[6] Niri M A, Zainuddin M Z, Man S, Nawawi M S A M, Wahab R A, Ismail K, Zaki N H A, Ghani A A and Lokman M A A *Middle-East J. Sci. Res. 12 (1) 101*

Impacts of the total solar eclipse of 9 March 2016 on meteorological parameters in Ternate

R Satyaningsih, E Heriyanto, Kadarsah, TA Nuraini, J Rizal, A Sopaheluwakan, and E Aldrian

Center for Research and Development, Indonesian Agency for Meteorology Climatology and Geophysics (BMKG), Jl. Angkasa I No. 2 Kemayoran, Jakarta 10720, Indonesia

E-mail: ratna.satyaningsih@bmkg.go.id

Abstract. The total solar eclipse on 9th March 2016 has given a unique opportunity to assess impacts of the eclipse on various meteorological parameters in Ternate. A weather transmitter, a net radiometer, and a heat flux plate for measuring soil heat flux in bare soil were deployed during this special observation in Madaha, Ternate (0°51′17.5″ N, 127°20′56.8″ E). The measured meteorological parameters presented here are net radiation, air temperature, relative humidity, and soil heat flux. The results show that those parameters are significantly affected by the eclipse. The percentage of the observed decrease of net radiation during the eclipse was proportional to the obscuration percentage. The air temperature measured at 1.5 m declined around 1.9 °C with approximately 11 minutes lag between the mid-eclipse and minimum temperature during eclipse. Influences from the surrounding environment in the measurement site minimize the temperature response. As a consequence of temperature decrease, relative humidity increased. A drop in soil heat flux was also detected but its response was slower.

1. Introduction

Although eclipses are astronomical phenomena, they also draw considerable interest from atmospheric scientists because solar radiation is the main source of energy to the atmosphere. Solar eclipses provide a natural 'laboratory' for studying the Earth's environment response to the abrupt change in radiation. The total solar eclipse event on 9th March 2016 started over the Indian Ocean, passed across the Indonesian region (Sumatra, Kalimantan, Sulawesi, North Maluku), and then ended near the Hawaiian islands. The greatest eclipse, the instance when the axis of the Moon's shadow cone passes closest to Earth's center, occurred over the Pacific Ocean. The variability of surface radiation during solar eclipses has been extensively studied [e.g. 1, 2, 3, 4, 5, and 6]. The results showed that all the radiation fluxes were significantly affected by eclipses. Zerefos et al. [2] and Kazadzis et al. [4] show that radiation decrease depends on its wavelength, which is more pronounced at the lower wavelength. There are a number of reports on impact of eclipse on meteorological parameters, such as air temperature, humidity, atmospheric pressure, and wind speed and direction [e.g. 1, 3, 6, 7, 8, 9, and literatures therein]. The range of temperature decrease during an eclipse varies from one location to the other. The existence of eclipse wind is debatable since for the same eclipse event, one location showed significant abrupt change in wind speed and direction, whereas others did not [9,10,11]. The fluctuation of relative humidity was largely reported as consequence of the temperature decrease.

Our measurements of meteorological parameters were carried out in Ternate (figure 1) before, during, and after the eclipse day. Ternate is one of the island in the North Maluku region and lies in the vicinity to totality path of the total solar eclipse of 9th March 2016 and experienced 100% obscuration. This measurement was aimed to investigate how the abrupt change in radiation, when the Moon covered the face of the Sun, affects net radiation, air temperature, relative humidity, and soil heat flux (SHF).

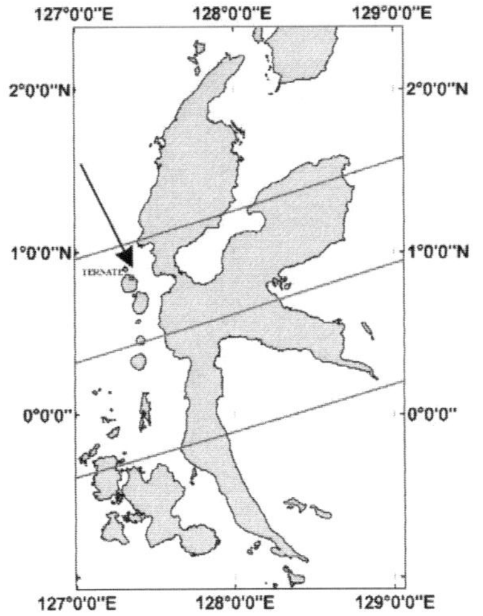

Figure 1. The 9th March 2016 total solar eclipse path over the North Maluku region. Regions inside the red curves experienced the total solar eclipse while those outside the curve had partial solar eclipse. The blue line indicates the central line. Red square shows the location of our measurement site.

2. Measurement: site, weather condition, and instrumentation

Our measurement site was located in the north-east part of Ternate Island. The eclipse at this location started at 08:36:05 LT, with mid-eclipse at 09:53:00 LT, and ended at 11:20:51 LT (the eclipse lasted for 2h 44m 46s and the duration of totality was 2m 21s). The Local Time is UT+9h. The instruments were installed in Madaha, Ternate (0°51′17.5″ N, 127°20′56.8″ E). The land where the instruments deployed has sandy soil and only small parts of it were covered with short grass.

A weather transmitter was used to monitor six essential weather parameters. This instrument consists of ultrasound sensor to measure wind speed and direction; separate capacitive sensors for barometric pressure, temperature, and relative humidity measurement; and precipitation sensor. Net radiation measurement was carried out using a net radiometer with thermopile sensor. The weather transmitter and the net radiometer were mounted 1.5 m above the ground surface on horizontal pole protruding around 50 cm and 100 cm respectively from the mast. The heat flux plate for soil heat flux measurement was submerged around 5 cm from the soil surface. Only measurement for the bare soil will be reported in this paper. All data were recorded every 20 seconds and its acquisition is acquired by connecting the instruments to a data logger. The measurements were conducted from 7-10 March 2016. Unfortunately, due to a technical problem on 8 March 2016, the data logger could not record the measurements on that day until early morning on the following day. The sky was partially cloudy for most of the days of the observation, but during the eclipse the sky around the sun was clear.

3. Results and discussion

The observed temporal variation of net radiation during 7-10 March 2016 is shown in figure 2a. On non-eclipse days, the net radiation started to increase in the morning, reached its peak around mid-day and decreased afterwards. During the night, as outgoing radiation leaving the Earth's surface is larger than the incoming radiation, the value of net radiation is negative. Frequent sudden drop in net radiation pattern were due to varying cloud coverage. It can be noticed clearly in figure 2a that during the eclipse net radiation decreased. Moreover, near mid-eclipse the value of net radiation is negative, which resembled night time condition. Thus the decrease of net radiation during the eclipse time was virtually caused by the eclipse, instead of cloud coverage. During the eclipse, the net radiation dropped from 341.87 Wm^{-2} to -40.21 Wm^{-2}. The net radiation reached its minimum shortly after the mid-eclipse.

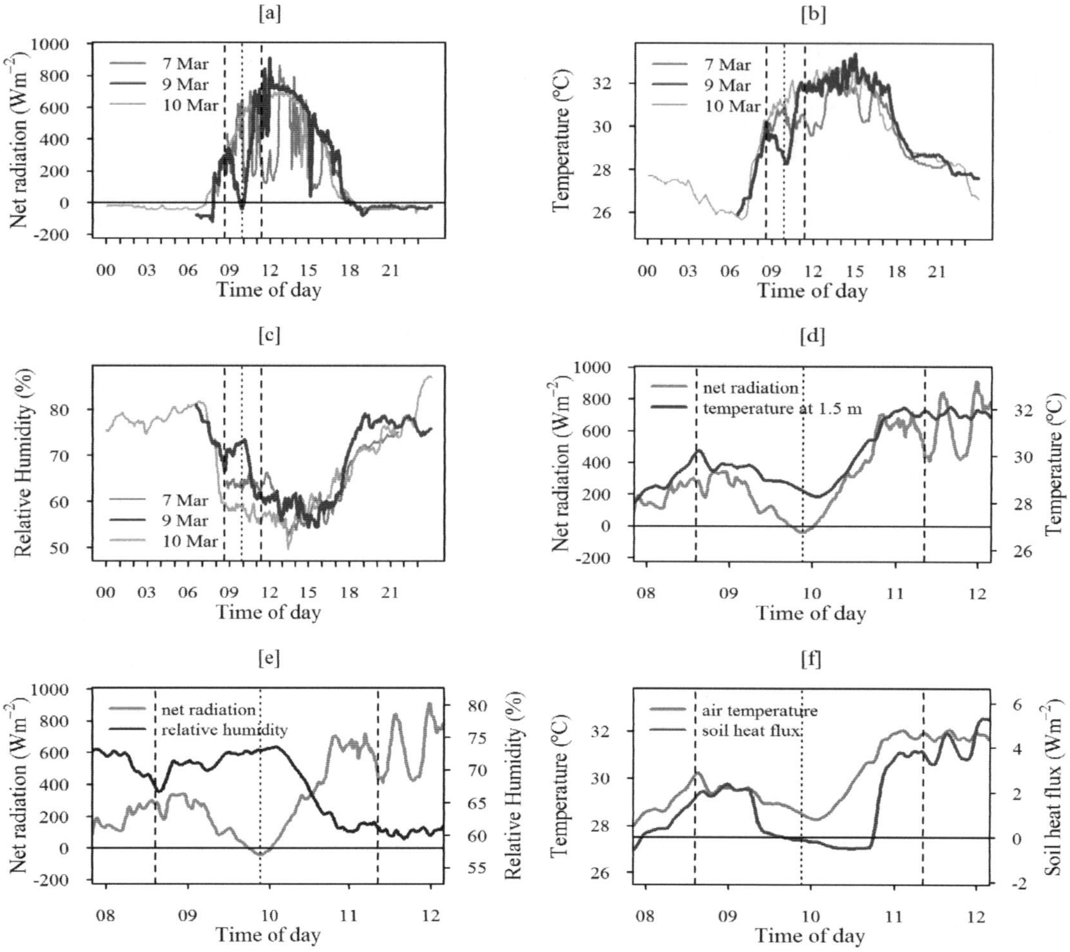

Figure 2. Temporal variation (5-minute moving average) of net radiation (a), temperature (b), and relative humidity (c) measured from 7th to 10th March 2016. Comparison of net radiation and air temperature (d), of net radiation and relative humidity (e), and of temperature and soil heat flux (f) during the eclipse. Vertical lines denote the onset, mid, and end of the eclipse respectively. Time is in local time.

Net radiation directly determines whether the surface temperature rises, falls, or remains the same. As the consequence, temperature pattern during the eclipse followed the net radiation pattern (figure 2b and 2d). A significant drop in from normal diurnal pattern for air temperature during the solar eclipse of 9th March 2016 is clearly seen. The temperature started to decrease from 30.2 °C at the beginning of the eclipse and reached its lowest temperature at 28.3 °C. This ~2 °C decrease is within the range already reported on other solar eclipse events [e.g. 8, 9, 10, 12, and 13]. However this decrease is lower than the temperature drop reported by Fernandez et al. [7] although both eclipses were total solar eclipses and occurred at similar time of day. Probably this is caused by the fact that the measurement site is located in an island surrounded by sea, which affected the temperature response (decrease) due to its larger heat capacity [10]. Figure 2d illustrates a comparison between net radiation and temperature during the eclipse. There was approximately 11 minutes lag between the net radiation and temperature minima during totality. The actual time lag is related to the thermal inertia of the surface layer [3].

The diurnal variation of relative humidity has opposite phase to that of temperature [14] during normal conditions as a consequence of temperature decrease during the eclipse, the relative humidity delayed its decrease until the temperature has reach its minima following the increase of net radiation after totality (figure 2c and 2e). Following the temperature, shortly after the eclipse began, usual decrease in relative humidity at that time of day was halted and started to increase. Relative humidity increased from 66.7% and reached its maximum at 73.5%. The maximum relative humidity was lagging to mid-eclipse by 12 minutes, which is comparable to time lag between mid-eclipse and minimum temperature. The relative humidity resumed its normal decrease afterwards. The increase in relative humidity did not surpass the highest relative humidity at nighttime as the temperature remained above the nighttime lowest temperature. The latter indicates that the sudden darkening during short eclipse events does not result in cooling at the same magnitude as during sunset and night [15].

SHF is affected by the surface temperature, soil moisture, and vegetation [5]. Because the temperature significantly decreased during the eclipse, it is expected that SHF will respond to this decrease proportionally. Figure 2f shows that the SHF obviously decreased in response to the eclipse but in slower response. The SHF took around 23 minutes to decrease from 2.351 Wm^{-2} after the eclipse started and the decrease persisted around 26 minutes after mid-eclipse until reaching a minimum value at -0.503 Wm^{-2}. A condition similar to night time during the eclipse is clearly seen in the SHF pattern, where the sign of the SHF values are negative.

4. Concluding remarks

Measurements of impact of the total solar eclipse of 9 March 2016 on net radiation, temperature, relative humidity, and soil heat flux showed that those parameters were significantly affected by the eclipse. The air temperature at 1.5 m declined 1.9 °C with around 11 minutes lag between the mid-eclipse and minimum temperature during eclipse. The percentage of the observed drop of net radiation during the eclipse was proportional to the obscuration percentage. The drop in temperature induced by the eclipse was initially expected to be even greater than the result of the measurements, since the eclipse occurred at late morning when temperature increases in normal days [10,13]. However the surrounding environment in the measurement site (particularly the sea) might affect in reducing the temperature response due to its larger heat capacity. Relative humidity increased from 66.7% and reached its maximum at 73.5% during the eclipse. The maximum relative humidity was lagging to mid-eclipse by 12 minutes. A drop in soil heat flux (SHF) was also detected although its response was much slower than other parameters.

Acknowledgments

The authors thank their colleagues from the Research and Development Division for Meteorology and Sultan Babullah Station in Ternate for their support during the eclipse field campaign.

References

[1] Foken T *et al.* 2001 *Meteorologische Zeitschrift* **10** 171-178
[2] Zerefos CS *et al.* 2001 *Adv. Space Res.* **27** (12) 1955-63
[3] Aplin KL and Harrison RG 2003 *Proc. R. Soc. Lond. A*, 459, pp. 353-371
[4] Kazadzis S *et al.* 2007 *Atmos. Chem. Phys.* **7** 5775–83
[5] Nymphas EF, Otunia TA, Adeniyi MO and Oladiran EO 2012 *J. Atmos. Terr. Phys* **80** 28-36
[6] Maturilli M and Ritters C 2016 *Earth Syst. Sci. Data Discuss.* doi:10.5194/essd-2016-2
[7] Fernandez W, Hidalgo H, Coronel G and Morales E 1996 *Earth, Moon and Planets* **74** 49-59
[8] Anderson J 1999 *Weather* **54** 207-215
[9] Nymphas EF, Adeniyi MO, Ayoda MA and Oladiran EO 2009 *J. Atmos. Terr. Phys* **71** 1245–53
[10] Founda D *et al.* 2007 *Atmos. Chem. Phys.* **7** 5543–53
[11] Tzanis C, Varotsos C and Viras L 2008 *Atmos. Chem. Phys.* **8** 425-430
[12] Segal M, Turner RW, Prusa J, Bitzer RJ and Finley SV 1996 *Bull. Am. Meteorol. Soc.* **77** 89-99
[13] Hanna E 2000 *Weather* **55** 430-446
[14] McIntosh DH 1972 *Meteorological Glossary* (London: Her Majesty's Stationery Office)
[15] Winkler P *et al.* 2001 *Meteorologische Zeitschrift* **10** 193-199

Journal of Physics: Conference Series **771** (2016) 012015 doi:10.1088/1742-6596/771/1/012015

Identification of moon craters and solar corona during total solar eclipse on 9th March 2016

Luthfiandari, N Ekawanti, F G Purwati and D Herdiwijaya

Astronomy Study Program, Institut Teknologi Bandung, Jalan Ganesha 10, Bandung 40132, Indonesia

E-mail: luthfiandari@s.itb.ac.id

Abstract. Total Solar Eclipse (TSE) is a rare natural event in which the positions of Sun, Moon, and Earth are perfectly aligned. In the past by using this phenomenon, many researches have been done to understand characteristic of the corona. In this paper we carried out the study of TSE which crossed over Indonesia from West to East on 9th March 2016. We observed TSE which occured in Palembang (2.9883° S 104.7513° E), Indonesia. The aim of this research is to understand the effect of moon craters on the appearance of solar corona and identification of solar active regions during TSE. This research was done using Canon SX170 IS camera with ND 5 sun-filter. Although the sky was cloudy during the totality of the phase, coronal video was still taken. Camera also took solar images of partial eclipse phase. Coronal images for every frame were then extracted from the video. Image processing of coronal images was done using RegiStax and PhotoScape freewares. To study solar corona, images from Virtual Moon Atlas, Hinode XRT, and SOHO-LASCO were compared with the result of oriented coronal image. Wider and many more moon craters were found having positive correlation with the brighter effect on solar corona as shown at westward coronal streamer. Those craters are represented by Bel'kovich crater, the biggest one. We also found that only the eastward coronal streamer was correlated with active region, sunspot number 12519, from behind solar limb.

1. Introduction

Total Solar Eclipse (TSE) occurs when the moon's shadow covers the entirety of the sun's disc. As the center of moon's shadow crosses the center of the sun's disc, so TSE becomes rare phenomenon. Observing TSE on the ground can pave the way to study coronal activitiy in cheap and valuable method [1]. This method can use digital camera for many exposures to cover many solar radii. Druckmüller did this method for discovering of a new class of coronal structure [2].

TSE which crossed over Indonesia on 9th March 2016 has saros series 130 . This series shows that the Moon was moving northward, and that the TSE occured at the Moon's descending node. This TSE started from West to East Indonesia in the morning, with different initial time at every place. By using this natural event, we studied coronal appearance during totality. The aim of this research is to investigate the effect of moon craters on the appearance of solar corona and the identification of solar active region during TSE. We observed the TSE that occured in Palembang (2.9883° S 104.7513° E), Indonesia, which started from 23:20:29.3 UT to 01:31:25.3 UT on 8th to 9th March 2016.

2. Methodology

We took both video and images of the eclipse phase which started from 6:20:29 to 8:31:26 Western Indonesian Time. Data were taken using Canon SX170 IS camera integrated with sun filter ND 5. Camera was attached to a tripod which could move in altitude and azimuth and took video of totality phase and image of partial phase. Due to an unfortunate circumstance, the sky was significantly covered with clouds, so that the Sun was barely visible, with only several phases of the TSE able to be observed. Allignment of solar coronal image usually can be done by a modified phase correlation method, based on Fourier Transform [3]. This method could be conducted if coronal images were clear from cloud. Since only several regions of corona appear during total eclipse phase, so another simple method is used on this research.

Video of the total phase were converted into images with 25 frames per second. Registax 6.0 was used for stacking images and improving image results. The images were then rotated using Photoscape v3.6.3, which was based on the orientation of the Sun at real time. Orientation of the Sun at totality phase was derived from orientation of the Sun at partial eclipse phase. Orientation of the partial eclipse was aligned using data from "https://www.solarmonitor.org/". After orientation of the total eclipse had been applied, the solar corona was analyzed using Virtual Moon Atlas and Solar Monitor web in order to identify moon craters and solar active region respectively. Westward of the solar disc was on the right side of the frame, whereas westward of the moon disc was on the left side.

3. Result and discussion

The clear images of partial eclipse were captured on 24 jpeg images. These data were then used to derive the rotation angle at total eclipse phase. Every rotated frame was built using Photoscape for the partial eclipse phase, both before and after the total eclipse phase. Orientation was based on the solar monitor web for Hinode XRT (X-Ray Telescope) data. Every rotation angle of all the partial eclipse phase images is shown on figure 1. The first clear image was created at 7:04:55 am and had a rotation angle of 74.5°. As time went on, the rotation angle was decreasing with 2 different regression types. The first fitting which occured before 7:33:25 am has a regression type of 2nd order polynomial. Rotation angle of its fitting dropped significantly with time. This was caused by a refraction of the thick atmosphere near the horizon. The secondary fitting which occured after 7:33:25 am has a linear regression, accounting for the slower decrease of the rotation angle than that of the first regression. Two kind of regressions are choosed by the biggest coefficient determination, R^2.

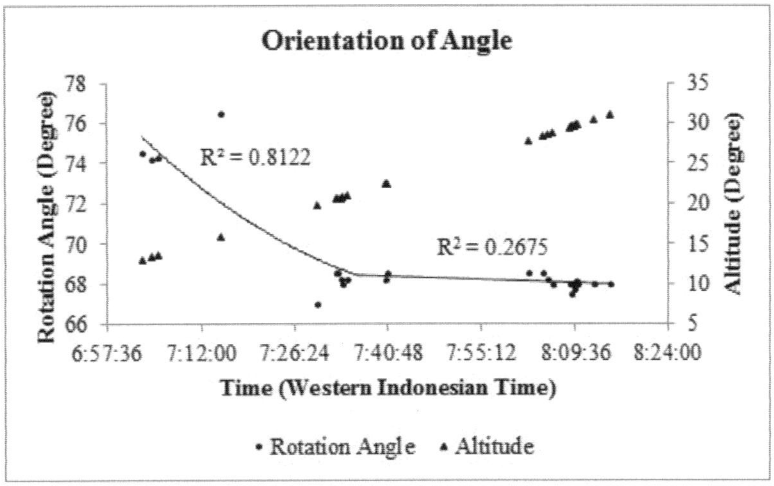

Figure 1. Angular orientation of each phase of the partial eclipse.

From the total eclipse phase MOV video, 275 jpeg images were generated. These images were stacked to become a single image, and then improved using Registax image processing. The total eclipse was captured at 7:21:23 am Western Indonesian Time. Based on this time, the total eclipse phase image has an average rotation angle of 71.05°. This angle was derived from the 2nd order polynomial regression as shown on figure 1.

3.1. Identification of moon craters

Orientation of the solar disc during total eclipse phase was stacked with lunar images at respective times of totality phase as shown on figure 2. Solar corona had been detected on the rightward and leftward of figure 2. Rightward corona is on latitude 45° to 82° and longitude (87±5)°. Then, leftward corona is on latitude 0° to 30° and longitude –(92±5)°. Craters in this region are chosen for craters which are on arbiter line which separates dark and bright sides of the Moon. By using these borders, the moon craters in this region are shown in table 1.

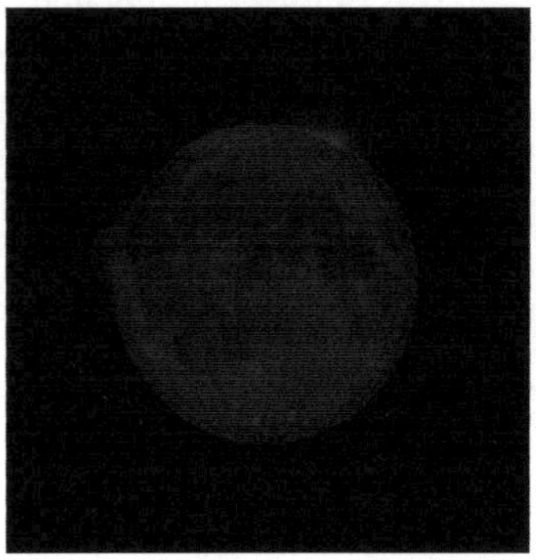

Figure 2. Stacked Lunar and Coronal Images at total eclipse phase.

Table 1. Moon Craters at Coronal Region

IAU Feature Name	Lat. (°)	Lon. (°)	IAU Diameter (km)
Petermann A	74.85	87.63	18.34
Hayn G	66.86	86.31	20.11
Bel'kovich A	58.65	87.41	57.26
Hayn F	68.01	85.75	62.54
Nansen	81.17	95.38	116.90
Bel'kovich	61.53	90.15	215.08
Einstein S	15.10	-91.67	19.63
Einstein R	13.83	-91.88	19.84
Sundman	10.76	-91.69	41.04
Moseley	20.95	-90.20	88.89

Data show only several region at solar limb which have coronal appearance. This appearance could be affected not only by clouds, but also by moon craters. As shown on figure 2, rightward/ eastward Moon has many more wide craters than leftward/ westward Moon has, so eastward region is brighter than westward region. Coronal region on eastward Moon is represented by Bel'kovich crater. On another side, coronal region is represented by Moseley crater.

3.2. Identification of solar active region

Other than moon craters, coronal appearance is also influenced by solar active regions. Therefore, solar corona can be identified by identifying solar active region. Images of solar active region and corona are shown on figure 3. Corona on eastward and westward solar disc were expected to have the same feature as Lasco C2 showed. However, stacked image between the data and image from Lasco

C2 shows differences in their coronal features. The eastward corona in Lasco C2 image is brighter than its corresponding westward corona and data show the opposite. Regions with brighter corona on Lasco C2 image show the effect of CME (Coronal Mass Ejection) which detected on eastward limb with a radial velocity of 688 375 km/s and angular width 6° ("https://www.heloviewer.org"). Based on Hinode XRT data, coronal appearance on eastward limb was affected by active region, sunspot number 12519, which is behind solar limb. This active region could be seen a couple of days after 9th March 2016, since it had direction of rotation from east to west. Active regions image from Hinode XRT shows loops above sunspots which represent strong magnetic field. Regions with strong magnetic field have high temperature, as magnetic energy is converted to thermal energy [4].

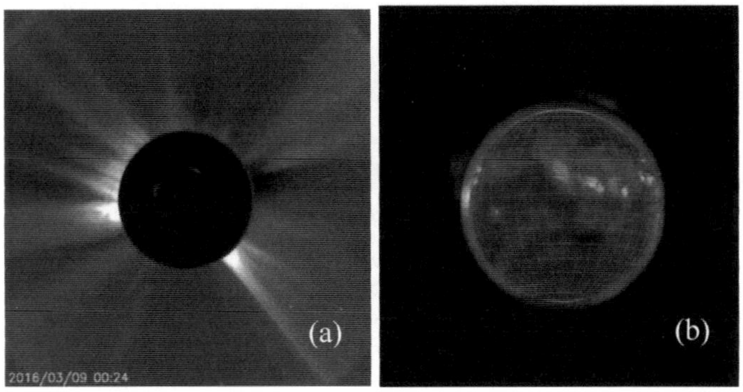

Figure 3. Corona from TSE stacked with solar active region image. (a) Stacked with SOHO Lasco C2 (b) Stacked with Hinode XRT.

4. Conclusion

By using a simple method which defining orientation based on partial eclipse phase, orientation of the Sun during TSE had been obtained. This method also showed an atmosphere refraction effect in the morning, before 8 am at local time. This simple method could be used for bad data, such as uncomplete corona because of clouds.

Coronal features during TSE were found having positive correlation with size and abundance of moon craters. On bright corona, coronal westward of solar disc, was correlated with wide and many of moon craters. On another side which had dim corona, coronal eastward of solar disc, was correlated with narrow and lack of moon craters. Both sides are represented by Bel'kovich and Moseley craters respectively. Coronal appearance was also influenced by solar active regions. Eastward coronal features had positive correlation with coronal appearance on Lasco C2, which is influenced by CME and active regions behind the solar limb, sunspot number 12519.

5. Acknowledgement

The SOHO/LASCO data used in this paper were produced by a consortium between the Naval Research Laboratory (USA), Max-Planck-Institut fuer Aeronomie (Germany), Laboratoire d'Astronomie (France) and the University of Birmingham (UK). SOHO is a project of international cooperation between ESA and NASA. *Hinode* is a Japanese mission developed and launched by ISAS/JAXA, with NAOJ as a domestic partner and NASA and STFC (UK) as international partners. It is operated by these agencies in co-operation with ESA and NSC (Norway).

References

[1] Pasachoff J M 2009 Scientific observations at total solar eclipses *Research in Astron. Astrophys.* **9** 613

[2] Druckmüller M, Habbal S F and Morgan H 2014 Discovery of a new class of coronal structures in white light eclipse images *The Astrophysical Journal* **785** 14

[3] Druckmüller M 2009 Phase correlation methode for the alignment of total solar eclipse images *The Astrophysical Journal* **706** 1605

[4] Phillips K J H, et.al 2000 SECIS: the solar eclipse coronal eclipse imaging system *Solar Physics* **193** 259

Investigation the effect of total solar eclipse March 9, 2016 on tidal elevation study cases: Bangka and Belitung islands

I M Radjawane1[1], E M Simanjuntak[1], A F Adziima[1] and I Sofian[2]

[1]Oceanography Study Program, Bandung Institute of Technology, Bandung, Indonesia
[2]Geospatial Information Agency, Cibinong, Indonesia

E-mail : ivonne@fitb.itb.ac.id

Abstract. Tidal elevation is generating primary by the gravitational forces between earth with moon and solar. Since the sun plays an important role on generating the tidal sea level then the effect of total solar eclipse (TSE) March 9, 2016 on surface water level has been analysed from real time one-minute tide gauge station from Geospatial Information Agency. Two stations was selected to observe the impact of TSE i.e. in coastal of Bangka Island where totally path of TSE was detected and the other is in the coastal Belitung Island, located near to Bangka Island. We compare the observed sea level with the predicted sea level from 9 constituents using Tidal Model Driver (TMD). The result shows that during the TSE in Bangka Island, the observed tidal phase is shifted and observed sea level is higher compare with the prediction one. But in Belitung Island, there is no pattern change of tidal phase and with small increasing sea level compare with the prediction.

1. Introduction

A total solar eclipse (TSE) took place on March 9, 2016 was visible along across 11 provinces of Indonesia. The TSE started in Western Indonesia and moved to Eastern Indonesia and ended in northern Pacific Ocean. TSE occurs when moon is directly in front of the earth and the moon's shadow falling on the earth completely hides the solar surface and thereby cuts off direct rays of sunlight from observer.

On the other hand we know tides or ocean tides is a phenomenon that affected by the gravitational attractive force of moon and sun. Tides is an alternating high and low sea level. Tides on the Earth are primary generating by the gravitational pull of both the moon and the sun with tides from the Moon being larger because the moon is so much closer. Tides occur in different positions and alignments of Sun, Earth and Moon. When the Earth, Moon and Sun are aligned then the resultant gravitational force exerted on Earth's water is strongest at these moments and we called the spring tides. Spring tides can only occur during full and new moon. Therefore, the TSE occurs during the new moon and the relation of on tidal elevation should be an interesting matter to study. So far there is no previous studies about the relation between eclipse and tide elevation that located in Indonesia area.

The aim of this research is to investigate the effect of TSE March 9, 2016 on the surface water level or tidal elevation from tidal movement point of view. In simple way, we will show the difference between the real time one minute tide data with the prediction one. We assumed the prediction one is a normal condition. We choose two tidal stations in the western part of Indonesia that passed by the TSE path.

2. Data and method

Two tide stations in western part of Indonesia were selected as the research area i.e. Muntok in Bangka Island and Tanjung Pandan in Belitung Island as seen in figure 1. Since we know that the path of TSE started at western part of Indonesia from Palembang and moved to eastern part of Indonesia. We choose the tidal station that passed by the TSE path. That is why both tide gauge stations is an ideal location. The tidal elevation data provided by the Geospatial Agency of Indonesia. The real time one-minute observed data was obtained from the tide gauge located in this two stations. We picked out the data from 1 to 16 March 2016, so we can cover the spring and neap tidal phases.

To see the effect of tide on TSE and normal events, we compare the observation results and prediction. We used the user friendly tidal model namely Tidal Model Driver (TMD) developed by [2] to predict the tidal elevation. There are 9 major tidal constituent considered in the TMD. The model was running with the same time with the observational one. We simulated also the tidal elevation in the same time in 2015.

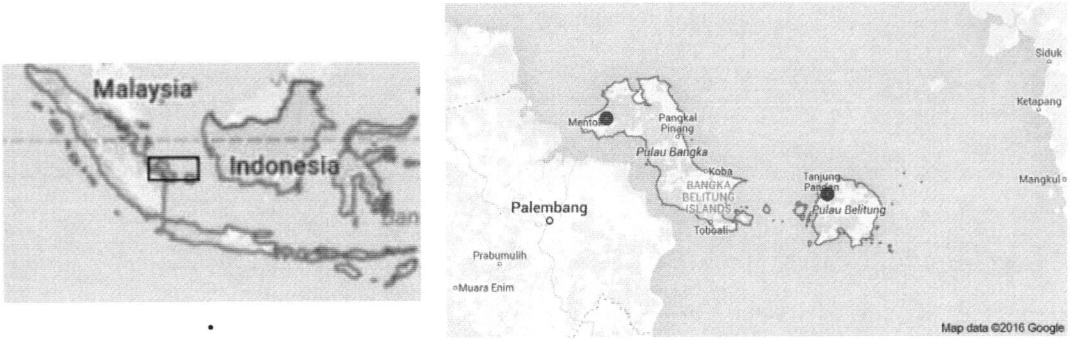

Figure 1. The location of two tide observation stations (Muntok and Tanjung Pandan). The station indicated by the blue circle. (Courtesy to Google Earth, 2016)

3. Results

The observation tidal elevation in Bangka and Belitung Islands on March 1-16, 2016 can be seen in figures 2 and 3. The tide type in both stations is fully diurnal tide, where in one day there is one time highest water level and one time lowest water level. This result confirmed with research from [3]. The spring tide is happened around the first week indicated by higher water level. After that around the second week, the neap tide which occurs indicated by the lower water level occurs. In both stations, it is clearly demonstrated that the TSE happened during spring in new moon phase. The observation tidal range in Bangka during spring tide and neap tide is about 2.4m and 0.5m respectively, while in Belitung is about 1.7m and 0.5m. If we compare the observational data with the prediction then we can find in general the same pattern between both results. In Belitung station, the observational and prediction tide range and phases slightly similar. But in Bangka the observational magnitude tide is bigger than the predictions, especially during spring tide, and also there is about one hour phase different, where the prediction one is earlier. After the TSE we can see the pattern is not similar for both station. The pattern and tidal elevation between the observation 2016 and prediction seem differs especially in the magnitude of tide. In both locations the predicted tidal range is smaller.

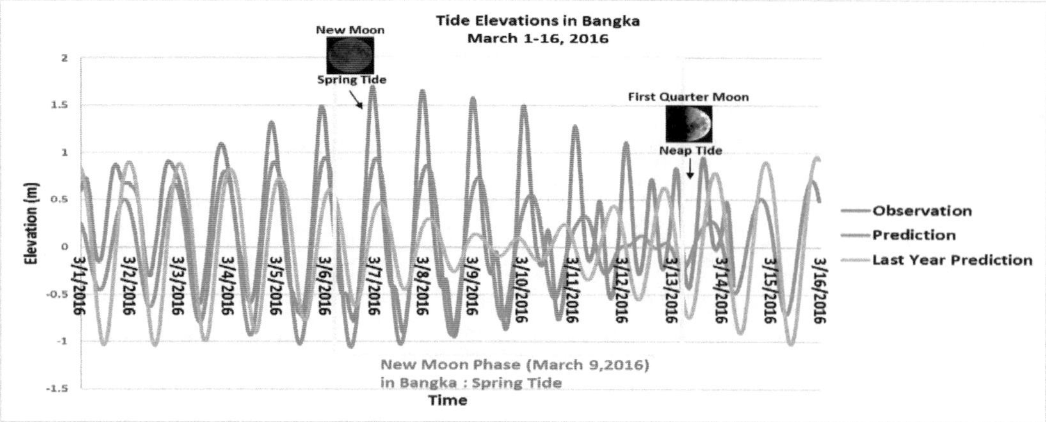

Figure 2. Tidal elevation on March 2016 from Observation (blue line), Prediction (red line) and Prediction in March 2015 (grey line) in Muntok, Bangka Island.

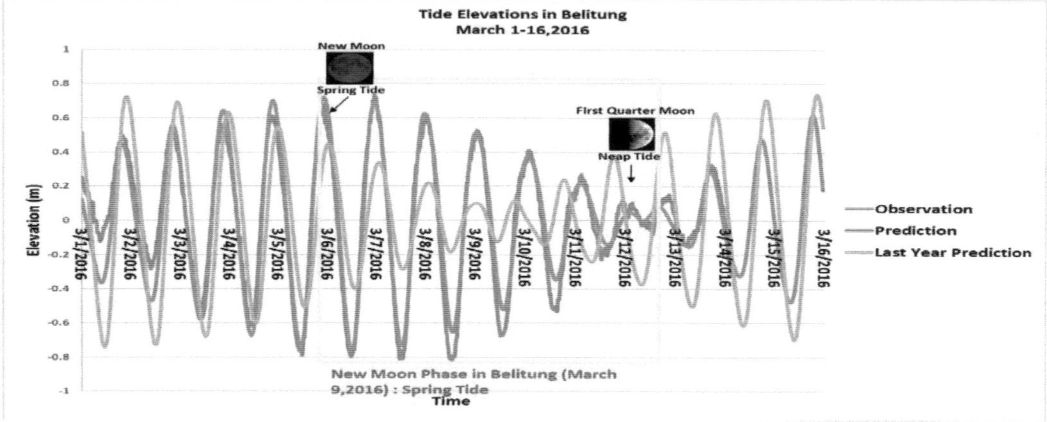

Figure 3. Same as Fig. 2 but at Tanjung Pandan, Belitung Island.

Further we will focus the analyses during the TSE phase in figures 4a and 4b. From the report it is known that the duration of maximum eclipse is just a short time i.e. 1-2 minutes (www.terangi.com). But the total process from sun rise is about 1-2 hour. So, we will investigate from 06.30 to 08.30. From figure 4a, in Bangka coastal, during this period the high water is going t to low water, but after that increase about 1 hour and go down again. We can assume that during TSE the low water becomes lower (the absolute magnitude is higher), it means the tide become stronger. That is why the curve is not smooth in this short interval time. The curve prediction in 2015 and 2016 is smooth but the magnitude is very small for 2015. There is also shifted phase between observation and prediction.

In Belitung station, the sea level is moving from high water to low water for both observation and prediction (figure 4b). Tide range and phase pattern slightly similar but the curve line of observation data is more negative (the magnitude is higher) compare to the prediction. It is maybe because of stronger force. But we need more detail and advanced investigation on math and physical analysis to conclude it.

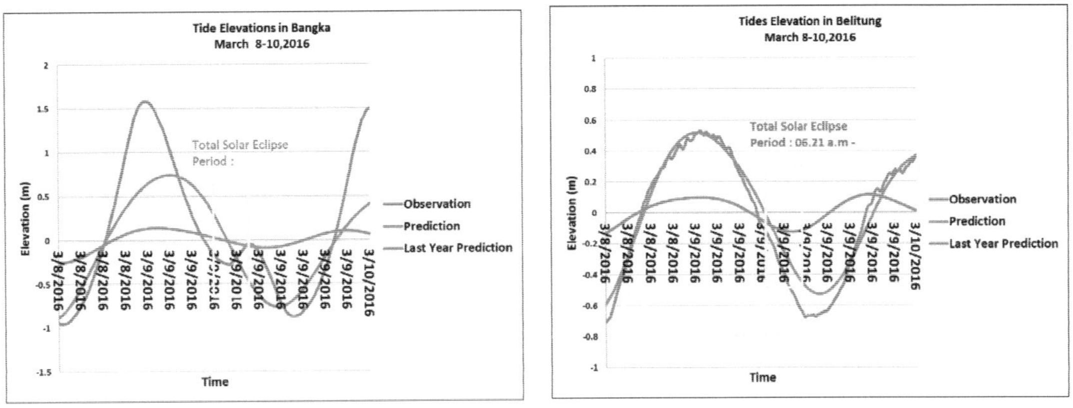

Figure 4. Tidal elevation on 8-10 March 2016 (before, occurrence, and after TSE) in Muntok Bangka Island (a) and Tanjung Pandan, Belitung Island (b). The yellow box indicating time during TSE happened.

4. Conclusions

The conclusions are concluded from a simple research method. We can conclude that during the TSE 9 March 2016, the water level is moving from high tide to low tide and tidal phase is in new moon phase in spring tide in both stations. The tidal level is higher and there is shifted tidal phase between observation and prediction in Bangka Island. The magnitude of tidal level is higher during TSE time compare with the prediction in Belitung Island.

This research gives just preliminary results. We know that there is still lot of remain questions that need to answer and debatable results to proofs in more detail point of views. Therefore, it should be some interesting multi-disciplinary research to be conducted.

Acknowledgment

We thank to Mr. Nabiel, Ms. Woro and Ms. Lina who support for the data processing. The idea of this research initiated by the National Strategic Research from Ministry of Research and Technology and Higher Education entitled "Database Arrangement of Coastal Vulnerability to Enhanced the Adaptive Capacity in Threat of Sea Level Rise in Coastal City and Small Island in Indonesia" for FY 2016. This research funded for attending the ISSEL conference.

References

[1] *Badan Informasi Geospatial* http://tides.big.go.id/ (accessed on May 20, 2016)
[2] Padman L and Erofeeva S 2005 Tide Model Driver (TMD Manual Readme file for Matlab ESR
[3] Wyrtki K 1961 Physical Oceanography of the Southeast Asian waters *Scripps Institution of Oceanography* (UC San Diego: Scripps Institution of Oceanography)
[4] Serm Murmson *What Tide Corresponds With a Solar Eclipse?* http://classroom.synonym.com/tide-corresponds-solar-eclipse-3392.html (accessed on July 12,2016)

The effect of total solar eclipse on the daily activities of *Nasalis larvatus* (Wurmb.) in Mangrove Center, Kariangau, East Kalimantan

Sya Sya Shanida[1], Tiffany Hanik Lestari[1], and Ruhyat Partasasmita[1,2]

[1]Department of Biology, Faculty of Mathematics and Natural Sciences, University of Padjadjaran. Jl. Raya Bandung-Sumedang Km 21, Jatinangor, Sumedang 45363, West Java, Indonesia. Tel. +62-22-7796412 line 104, Fax. +62-22-7794545
[2] Programme of Magister Biology, Faculty of Mathematics and Natural Sciences, University of Padjadjaran. Jl. Raya Bandung-Sumedang Km 21, Jatinangor, Sumedang 45363, West Java, Indonesia. Tel. +62-22-7796412 line 104, Fax. +62-22-7794545

E-mail: syasyashanida@gmail.com, tiffanyhaniklestari42@gmail.com, Ruhyat.partasasmita@unpad.ac.id

Abstract. The total solar eclipse is an interesting phenomenon because the sun is covered by the moon. This phenomenon is like a night deception for animals, humans, and plants. One of the animals is Bekantan (*Nasalis larvatus* (Wurmb.)). *Nasalis larvatus* change its activity when this phenomenon occurs. The aims of the present study are (1) daily activity of *Nasalis larvatus* on total solar eclipse on March 9th, 2016 and (2) the effect of total solar eclipse on its activity in Mangrove Center, Kariangau, East Kalimantan. The *adlibitum* method was used in this study on Bekantan's adult female. The result shows that the total solar eclipse has considerable effect on the daily activity of Bekantan. During total solar eclipse, the activity of Bekantan significantly stopped. When the total solar eclipse finished, Bekantan started its daily activity, and it was indicated by feeding activity which was led by alfa-male.

1. Introduction

The total solar eclipse is an attractive natural phenomenon. This phenomenon happens when the sunlight is not able to reach the earth since it is blocked by the moon. At this moment, the sky becomes dark as the dark in the night. Therefore, some animals, plants, and even humans, presumed the total solar eclipse as a night. Bekantan (*Nasalis larvatus*) is one of the animals that is deceived by total solar eclipse. Bekantan is an animal from Primate (ordo), Cercophitecidae (family), and Colobinae (subfamily) [5].

Bekantan's activity is affected by the intensity of the light (night to morning). Bekantan starts their activities at 07.00 a.m. The daily activities of Bekantan throughout the day (05.30 – 18.30 a.m.) are walking, playing, resting, running from one tree to another, and feeding. The highest peak of feeding activity was observed in the afternoon. In fact, it is difficult to distinguish the feeding activity and the movement activity of Bekantan. This is because Bekantan did both the feeding and the movement activities within a short period of time (in minutes) since Bekantan forages for food by moving in the trees. This results the fluctuation of the frequency of movement activity is nearly similar to that of feeding activity. Around 12.00 – 12.30 p.m., the intensity of feeding activity increased. After that, the feeding activity decreased and Bekantan start preparing to rest and sleep until 14.00 p.m.[4].

It was observed that the frequency of the feeding activity of Bekantan is not fixed only on one tree. The feeding activity of Bekantan was always followed by movement activity such as jumping between branches in one tree and/or running to another tree. This observation also exhibited that the movement of Bekantan from one tree to another for feeding occurs in every 10 – 15 minutes. It is worth mentioning that both the number and the distribution of trees are sufficient to supply the food demand of Bekantan. The purpose of the present study is to investigate the Bekantan's daily activity during total solar eclipse on March 9th, 2016.

2. Method

This study was conducted using the *adlibitum* method on one Bekantan's adult female in Somber River, Mangrove Center, Karingau, East Borneo (as shown in Figure 1) on March 9th, 2016 at 06.00 to 11.00 a.m.

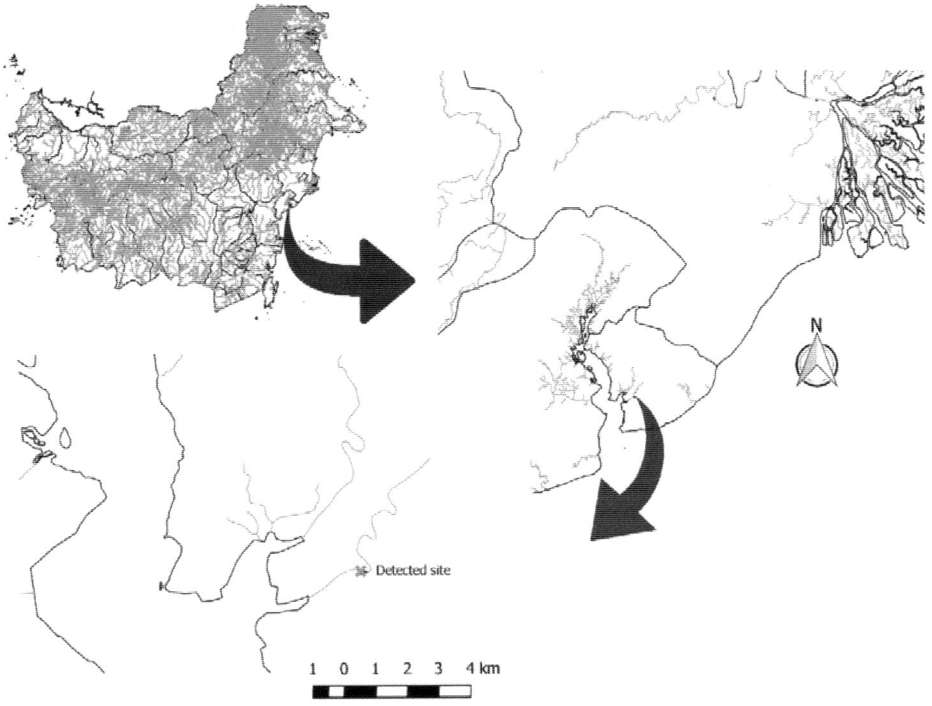

Figure 1. Area Study in Mangrove Center Kariangau, Somber River, Balikpapan, East Kalimantan ($1^0 12'28.25''$S and $116^0 50'18.3\ 0''$E)

3. Data analysis

The result of the present study is a graphic of the frequency of the activities of the Bekantan such as feeding, resting, and moving, as a function of time.

4. Result and discussion

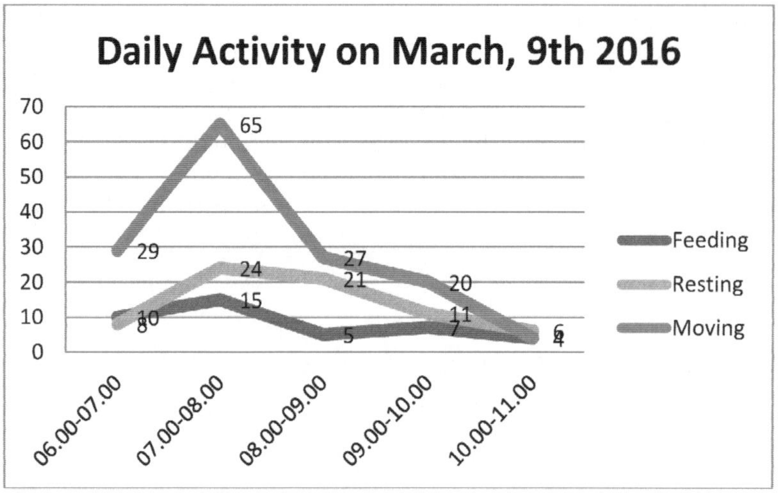

Figure 2. Graphic of the daily activities of Bekantan on March 9th, 2016

4.1. Daily activity

It can be seen from Figure 2 that the moving activities increased from 29 to 65 at 06.00 to 08.00 a.m, while the duration of feeding activity increased up to 5 minutes followed by movement from one tree to another to eat leaves. It is also observed that the duration of feeding activity of Bekantan at one tree depends on both the type of the tree and the amount of the food in the tree. On *Ganua motleyana*, Bekantan could stay around 20 – 50 minutes [2]. While for doing the movement, at the same range of time the feeding activities increased from 10 to 15 times.

[6] Based on the mealtime, Bekantan used around 66% of their times to eat leaves and 26% to eat fruits [7]. [3] The main main food of Bekantan in Mangrove Center at Kutai National Park is *Avicennia alba, Rhizophora apiculata,* and *Bruguiera sexangula.*

The intensity of resting activity increased from 8 to 24 times at 06.00 to 08.00 a.m., and usually done when the Bekantan sat watching the neighborhood. Long resting activities carried out up to 25 minutes. This animal is choosing resting place (tree) as well as providing food, such as buds, flowers, fruits, and leaves, for example *S. Caseolaris*, *Ilex cymosa*, dan *Ficus* sp [1].

Contact 1, the first phase of partial solar eclipse, occurred at 07.25 a.m. Interestingly, around 35 minutes after Contact 1, it was observed that the moving activity, including the movement of Bekantan to the resting place, decreased to 27 times (at 08.00 to 09.00 a.m.). The same fashion also was observed on the feeding activity. The feeding activity decreased to 5 times at 08.00 to 09.00 a.m. The decrease of both moving activity and feeding activity is due to the Contact 2, the phase of total solar eclipse, which occurred at 08.33 a.m. The peak of total solar eclipse happened one minute after Contact 2. It is worth noting that the activities of Bekantan during Contact 1, particularly when Bekantan start moving to resting place, was led by alfa male.

When the the last phase of total solar eclipse (contact 3) occurred (at 08.34 a.m.), it is observed that Bekantan start resuming their normal daily activities (see the activities at 09.00 to 11.00 a.m. in Figure 2), and it is indicated by feeding activity of alfa male. However, it was observed that the feeding activity of Bekantan after total solar eclipse is not as high as before total eclipse. The same pattern also was observed on both moving activity and resting activity. At 10.00 – 11.00 a.m. it was shown that the feeding, resting, and moving activity decreased to 4, 6, and 4 times, respectively. [8] It is worth mentioning that the feeding activity of the Bekantan is also influenced by the weather. The feeding activity of Bekantan is higher at higher light intensity. Therefore, the feeding activity of Bekantan in the morning is higher than the feeding activity in the afternoon or gloomy weather (in case of raining). Indeed, the gloomy weather and/or raining is mostly happen at noon or afternoon.

4.2. The activity alteration on total solar eclipse

During Contact 1, we observed no considerable effect of solar eclipse on the daily activities of Bekantan such as feeding, playing, and moving. However, it was exhibited that the feeding activity of Bekantan start decreasing at 08.15 a.m., when the sky brightness becomes similar to the sky brightness in the afternoon. The activities of Bekantan such as feeding, playing, and moving, significantly decreased at 08.30 a.m. In other hand, it was observed that Bekantan moves around the tree to look for the comfortable place for resting. At that time, the sky over mangrove forest started going to be dark. Interestingly, the total solar eclipse had many significant effects on other animals such as chicken and bird. During the total solar eclipse, it was shown that the chicken go into its cage and stop crowing, and birds perched on the tree and stop chirping. After several minutes of total solar eclipse, the brightness of the sky increased (Contact 3). It was observed that the environmental condition is similar to the condition in the morning. Interestingly, the activities of the animals also was observed similar to their normal activities in the morning. For example, the chicken was crowing, the birds started flying and chirping. Indeed, Bekantan also started its normal morning activities which is feeding. It is worth noticing that all activities of Bekantan is always led by alfa male. When the total solar eclipse finished, Bekantan resumed to its normal daily activities such as feeding, playing, and moving.

Acknowledgement

We give our best appreciation to GMT-BALIKPAPAN 2016's team, and Mr. Irfan's family who have contributed us on facilitating the research. We also thank Mr. Agus and his staff as a manager of Mangrove Center Kariangau Balikpapan.

References

[1] Alikodra H S 1997 Populasi dan Perilaku Bekantan (*Nasalis larvatus*) di Samboja Koala, Kalimantan Timur *Media Konservasi* **5** (2) 67-72

[2] Bismark M 1984 *Biologi dan Konservasi Primata di Indonesia* (Bogor: Makalah Fakultas Pascasarjana Institut Pertanian Bogor)

[3] Bismark M 1986 *Perilaku Bekantan (Nasalis larvatus) dalam Memanfaatkan Lingkungan Hutan Bakau di Taman Nasional Kutai Kalimantan Timur* (Bogor: Thesis of Magister Science Program of IPB)

[4] Bismark M 2009 *Biologi Konservasi Bekantan (Nasalis larvatus)* (Pusat Penelitian dan Pengembangan Hutan dan Konservasi Alam Bogor)

[5] Jolly A 1972 *The evolution of Primate Behavior* (New York: Mac-Millan Publishing Co. Inc.)

[6] Matsuda I Tuuga A and Higashi S 2009 The Feeding Ecology and Activity Budget of Bekantans *American Journal of Primatology* **7** (1) 478–492

[7] Matsuda I 2008 *Feeding and Ranging Behaviors of Bekantan (Nasalis larvatus) in Sabah Malaysia* (Hokkaido: Dissertation Graduate School of Environmental Earth Science Hokkaido University Hokkaido)

[8] Purba E F B 2009 *Studi Keanekaragaman Jenis Tumbuhan Pakan Bekantan (Nasalis larvatus) di Taman* Nasional *Tanjung Puting Kalimantan Tengah* (Bogor: Thesis of Under Graduate Program of IPB)

Astronomy in Buginese–Makassarese culture based on historical and ethnographical sources

N Hasanah[1] and D A Suriamihardja[1]

[1] Geophysics Study Program, Faculty of Mathematics and Natural Science, Hasanuddin University, Makassar, Indonesia

E-mail: n.hasanah@unhas.ac.id

Abstract. Our research has been looking for astronomical aspects in Buginese-Makassarese culture from historical to ethnographical sources. We found from history that astronomy had been used by Buginese – Makassar society long before Islam came to South Sulawesi at early 17th century and had their own first telescope at the time of Gowa's Prime Minister, Karaeng Pattingaloang. Meanwhile from ethnographical source, utilization of astronomy penetrated not only at Islamic calendar and worship time, but also at agriculture, sailing navigation, and weather prediction. Some of our literature were taken from *Lontaraq* (ancient manuscript) which now the experts are not many and old, especially in the field of astronomy. It was found that *Lontaraq* was using lunar period to characterize local weather [called: *Pananrang*]. Therefore, we need to excavate immediately more about this matter before the experts reduced in number.

1. Introduction

In 17th century, Makassar was the busiest port in entire region of eastern among seafarers. Since the time as famous city with spice trade, Makassar has been inhabited by multi ethnics (indigenous people, Malay, Arabs, Persian, Indian, Chinese), although most of them are still dominated by two the largest ethnics in South Sulawesi, i.e. Buginese and Makassarese.

In spite of having different words in use, two ethnics Buginese-Makassarese have been sharing the same traditions and culture, particularly in their script called *Lontaraq*. The named *Lontaraq* to their scripts was due to number of records written on the leaves of palm trees (*Borassus Flabellifera*) [1]. Based on earlier sources, Kathryn G. Anderson [2] explained that: "*The Buginese script is derived from an Indian prototype, the Pallawa script of southern India that spread to Southeast Asia in the fourth or fifth century.*" *Lontaraq* contained a wide range of knowledge that govern their social life, such as agriculture, commerce, and weather forecasting [3]. In Javanese society, weather forecasting was known as *Pranatamangsa* [4], whilst in *Lontaraq* Buginese-Makassarese was called *Pananrang* or *Kutika*. This *Pananrang* explains the characterization of annual local weather over period of eight years which is called *Sipariyama* period [3]. Several studies had tried to analyze the relationship between *Sipariyama* period with Dipole Mode [5], ENSO [6], and Lunar motion period [7]. On the other hand, Ammarel [8] had described astronomical aspects contained in Buginese navigation system by tabulating some local stars and asterisms. The story behind some of the stars (stars in Buginese and Makassarese word is '*bintoéng*') had similarity with Greek myths [9]. Perhaps it was due to universal information from Greek myths about asterisms that got into the thought of Buginese sailor and eventually grown into local myths. However, to adapt the paradigm of universal science needed not only acceptance, but also needed a very critical look at mathematical elaboration and physical background [10].

Until recently, number of researchers who have tried to analyze scientifically the background knowledges of *Lontaraq* was still few, particularly in traditional astronomy. Quite different to Javanese society, the *Pranatamangsa* had been widely studied for its usage as guidance for agriculture activities [11]; cosmological, meteorological, and bio-climatological aspects [4]; and its dependency to the presence of certain stars [12]. Whilst from archeology side, Javanese had solar gnomon called *Bencet* [13] and *Lingga* [14] that been used as solar time marker. Through simple comparison that had been mentioned above, it obviously needs to be considered that astronomical aspects in Buginese-Makassarese culture should be explored more widely as interdisciplinary studies. Therefore, this paper would like to write some historical and ethnographical aspects of astronomy in Buginese-Makassar culture.

2. Historical aspects

The existence of *Lontaraq* was already began in South Sulawesi long before Islam arrived in early 17th century. In this pre-Islamic times, *Lontaraq* manuscript never mentioned the dates for events [1]. Kathryn G. Anderson [2] also said that: "*... the precise date when writing reached South Sulawesi cannot be determined with certainty...*" Basically, pre-Islamic *Lontaraq* only reckoned the length of events. For example, the length of the first ruler's reign was lasted for four *pariyamas*. Owing to one *pariyama* was equal to eight years, so that the ruler's reign was 32 years. It didn't mean that Buginese-Makassarese society do not own calendar system, but caring to the time of events was assumed to be not important. Abidin [1] in his paper stated that Buginese calendar consisted of 12 solar months with Sanskrit names but had different arrangement sequence with Indian and Javanese calendars. They divided one month into weeks of five days. This five-day 'week' was called '*sipasa*', means one market cycle (Javanese and Balinese society called it *Pancawara*: *Legi*, *Pahing*, *Pon*, *Wage*, and *Kliwon*). Apart of five-day 'week', seven-day week also existed in Lontaraq, but it did not had name (Javanese and Balinese society called it *Saptawara*). When Islam came, *Lontaraq* had written the dates in Islamic calendar and used Arabic to write Arab-Malay language.

Lontaraq itself consists of a wide variety of documents. The one which included astronomy was Lontaraq Pananrang. Figure 1 depicts a sample of the *Lontaraq Pananrang*'s tabulation which is written in 1907 [15]. Although the original *Lontaraq* had been rewritten in Latin alphabet, it was still very difficult to translate the used Buginese language into Indonesian, because some of the words are not in use anymore by Buginese modern society today.

There were some points that can be noticed from figure 1. (1) Firstly, column-1 can be translated as the Name of Seven Guidance (N7G). Looking at Ammarel's table of Buginese stars and asterisms [8], four of it were mentioned in N7G, which is {a} *Oromporong* (Pleiades), {b} *Warawarae* (Venus: morning), {c} *Tanratellué* (Delta, Epsilon, and Zeta Orionis), and {g} *Walue* (Alpha and Beta Centauri). So it means that the guidance in N7G was celestial objects. Meanwhile, {d} *Manu'e* and {f} *Empang* were mentioned by Saleh [16]. He stated that *Manu'e* was an asterism that visualized as a chicken, but he did not tell the shape of *Empang*. Only {e} *Etapatae* did not have a reference. (2) Secondly, it seems that the table in figure 1 was used to determine the name of a day based on N7G's position at a certain time. Column 2 – 6 respectively meant afternoon (*asara*), sunset (*orowe*), midday (*tengasso*), sunrise (*abuwe*), and morning (*ele*). The last column gave the name of the day in Arabic alphabet: Friday – Thursday (up to down). This day determination probably had intention to be used as one of indicator which day was considered auspicious. This idea was still uncertain because we could not understand the meaning of signs in column 2 – 6. It needs some epoch time to confirm the position of N7G. This works need further research.

Aside from *Lontaraq*, Arsuka [17] explained that in the mid of 17th century, a Galilean telescope had been possessed by Gowa's Kingdom as a tool to observe stars in Makassar land's sky. Their Prime Minister, named Karaeng Pattingalloang (Karaeng mean Prince), used it to explore Moon's position and evolution. It was so unfortunate that the story about the telescope existence could not be found in any museum in Makassar. We assumed that there was likely other party who took over and look after the telescope during the colonial period, as for the original manuscripts of *Lontaraq* which is now widely claimed to be in Leiden, Netherlands.

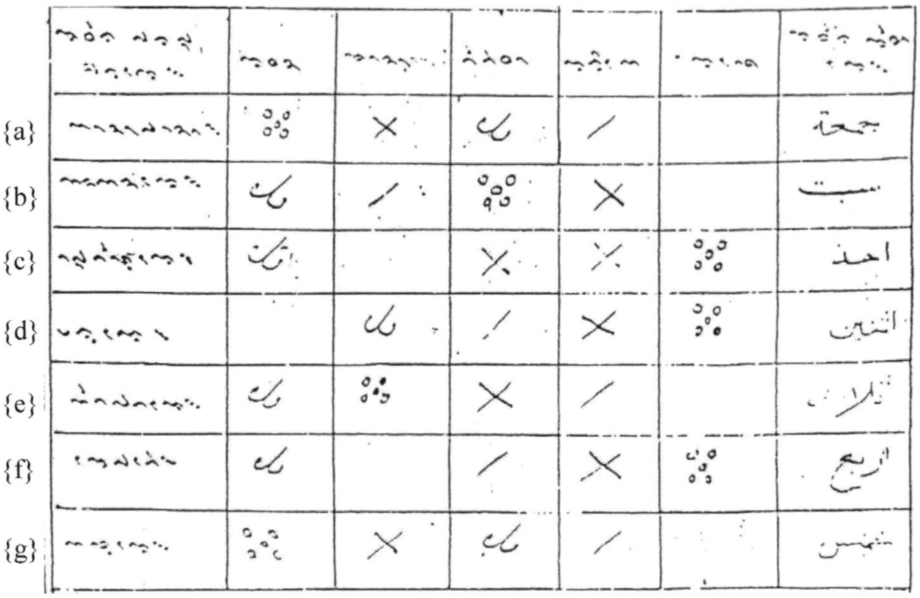

Figure 1. An example of *Lontaraq Pananrang* manuscript [15].

3. Ethnography aspects

Astronomy in Buginese-Makassarese had already established in navigation sector. Ammarel [8] had described some of it that been used by the Buginese in Balobaloang (small island located midway between Makassar and Bima of Sumbawa). Buginese - Makassarese had a sharing view that the universe was not an empty space, but all the contents and the space were proceeded naturally, while oceans was part of the cosmos that fulfilled with magicians and wonders [16].

Besides being used for navigation, the Buginese - Makassarese also used celestial objects to predict the weather. *Lontaraq Pananrang* accommodated a method to predict the weather for the purposes of agriculture, travel, trade, and even in building houses. The Forecasting was based on characterization of local weather within eight years period (*Sipariyama*). All of the years in *Sipariyama* period was named with Arabic letters. It was tailored from Islamic calendar which used Lunar's motion. The eight-year term was also used in Javanese Islamic calendar which one year had 354.375 days and became full cycle after eight years (2835 days) [18]. *Sipariyama* adopted this eight-year period probably because it had same background with Javanese Islamic calendar which is originated from Islamic culture, even though the name of the year and the number of days in some years had little difference. However, retrospect from history, the eight years period had been used by Buginese - Makassarese long before Islam arrived. This fact indicated that they had been aware with the regularity of celestial bodies.

It should be recognized that the results from study to evaluate the *Sipariyama* as a scientific term was still inadequate. This possibly was due to the oldness of *Lontaraq* manuscript so that the letters were hard to read and to interpret. It also needed a wider interdisciplinary studies that involved all expert in natural science, social science, and humanities. Therefore, more study to assess *Lontaraq* still needs to be improved so that the knowledge contained in *Lontaraq* as cultural heritage is maintained its glory.

4. Summary

This paper has presented some usages of astronomical aspect in Buginese-Makassarese culture based on historical and ethnographical sources. *Lontaraq* manuscript is one of the authentic evidence at how

Buginese- Makassarese society understood and used it in their daily life. *Oromporong*, *warawarae*, *tanratellue*, *walue*, and *Manu'e* were some of examples of celestial object's name that flourished the record of Buginese – Makassarese culture. The benefit of these celestial objects are not only for navigation purpose, but also for predicting weather to begin agricultural activities. Discussing the weather for the next 4 to 6 months is very beneficial to reduce some risk economically in calculating a cost of production. The discussion itself becomes a traditional events called '*tudang sipulung*' attending by farmers, climatological experts, and readers of *Lontaraq* (*palontaraq*) to decide when the agriculture activities (particularly to grow rice in paddy fields) should begin and how to manage collectively water in watering paddy fields. Using lunar calendar, *Lontaraq* characterized the local weather within eight years period called *Sipariyama*. It was little different from *Pranatamangsa* in Javanese that used solar calendar with one year period. Now the existence of *Lontaraq* and its knowledge are endangered because of the rareness. Therefore, from our consideration a wide research on *Lontaraq* is still necessary to be performed.

Acknowledgments

The authors feel indebtedness to the Hasanuddin University for supporting all the expense to present this paper in the International Symposium on Sun, Earth, and Life (ISSEL) 2016 in Bandung, on June, 3-4, 2016. The authors also thank to all colleagues from Department of Physics for the assistance in interpreting Buginese words written in *Lontaraq*.

References

[1] Abidin A Z 1971 *Indonesia* **12** 159-172
[2] Anderson K G 2003 *The Open Door: Early Modern Wajorese Statecraft and Diaspora* (Hawai'i: Dissertation of Doctor of Philosophy, University of Hawai'i)
[3] Hamid P, Mappasere and Batong H 1990 *Traditional Astronomy and Meteorology in the Province of South Sulawesi* (Jakarta: Department of Education and Culture, Project of Inventory and Development of Cultural Values)
[4] Daldjoeni N 1984 *Environmentalist* **4** (S7) 15-18
[5] Jafar, F 2005 *Dipole Mode Interaction with Monsoon According to Presipitation Variability in Province of South Sulawesi* (Makassar: Essay of Physics Bachelor Degree, Physics Department, Hasanuddin University)
[6] Prihatin L 2005 *ENSO (El Nino Southern Oscillation) Interaction with Monsoon According to Presipitation Variability in Province of South Sulawesi* (Makassar: Essay of Physics Bachelor Degree. Physics Department, Hasanuddin University)
[7] Hasanah N, Suriamihardja D A and Abdulsamad B 2015 *Proc. Makassar National Seminar of Physics 2015* (Makassar: Department of Physics, Hasanuddin University) p 224
[8] Ammarel G 2008 *Encyclopaedia of the History of Science, Technology, and Medicine in Non-Western Culture*, ed H Selin (USA: Springer) pp 324-333
[9] Hasanah N and Suriamihardja D A 2013 *Proc. Makassar National Seminar of Physics 2013* (Makassar: Department of Physics, Hasanuddin University) p 55
[10] Hidayat B 2000 *Astronomy Across Cultures: The History of Non-Western Astronomy* vol 1, ed H Selin (USA: Springer Science+Business Media Dordrecht) pp 371-384
[11] Daljoeni N and Hidayat B 1986 *In History of Oriental Astronomy*, ed G Swarup, A K Bag and K S Shukla (Cambridge: Cambridge University Press) pp 249-252
[12] Radiman I 1990 *Seminar of History and Culture of Galuh* (Tasikmalaya, Java)
[13] Aveni, Anthony. 1981. *Tropical Archaeoastronomy*. Science 213: 161 – 171.
[14] Wiramihardja S D 2011 *Proc. Seminar of Indonesian Astronomy Association*, ed. Dermawan et al. (Bandung: ITB) p 9
[15] *Lontaraq Pananrang* 1981 Published by Regional Government of South Sulawesi
[16] Saleh N A 1997 *Bosara* **7** (III) 43-50
[17] Arsuka N A 2002 *Inter-Asia Cultural Studies* **3** (2) 309-319
[18] Khazin M 2004 *Falaq in Theory and Practice: Calculation of Qibla, Prayer Times, Beginning of Lunar Calendar and Eclipses* (Yogyakarta: Buana Pustaka) p 119

Pawukon: from incest, calendar, to horoscope

Agustinus Gunawan Admiranto

Space Science Center, LAPAN, Jl. Dr Djundjunan 133 Bandung 40173, Indonesia

E-mail: gunawan.admiranto@lapan.go.id

Abstract. Javanese calendar has several cycles, i.e. 5 days (pasaran), 6 days (paringkelan), 7 days (week), 8 days (padangon and padewan), 30 days (month), and 365 days (year). There is another 210- day cycle caled pawukon which divided into 30 part caled wuku. This cycle originated from an incest tale about a king named Prabu Watugunung which married his mother named Dewi Sinta and his aunt named Dewi Landep. In this marriage they had 27 sons and all of them are called wukus. In this tale it was told that this incestuous relationship caused some havoc in the world and the gods decided to kill this family. After some struggle, all of them are killed and then the gods brought them up to paradise one by one starting from Dewi Sinta and ended with Prabu Watugunung. This ascencion needs 30 weeks (210 days) because to be ascended one wuku had to wait for 7 days, and after one cycle is finished the cycle starts all over again. The establishment of a cycle of pawukon is regarded as an effort to create a cosmos out of chaos (incestuous relationship), and furthermore pawukon is used as a kind of horoscope to determine one"s fate in the future. It is because the cosmos is regarded as a clockwork in which each element of this clockwork works in a predetermined fashion.

1. Introduction

Mankind are heavily influenced by cyclical patterns which occur in nature such as diurnal cycle as manifested in day and night, monthly cycle which can be seen in the changing phases of the moon, and annual cycle of stars. These cycles were then codified and quantified into a system which we call as calendars. The calendar is very useful for the working of the society and it also gives some certainty and standard about some elapsed time and a sense of security to man.

Calendar is a part of human culture that has a very important role in their lives. The existence of calendars is one of the sign of how advanced an ancient civilization is. Calendars are used as a means of calculating the time and the pattern of 24 hours in a day. The division of hour into minutes and seconds is rooted in the culture of Ancient Egypt and Babylonia [2]. The word calendar rooted from Latin word *calare* which means call or proclaim. In ancient Roman times, every new moon the priests proclaimed (calend) the arrival of the new moon through various rituals and trumpet [3].

In addition to providing a clear benchmark of the passing of an era, calendar also provides a kind of security to human because it provides certainty about time. Therefore, since time immemorial man has always interpreted the passage of time as a cyclical one. The linear passage of time is an invention of modern times, which got influenced by monotheistic religions with their eschatological concepts.

This short paper describes a cyclical Javanese calendar, especially to a cycle called Pawukon which is connected with a taboo phenomena in Javanese society i.e. incest. Furthermore, it will be described here how Pawukon developed into a kind of horoscope which described one's fate based on his or her day of birth.

2. Some cycles in Javanese calendars

Our calendar has only seven-day cycle (week) and thirty-day cycle (month), but Javanese calendar has several cycles, i.e.:

a. Five-day cycle (*pancawara*)
    Paing, Pon, Wage, Kliwon, Legi
b. Six-day cycle (*Paringkelan* or *sadwara*)
    Tungle, Aryang, Wurukung, Paningron, Uwas, Mawulu
c. Seven-day cycle (*saptawara*)
Dite, Soma, Anggara, Buda, Respati, Sukra, Tumpak
d. Eight-day cycle (*Padewan* or *astawara*)
    Sri, Endra, Guru, Yama, Lodra, Brama, Kala, Uma
e. Nine-day cycle (*Padangon* or *sangawara*)
    Dangu, Jagur, Gigis, Kerangan, Nohan, Wogan, Tulus, Wurung, Dadi.

In longer time scales there are monthly cycle and cycles called Pranatamangsa and Pawukon. In this respect, the Pawukon cycle will be discussed further.

3. Pawukon

Figure 1. Prabu Watugunung

Pawukon originated from a tale of a king named Prabu Watugunung which married his mother (Dewi Sinta) and his aunt (Dewi Landep) out of his awareness. This incestuous marriage gave them 27 children. Initially, Dewi Sinta was not aware of her incestuous relationship with his son, and when she realized this, she decided to end their marriage.

Because she didn't dare to ask Prabu Watugunung to end their marriage, Dewi Sinta then asked his husband (his son also) to attack the domain of the gods (*Kahyangan*). She told Prabu Watugunung that she want the wife of Batara Indra to be the wife of Prabu Watugunung. She asked his husband to attack the Kahyangan with the hope that they will perish during their attack.

The expectation of Dewi Sinta was realized when her husband and their children were perished during the battle with Batara Wisnu. Afterwards, Dewi Sinta asked the gods to take the spirit of her, her sister (Dewi Landep), Prabu Watugunung and their children to Kahyangan.

The gods granted her wish, and all of them were ascended to Kahyangan starting from Dewi Sinta, Dewi Landep, their sons, and this sequence ended with the ascencion of Prabu Watugunung. The ascension of each of them was done each week so this process needed 30 weeks (210 days) to finish. This ascension process is the origin of a cycle in Javanese calendar which called Pawukon in which the date after the ascension of Prabu Watugunung the process was started all over again from Dewi Sinta to repeat the cycle.

4. Closer investigation on Pawukon

In Javanese culture the marriage of Dewi Sinta and Prabu Watugunung is a taboo event and considered as an effort to destroy the existing social structure. The folklore which describe incestuous phenomena is quite common and can be found on other cultures, like Oedipus (Greek), Sangkuriang (West Java), Lumaluindung (North Sulawesi).

There are common ideas in these folklore. The incestuous relationship are established with no awareness of the parties involved, these relationships ended in tragedies, and orders were restored after the ends of these tragedies. In short, chaos which were started during the incestuous relationships become cosmos after the relationships were consummated.

What is cosmos? Cosmos is a state of being in which all events and circumstances are in ordered state and predictable. The predictability of a cosmos can be inferred from its cyclical nature, so this can explain the nature of the ascencion of the family of Prabu Watugunung to heaven which was executed one of a time every week so the cycle of ascension could be established from this process and a cosmos was restored. A cosmos is a system which works like a clockwork so the future events in this cosmos can be expected to occur bssed of the observed events in the past.

5. Pawukon as a horoscope

Because of its cyclical nature and a part of cosmos in Javanese culture, some people regard that Pawukon can be used to predict one's future fate based on his or her date of birth. This results in a building of a horoscope no different from western hosorscope which based on the passage of the Sun in the stellar constellations.

Pawukon horoscope is a part of a complicated system in Javanese culture called *petung* to calculate one's fate, the determination of important events (marriages, house moving, journeys, commemoration of deceased persons, etc), and one's general predisposition so these events can be conducted properly without any unwanted results caused by the obstruction of predetermined order which exists inherently in the Javanese world.

6. Conclusion

One of the purposes of the establishment of of a calendrical system is to get a sense of security and assurance in the world they inhabit. The cyclical pattern of the calendar give them the assurance that the pattern of all existing events and circumstances are predetermined from the beginning and in order to survive in the world people have to conform with this pattern.

In Pawukon case, the consecutive ascension of the family of Prabu Watugunung which then repeated continuously to be a cycle is aneffort to build a cosmos which was previously destroyed in an incestous relationship between Prabu Watugunung and Dewi Sinta.

In this cosmos, in order to survive, people have to adjust their lives with the life of the world, the microcosmos has to be adjusted with the macrocosmos. This adjustment is accomplished using *petung* mechanism as described above.

References

[1] Nagara 1957 *Almanak Waspada* (Jogjakarta: PESAT)
[2] Kelley D H and Milone E F 2011 *Exploring Ancient Skies: An Encyclopedic Survey of Archaeoastronomy* (Springer Science+Business Media, LLC)
[3] Langone J 2000 *The Mystery of Time* (National Geographic Publisher)

The Assembled Solar Eclipse Package (ASEP) in Bangka Indonesia during the total solar eclipse on March 9, 2016

Cahyo Puji Asmoro[1], Agus Fany Chandra Wijaya[1], Nanang Dwi Ardi[1], Arman Abdurrohman[1], Judhistira Aria Utama[1], Asep Sutiadi[1], Hikmat[1], Taufik Ramlan Ramalis[1] and Bintang Suyardi[2]

[1]Earth and Space Laboratory, Physics Department, Universitas Pendidikan Indonesia, Jl. Dr. Setiabudhi No 229 Bandung 40154
[2]Kementrian Komunikasi dan Informatika (KOMINFO), Jl. Medan Merdeka Barat No 9 Jakarta 10110

E-mail: cp.asmoro@upi.edu

Abstract. The Assembled Solar Eclipse Package (ASEP) is not only an integrated apparatus constructed to obtain imaging data during solar eclipse, but also it involved sky brightness and live streaming requirement. Main four parts of ASEP are composed by two imaging data recorders, one high definition video streaming camera, and a sky quality meter instrument (SQM) linked by a personal computer and motorized mounting. The parts are common instruments which are used for education or personal use. The first part is used to capture corona and prominence image during totality. For the second part, video is powerful data in order to educate public through web streaming lively. The last part, SQM is used to confirm our imaging data during obscuration. The perfect prominence picture was obtained by one of the data capture using William-Optics F=388mm with Nikon DSLR D3100. In addition, the diamond ring and corona were recorded by the second imaging tool using Sky Watcher F=910mm with Canon DSLR 60D. The third instrument is the Sony HXR MC5 streaming set to be able to broadcast to public domain area via official website. From the SQM, the value of the darkness during totality is quiet similar as a dawn condition. Finally, ASEP was entirely successful and be able to fulfil our competency as educational researcher in university.

1. Introduction

Total Solar Eclipse (TSE) has become a magnet for all people, especially for researchers. Not only the beauty of a rare phenomenon and fascinating but also the amount of information that can be extracted related to science. Moreover TSE March 9, 2016 was occurred in Indonesia, was strong circulated by TV and social media. The formation of a national committee of a total solar eclipse in 2016 became an important part in the success of the moment of the eclipse this time, learning from the experience of 1983. TSE committee was able to become a center of information to the public and reporters so that the moment TSE can be enjoyed.

Universitas Pendidikan Indonesia, one of the leading universities in Indonesia, under department of Earth and Space Laboratory, has participated actively to oberseve 9 March TSE. Following [1, 2, 3, 4] about solar eclipse instrument and scientific observations, and solar photographic techniques and image processing, also in order to fulfill Tridharma of university (eduction, research, and community service), we created a package of a total solar eclipse observation assembly consisting of four instruments commonly used by individuals, schools or clubs. We give a name The Assembled Solar Eclipse Package (ASEP), an integrated apparatus constructed to satisfy in getting imaging data during solar eclipse and also it involved sky brightness and live streaming requirement.

2. Methods

The ASEP instruments are Sky Watcher Telescope F 910 mm, William Optics Telescopes F 388 mm, Video Camera Sony HCR MC5, and Sky Quality Meter (SQM) linked by a Laptop and motorized mounting.

Figure 1. The ASEP controled by three persons, the first person controlled the Mounting Vixen Spinx and SkyWatcher 910mm + Canon 60D, the second person controlled William Optics + Nikon D3300 and SQM, and the third person controlled Sony HCR MC5 and release-installing the Filter ND5.

The role of both telescopes is to acquire the image of a great total solar eclipse, SkyWatcher telescope was running on video mode when it would enter the second contact and after the third contact. The William Optics telescope was used on image mode. Data from both telescopes are a learning media in education.

Sky Quality Meter instruments that was stored at the upper end of tube Skywatcher, was to investigate sky brightness during the daytime when the total solar eclipse occurred. Video cameras Sony HCR MC5 were attached to Sky Watcher telescope for video capture and were broadcasted live via the portal Ministry of Communications and Information Technology Republic Indonesia and www.gerhana-indonesia.id so they can be seen by the public, this is a form of community service.

3. Result and discussion

The parts from William Optics F=388 mm with Nikon DSLR D3300 captured TSE with resolution 300 dpi, 24 bit, ISO 100 and with a different exposure time because no detector can satisfactorily cover whole brigthness range on single image [2]. At exposure time 1/2500 sec we got the briliant reddish prominence (validate with Solar Dinamic Observatory) [5] and at exposure time 1/100 sec we could see the crown majesty of corona. So after combined two image in different exposure time with software Photomatix Pro Ver.5 [6] we obtained prominence and corona at the one single image.

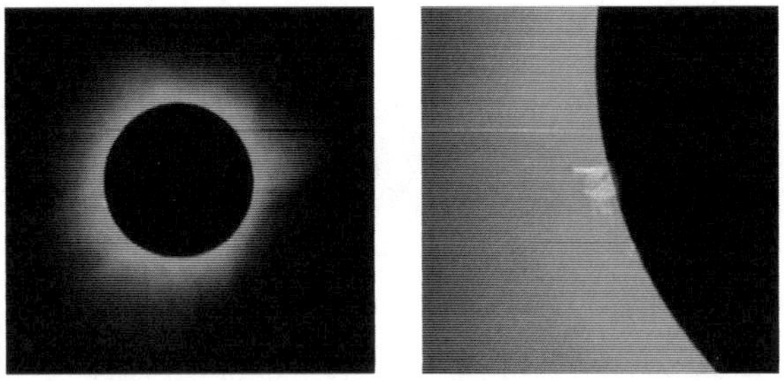

Figure 2. Image data from William Optics F=388mm

SkyWatcher F=910 mm with Canon DSLR 60D was used to record the zoom of TSE. With resolution frame 1920 x 1088 and frame rate 25 frames/second. The video was taken at Sport Mode, to track the Sun. So the image a little bit over exposure for the corona. But, fortunately we can see "The Diamond Ring" before and after totality. Size of the diamond before totality was smaller because of the cloudy weather.

Figure 3. Video capture from SkyWatcher F=910mm

The sky brightness could not be determined at the beginning of the eclipse due to the high intensity of the sky i.e., too high for SQM to be functioned normally. So we installed ND5 filter to provide two functions, the first is to avoid the sensor to be damaged due to continue pointing to the Sun and it can reduce the intensity of incoming light so that it can be read by SQM. We released ND5 filter before the second contact and re-instaled it after the third contact. The sky brightness when maximun totality is 12.21 mag/arcsec^2. It means that the darkness of the totality of the eclipse this time is roughly similar to a half hour after sunset or before sunrise [7]. We can see the "valley" at a temperature (figure 4), this means that there is an influence when the moon covered the solar disc. The decline begins at 7:11:22 until 7:43:11 local time. Temperature decreased gradually to 1.3 °C and then rise over time.

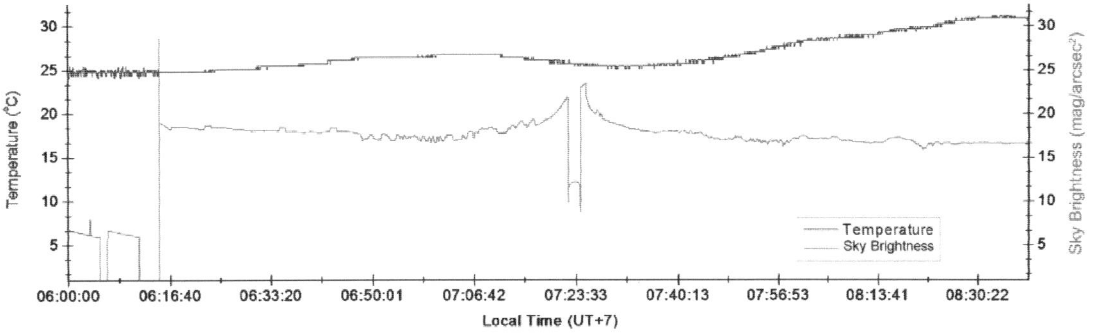

Figure 4. Graphics from Sky Quality Meter

Video streaming of the eclipse can be accessed at http://gerhana-indonesia.id/id/live-streaming/. Initially the process stream has no problem, but after 45 minutes there was a problem on the Internet because too many people accessed the internet in these locations. But all the data stored on the internal hard drive. After process of editing, the video was uploaded on Youtube to be accessed by the public.

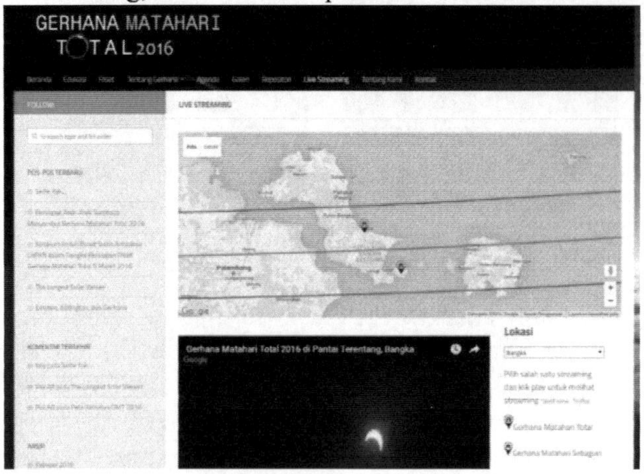

Figure 5. Live Streaming at website http://gerhana-indonesia.id/id/live-streaming

4. Conclusions

Finally, The ASEP was able to fulfill our competency as educational researcher in university. For education and teaching, picture and video will be a learning media for the phenomenon of eclipse and the subject of literacy in earth and space science. For research and development, sky brightness data from this eclipse can be advanced research material for the next eclipse. For public services, via video streaming we want to give confidence to enjoy the eclipse and not to scare of it. The results obtained from The ASEP can be seen on Youtube Channel "Laboratorium Bumi & Antariksa" (link Video : https://youtu.be/iri-wqy_PSo).

References

[1] Phillips K J H, et al 2000 *Solar Phys.* vol 193 (German: Kluwer Academic Publishers) p 259
[2] Pasachoff J M 2009 *Research in Astronomy and Astrophysics* **9** 613-634
[3] Littmann M, Espenak F and Willcox K 2008 *Totality : Eclipses of the Sun* (New York : Oxford University Press)
[4] Yudhoyono S B 2012 Undang-Undang Nomor 12 tentang Pendidikan Tinggi (Indonesia : Kementrian Sekretariat Negara)
[5] Pesnell W D 2016 *Solar Dinamic Observatory Data* (Maryland : NASA Goddard)
[6] HDRSoft 2016 *Photomatix Pro version 5.0 User Manual* (United Kingdom : HDRSoft)
[7] Flanders T 2008 *Rate Your Skyglow* (Cambridge : Sky&Telescope)

Tutulemma of near equator Partial Solar Eclipse 2016

F Mumtahana[1], Sartika[2], A G Admiranto[1], E Sungging[1], M Z Nurzaman [1], R Priyatikanto[1,3], T Dani[1]

[1]Space Science Center, LAPAN, Bandung, Indonesia
[2]Atmospheric Science and Technology, LAPAN, Bandung, Indonesia
[3]Doctoral Program in Astronomy, ITB, Bandung, Indonesia

E-mail: farahhati.mumtahana@lapan.go.id

Abstract. Tutulemma which stands for tutulma (eclipse) and analemma is a Turkish name to describe analemma which contains solar eclipse. The photographs of analemma which looks like a figure 8 pattern of the solar path in the sky throughout a year is very common. However, this observation is rarely done in low latitude countries, especially tutulemma's plot. Related to the eclipse event on 9 March 2016, we built the first tutulemma in Indonesia which cointains partial solar eclipse observed from the rooftop of LAPAN Bandung office. The purpose of this endeavor is also to understand how the sun moves in one year, particularly at low latitude region based on observation data. We took the picture of the Sun every week at the same time of the day (the time of the eclipse). In this observation, we used a tripod and DSLR camera with a variable ND filter to take the data. To obtain the pattern, the weekly data were combined by image stacking using the same foreground. Finally, we got a figure 8 shaped pattern tutulemma which a bit different from high latitude country ones.

1. Introduction

As it is described in the book Sun Heavenly Mathematics and Architecture [1], analemma is the trajectory traversed by the Sun across the sky in a certain elapsed time when the picture of the Sun is taken from the Earth surface at the same position and time along a year. The combination effect of 23.5° obliquity of Earth's axis and the elliptical orbit of the Earth causes the Sun seems to move up and down in the sky and generates two loops like figure 8. In Northern hemisphere, the highest point of apparent position of the Sun in figure 8 is generated around the Summer Solstice [2]. Then, it starts moving down the sky, generating the first loop of the figure 8 and it repeats similarly during the winter months to generate the second loop of the figure 8 curve. Other than the elliptical Earth's orbital path, the Sun is not in the center of the Earth's orbit. It means that there are one part of the orbital path, called Perihelion, which is closer to the Sun and the farther one, Aphelion. Consequently, the Earth seems like move faster around the Sun when it is at Perihelion in Winter Solstice than when it is at Aphelion. This has the effect of flattening out the bottom half of the curve.

The first successful single photo frame of an analemma was taken from February 27th, 1978 to February 17th the next year by Dennis di Cicco [2], an amateur astronomer from New England. The analemma's position is determined not only by the latitude at which the images are taken, but also by the time of day. The direction of the analemma also differs depending on the observer location on Earth. [2] Observers in the Northern hemisphere will obtain an analemma curve with the broader loop at the bottom and observers at the North Pole will observe only the top loop of the analemma. Meanwhile,

Journal of Physics: Conference Series **771** (2016) 012021 doi:10.1088/1742-6596/771/1/012021

observers in the Southern hemisphere will see otherwise. Higher or lower latitude will give more vertical photographs of figure 8 analemma. The apparent variation position of the Sun has been intensively studied by astronomers as it is well understood and mathematically calculated [3][4]. Based on theory, the analemma at the equator should look like completely horizontal 8-shaped even though probably no one has ever made equatorial analemma's photographs.

There are some analemma portraits, which contains the image of a solar eclipse known as Tutulemma. Tutulemma derived from the Turkish word tutulema (eclipse) and analemma which means analemma containing eclipse. The first tutulemma photograph published by Tunç Tezel from Turkey who succeed taking the pictures with the background of the sun's path in one year and total solar eclipse occurring in March 29, 2006 using a playground as its foreground [5]. In Asia, particularly Indonesia, analemma potrait has not been quite popular or perhaps yet unpublished. Therefore, Tutulemma team consisting 7 members in LAPAN has been formed with three main aims. First, we want to make the first analemma photograph in Indonesia which contains imagery of solar eclipse 2016. As it was conducted in the office rooftop of LAPAN Bandung, the eclipse image was Partial Solar Eclipse (PSE) only. Second, we want to proof how the analema's figure in latitude -6° is consistent with the theory. The trajectory traversed by the Sun in Indonesia which located near the equator would be very unique and different with those from Turkey or high latitude countries which had already made a lot of analemma photographs. Theoretically, analemma at near equator will shaped like horizontally tilting figure 8. Third, for educational purpose, such as understanding how the sun moves in one year particularly at low latitude region based on observation data, studying equation of time and coordinate system. There are some variations in the length of a solar day and the difference between solar time and mean time. The variations are encapsulated in an expression called the Equation of Time (the term 'equation' is used here in a historical sense, meaning a correction or adjustment) [6].

2. Data and method

Making analemma's photograph might look simple, however, it needs consistency, patience, and teamwork of a weekly observation along a year. We were divided in two small team consisting three up to four people which responsible taking the data according to the schedule. Data were collected every Wednesday at around 07:21 LT when maximum PSE occurs in Bandung. The shooting position and instruments must be the same throughout the year of data collection. We used DSLR camera and tripod configured to be fixed for the weekly observation shown in figure1.

Figure 1. Default position of the camera and tripod.

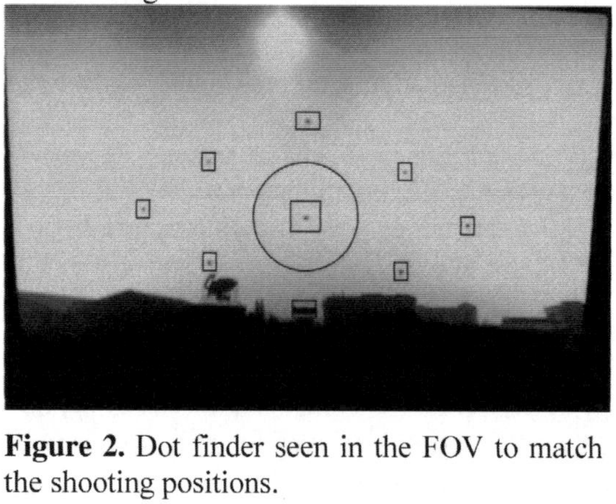

Figure 2. Dot finder seen in the FOV to match the shooting positions.

Our procedures of taking analemma data that was written in the Analemma team's rules or Standard Operating Procedure (SOP) are as follows: image acquisition was conducted every Wednesday morning at the rooftop of LAPAN Bandung office, started from March 2015 until March 2016. There are 48 observation days in one year, which is enough to construct analemma's path. The camera used was CANON EOS 700D, with Natural Density (ND) variable filter up to ~ 400 for reducing sunlight. Filter was set to see the homogeneous (at certain points around the word "MAX") in which also depends on the weather. The shinier day, higher number in variable filter should be set, yet in cloudy day less number was better as the clouds also performs as filter. Camera setting were 1/60s exposure time, focal ratio f/9, ISO 100, focal 18mm (maximum zoom out), while the focus setting was adjusted using the Auto Focus (AF) or manual. For the purposes of better processing, images were taken in RAW + best JPEG mode. The position of the tripod and camera were arranged so that the field of view (FOV) finder of camera may look the same for each week (figure 2). We used several objects seen from roof top LAPAN Bandung office to be marked in camera finder in order to make sure, if the camera was always set at the same position in every data acquisition.

The method used in final processing was the image combining and compositing. Before that, every weekly observation images were categorized as good and bad data and were processed individually. Ideally, to get a portrait which is consistent and has equal position, it requires a fixed monopod or a fixed camera attachment like the others did. As our team only used a portable tripod, weekly images sometimes were captured in different angle and size. To get the exact sun's position in individual frame, the appearance of foreground or scenery is very important to be used as reference. Even though in some cases, the portrait of the sun with maximum ND filter will remove the entire foreground and only leave small yellow sun's image in dark frame. Therefore, all the best weekly images were processed separately first in order to emerge both the reference foreground and the best sun's image in the same frame. Then, to get the desired result, after a year passing at around PSE 2016, each of those processed weekly image layers were combined one by one starting from the oldest to the newest. When the foreground of every frames were stacked precisely, we finally obtained the figure like 8-shaped from one-year solar pattern.

3. Result

During a year observations, we obtained various quality of data: sometimes good in clear sky, sometimes bad due to cloudy weather and problem in instrument even the slight observer's error. Figure 3 and figure 4 shows the example of bad data due to the weather. On February, 10th 2016 Sun's image was taken too bright so this kind of pictures need further processed (figure 3). On February 3rd 2016, the weather was too cloudy and unfortunately, no data was obtained. (figure 4). A whole one year, there were several weeks with these kind of bad data and two weeks of no data. However weekly data mostly look like in figure 5, which were very dark with perfect small yellow sun.

Figure 3. Image of the Sun taken on February, 10th 2016 which is too bright.

Figure 4. Image of cloudy day taken on February, 3rd 2016.

Figure 5. Example of good data with maximum ND Filter taken on January, 13th 2016

To complete this project, time lapse of Partial Solar Eclipse which is shown in figure 6 was taken with same instrument with maximum zoom in between 06:20 LT – 08:32 LT at rooftop office of LAPAN Bandung. In this place, the Moon covered up 88% of Sun's disk at 07:21 LT on March, 9th 2016 Combining the maximum PSE image, which is marked in circle, with images obtained during one year around before and after PSE, finally we made a figure 8-shaped pattern as it is shown in figure 7.

It can be seen in our analemma photograph, there are some blank space around February and March due to the bad weather. We can also recognize different shape of sun's images for each week due to the different style of the observers, instrument problem and more importantly the morning local weather. For example, the dots look bigger around summer solstice than around winter solstice, due to cloudier weather by the time images were taken. The morning weather was getting cloudier around August, so one of the sun's image appears not in full circle shape because the sun was hiding between the gaps of clouds.

Figure 6. Time Lapse of PSE 9 March 2016 observed in Bandung

Figure 7. Analemma picture taken from March 2015 until March 2016 with PSE in circle mark.

This nearly horizontal analemma describes typical sun's path from near equator, particularly in latitude -6^0 South at 7:21 am. During summer solstice around June, the sun was at Aphelion that can be considered like it moves more slowly, therefore the gaps between the dots look narrower. Meanwhile, during winter solstice around December, the gaps became wider because the earth moves faster at perihelion. The 8-shaped loop is broader at the top-right in winter solstice, because we observed from southern hemisphere. Hence, as it was calculated and discussed by the old astronomers, the shape of our one year analemma observation is certainly look like the theory we mentioned above.

4. Summary and discussion

We have briefly show our result of one year analemma's observation which is consistent with the theory and calculation precisely for low latitude region. In reality, along a year, weekly images vary widely with the technical problems (change SOP, instrument problem), observers' error (different angle, zoom in/out), and weather (clouds, haze). Therefore, Ideal Analemma's photography should be conducted exactly not only with the same instrument and setup, but also with the fixed tripod/monopod and with the same person if possible (max 2). Finally, Tutulemma's PSE in Lapan can be useful as original reference for education purpose. Indonesia is situated around the equator, so this will definitely have the distinctive characteristics of its analemma shape. Understanding the analemma can be the study of the equation of time, the coordinate system of the earth and celestial and how they move, calculation time the sun rises and sets, and some other knowledges.

Acknowledgments

We thank our colleagues in sun and space group of Space Science Center for the kind assistant during our weekly observation along one year. We also thank to Space Science Center LAPAN for giving support to conduct this activity and allowing us joining ISSEL 2016 to publish our satisfying result.

References

[1] Lee Jin You, Roger, Lee Ji Hao, Theophilus, Lim Guang Yong, Lim Ghim Hui, Lim ShuEn Adele and Lim Wee Kee Sun and Architecture *Heavenly Mathematics GEK 1506*

[2] diCicco D 1979 *The First Ever Analemma* http://twanight.org/newTWAN/photos.asp?ID=3001422
[3] Lynch P 2012 *Irish Math. Soc.* **69** pp 47-56
[4] Kittler R and Darula S 2004 *Architectural Sci. Rev.* **47** Issue 2 pp 141-144
[5] Tunc T 2007 Sky & Telescope **113** no 1 p 134
[6] Hughes D W, B D Yallop and C Y Hohenkerk 1989 *Mon. Not. Roy. Astr. Soc.* **238** 1529–1535.

International Symposium on Sun, Earth, and Life (ISSEL) IOP Publishing
Journal of Physics: Conference Series **771** (2016) 012022 doi:10.1088/1742-6596/771/1/012022

UNAWE Indonesia project: raising total solar eclipse 2016 awareness through educational packages

A. T. Handini[1], Y. Yulianty[1,2], P. W. Premadi[1,2,3], A. Annafi[1]

[1]UNAWE Indonesia
[2]Bosscha Observatory, Lembang, Indonesia.
[3]Astronomy Department of ITB, Bandung, Indonesia

Abstract: On March 9th 2016, some regions in Indonesia witnessed the Total Solar Eclipse (TSE). At that time, Indonesia was the only mainland in the world that could observe TSE. This moment is extraordinary because its probability of being observed at the same point happens every 350 years. On a purpose of raising public awareness and engaging more participations, UNAWE Indonesia developed a handy material and sent the so called educational packages which provided essential information about eclipse, especially TSE. Each package contains of modul of TSE study and materials for simple TSE activity so people who received the package could observe TSE safely. In the delivery of this program, we established a collaboration partner, whom are teachers, as an 'eclipse ambassador'. The local partners were responsible for socializing the information for school or their community. We encouraged them to conduct an observation activity and ask them to fill an evaluation report. As a result, 48 out of 59 partners or equal to 81.3% of all partners has sent the evaluation of the package and documentation of the activity during TSE. They responded positively as it was easy to understand, functional, and convenient.

1. Introduction

The trajectory of Total Solar Eclipse (TSE) in Indonesia was a narrow and long path. It extended from west to east Indonesia. It passed through not only at big cities such as Palembang, Palu, Ternate and Balikpapan, but also other remote districts across the archipelago. These areas are known for having difficulty with comunication and transportation access such as Muko-Muko in Bengkulu, Kapuas in West Borneo, also North Halmahera in North Mollucas.

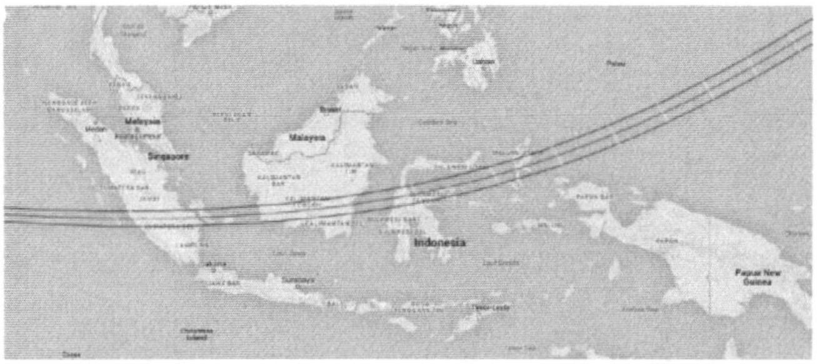

Figure 1. The trajectory of TSE 2016 in Indonesia

As a member of UNAWE international, UNAWE Indonesia has a mission to introduce beautiful cosmos for children, especially children who are in unfortunate condition in socio-economically as well as geographically. We realized how it is became very important for the public to get the need of correct information of TSE and how to observe it. We were conducting the project by developing simple teaching material that provide resource to educate student the fundamental principle of science behind the eclipse that is easy to understand and deliver. These material will helped teacher (especially science teacher) for conducting a selected topic science in a guided. The material also give information when TSE occurs and how they can safely observe in a right method.

Moreover, with the limited number of big expedition by both professional and amateur astronomers to observe TSE across the islands, we wanted to share knowledge by ensuring a wider audience of Solar Eclipse observation in rural area with this first hands experience. The distribution points of our package not only targeted areas in the totality path, but also it wider than that.

2. Method

2.1 Gathering eclipse ambassadors

Local school teachers and educator were our main target to be given the educational package. These teacher would be asked to conduct socialisation program, workshop activities and arrange observation on 9th of March. One of main network we got to get the teacher was Pengajar Muda from Indonesia Mengajar.

We were choosing the person in local area who have the same spirit in educating the children. Pengajar Muda has one year teaching placement at elementary school in remote area with difficult access of transportation and communication. Their presence in those areas not only for teaching, but also they have to convince parents and local teachers how important education is for their children or student. For the local teacher and educator, we have to find their contact with our networking. After we know their contact, we personally asked them by phone call or text whether they are willing to be our ambassadors. When they were agreed to conduct the program, they received our package.

2.2 Developing material study and arranging educational package

We set up a simple package with all our materials inside it. Teachers would be able to not only deliver the information but also conducting the observation with ready to use eclipse glasses. We also selected material study, topic and activities highly on TSE topic. Material that we provide educational packages included some of material things and simple learning module to execute in school and communities. Here are materials in our educational package:

a. *Booklet*
Booklet provided contains essential and brief information of TSE on the March 9th, 2016. The booklet gives information about the solar eclipse: what eclipse is, how eclipse happens, as well as when and where the eclipse occurs, how to observe eclipse safely.

b. *Activity material*
The activity material consist of aluminium foil and compact disc (CD). We sent aluminium foil to make the pinhole box because of, how it is considered as a hard to get material in those remote areas. And CD contains of activities modules and videos about various astronomy topics especially eclipse. We put also all material study in our website as an open resources to ensuring a larger audiences.

c. *Teacher paper work*
The papers work consist of Term of Reference(ToR) and evaluation paper. ToR as a guide on how to conducting the programs that are socializing the TSE and plan observing activity in the day of eclipse with children, teacher and communities. And for evaluation paper, our ambassadors need to fill the question and sent it back to us along with the activity documentations (if it possible)

d. *Other materials*
We sent the eclipse data location for spesific area that received our educational packages. We also sent 5 eclipse glasses for each educational packages. These eclipse glasses can help them to observe TSE safely.

Figure 2. Our materials and the packages

3. Result

Figure 3. The distribution map of recipients locations that received our educational packages.

In collaboration with Indonesian Postal Office (*PT Pos Indonesia*), our educational packages have reached 59 locations (schools) across 14 provinces in Indonesia within 2 months of delivering period (Jan-Feb 2016) . We have successfully delivered all way through from the Western tip point of Indonesia in Muko-Muko district to East Indonesia in East Halmahera district, both ends of the North in Natuna Islands district and south of Indonesia, Kota Kupang. The number in the circle of figure 3 represents the number of packages that received in the locations. And we also have successfully sent 42 packages to 42 locations in totality path from all of the packages.

We have received 48 evaluations of 59 packages that we distributed or 81.3% from all packages. The evaluation papers sent back to us for our evaluation after the project. 45 evaluation papers have been received along with the activity documentations, 3 evaluations were carried out by phone and 11 evaluations can not sent back to us due to difficulties of transportation and communication access.

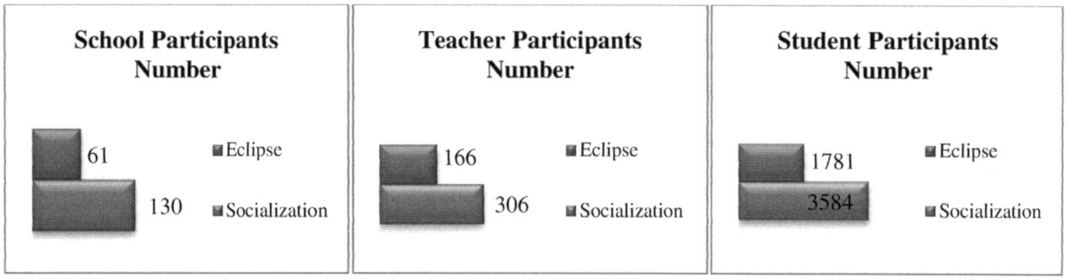

Figure 4. Participants number at all locations during socialization and eclipse day.

From the evaluation papers, the participants reached more than 3000 that involved socialization and eclipse observation arranged by our ambassadors. From the figure 3, the number on socializiation is more than the number on the day of eclipse. The decrease number happens because of several factors which are bad weather in the day of eclipse particularly in West Indonesia, while the national holiday when eclipse day so many people choose to observe at their house, and because they come to several different schools, but when the observation day they just organize eclipse observation in one location. It was a well performed project that shown how engaging teacher to

Figure 5 derived from the evaluation paper that 96.5% of our ambassadors can implement the project. It is equal with 44 ambassadors had succesfully implemented the eclipse project from 48 ambassadors who sent back the evaluation papers. This show that our project as deliverabled. There were 4 ambassadors that had not successfully conduct the eclipse observation. It is due to: bad weather in Borneo and Sumatra, parents prohibition in Musi Banyuasin district, and moslem pray in Muko-muko distrct.

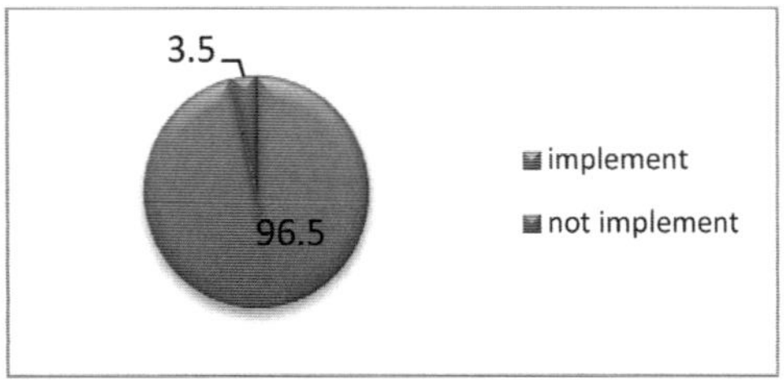

Figure 5. The deliverability of eclipse project for ambassadors

4. Summary and conclusions

The conclusions reached in this project can be summarized as follows:

a. For some locations, educational packages program is the main program, even the only program, related TSE. Beyond TSE, the program could be an inspiration for teachers to teach or deliver about science in class.

b. From evaluation reports of Eclipse Ambassadors, we seen that the packages were very useful, easy to understand and implemented in their schools. Indeed, our program is an educational program that is new to be implemented along with a phenomenon that will happen. We also face the challenges of both technical and non-technical when prepare this program, such as making the module that easy to read and understand, setting up the packages within a very short time, and find contact of Eclipse Ambassadors also ask them if they want to organize the program in their location or school.

Figure 6. The activities in locations that received our packages.

Acknowledgments

This project was supported by PT Pos Indonesia for distributing and guarding our packages until safely arrived at the locations. We thank our colleagues from Yayasan Ganesha ITB83 for their contributing to our project. Thanks to Yayasan Gerakan Indonesia Mengajar and Bosscha Observatory for our collaboration in this project. We also thank to Sahabat Gerhana Matahari for their donate.

Public outreach and education during the 2016 total solar eclipse in Palu and Malang

A P Rachmadian[1], C Kunjaya[2], W Wahono[3], A A Anugrah[1]

[1] Universitas Ma Chung, Villa Puncak Tidar N-01 Malang 65151, Indonesia
[2] Institut Teknologi Bandung, Ganesha 10, Bandung 40132, Indonesia
[3] Wow Photography, Mangga Besar 8 no 11, Jakarta 11180, Indonesia

E-mail: almirapraza19@gmail.com

Abstract. MAGIC (Ma Chung Galactic Club) of Ma Chung University, Malang, is one of the amateur astronomers club who did public outreach and education during the Total Solar Eclipse March 9, 2016. The motivation for doing this was the bad experience during Total Solar Eclipse 1983. At that time the Indonesian government forbid the people to observe the Total Solar Eclipse in a way to avoid blindness. We try to fix this misunderstanding by educating people the safe way to enjoy the partial and total solar eclipse. MAGIC team was divided into two teams, one team went to Palu and did the solar eclipse related education in six high schools before and during the eclipse. The other team did the observation on Ma Chung University campus, Malang, to accompany people who want to see the partial solar eclipse through filtered telescopes. The sky during the solar eclipse was clear both in Malang and Palu. People were very excited and satisfied with the solar eclipse, and their interest to astronomy is increased.

1. Background

Amateur astronomy is a hobby where participants enjoy watching the sky and the abundance of objects as seen with the naked eye, binoculars, or telescopes. Even though scientific research is not the main goal, generally many amateur astronomers could still make a contribution to astronomy.

MAGIC (Ma Chung Galactic Club) is one of the amateur astronomer clubs in Malang founded since September 2015. The purpose of this club is not just to provide an experience as a hobby but also, in part, for education. There are approximately 50 members in this club, many of whom are students of Ma Chung University. MAGIC sometimes partners with MAC (Malang Astronomy Club) and works in collaborations in some projects and observations. There were many activities held collaboratively such as moon observation during the International Observe the Moon night, planet observation, stargaze, scientific discussions, Galaxy Forum and the most significant one is the Total Solar Eclipse. The information of the eclipse is available on NASA website [1].

The Total Solar Eclipse observation was done in Malang (East Java) and Palu (Central Sulawesi). It was a great opportunity to do a public outreach for astronomy, to raise the general awareness about the universe. We learned that during the Total Solar Eclipse of 1983, many locals were exposed to incorrect information based on traditional mysticism and incomplete education of the solar eclipse, afraid of blindness *et cetera*, making them afraid of observing this remarkable event. Many printed source are available regarding this, for example [3].

In 2016, we wanted to work together with the astronomers from LAPAN, ITB, Planetarium and other astronomy clubs to reach out to as many people as possible to give them more correct information on the Total Solar Eclipse, in the hope that many could then enjoy the Solar Eclipse moments in the correct and safe way. There are many references connected to the learning of safe observation and practices during the Solar Eclipse, see for example [2].

2. Total solar eclipse

The total solar eclipse took place on 9 March 2016 [1]. Indonesia was the only country which was traversed by the totality, so that it is a great opportunity to do public education in astronomy and science for Indonesian people. The total eclipse was expected to be obviously visible in many parts of Indonesia, including Palembang, Bangka Belitung, Palu and Ternate. As the solar eclipse arched across the skies of Indonesia, millions gazed upwards hoping for a glimpse of the rare natural phenomenon. Unfortunately, observers in Malang, like most of the other regions of Indonesia, could only enjoy a partial eclipse, and not in its totality.

3. Eclipse related activities

Figure 1. Students in Palu tried the telescope

Figure 2. Preparation in Malang

Before the many planned Solar Eclipse events, several meetings and observational trials were executed by MAGIC in collaboration with MAC. During the 2016 event, the MAGIC team was divided into two – one team went to Palu and did a series of solar eclipse-related education seminars, a week prior to the eclipse day, and a mass eclipse observation together. The other team did the observation in Ma Chung University, Malang, to accompany those who had wanted to see the partial solar eclipse through filtered telescopes.

In Palu our team visited 6 high schools a week before the eclipse. We delivered interactive lectures on astronomy, and also invited students and teachers for a solar observation through a telescope equipped with a solar filter. We informed students on how to observe the solar eclipse safely and invoked more interest in astronomy especially during the precious moment of Total Solar Eclipse. Most of the students were very enthusiastic, because it was their first experience to see the Sun through a telescope.

On 9th March, from 07.27 a.m WITA to 10.00 a.m WITA, we observed the Total Solar Eclipse together with about 100 observers from all walks of life in a private villa in Donggala Kodi, half way up a mountain, west of Palu City. Although the main purpose of the observation was public outreach, we also did eclipse astro-photography. See the result in figure 6. All the visitors could see the sun through our telescope, mostly with solar filter attached, or through the eclipse glasses particularly before the first until the fourth contact. During the first until the second contact, the sky slowly became darker, as the temperature dropped. The Sun was totally covered by the Moon for 2 minutes, from 08.37 to 08.39.

Between the second and third contact, the solar disk was totally obscured by the Moon. The solar filter was removed from the telescope and everyone could take off their eclipse glasses so they could see the total eclipse directly. However, a few people were lucky enough to see the totality through our telescope, because the totality was so brief, only about two minutes. At that moment, the sky was dark enough to make Venus reappear in the sky. Most of the people were not aware of this, until it was announced that Venus had appeared.

Figure 3. Give instruction to public

Figure 4. The atmosphere during the The Solar Eclipse in Palu

In Malang, the observation was held at the plaza of Student Center of Ma Chung University. Observation preparation started from 05.30. Guests and reporters came as early as around 06.00 a.m. because they did not want to miss the moment. A total of approximately 100 visitors, including foreigners, and 10 reporters from different media came. Although it was not a total eclipse in Malang, everyone could still feel the brightness of the morning sky dimmed and a little drop in temperature and they were very satisfied in watching the partial eclipse through the telescope or using eclipse glasses, which lasted more than 2 hours, from 06.20 to 08.40. This is a good opportunity to spread common astronomical knowledge and more importantly, to correct some common misconception among the locals.

The enthusiasm of Total Solar Eclipse had inevitably increased the interest in astronomy, worldwide. We know this from the large number of news articles and open discussion circulated in newspapers, television, websites, and social media. Hopefully, it will increase the interest of studying astronomy among young people.

Figure 5. The atmosphere in Ma Chung University, Malang

Figure 6. The solar eclipse.

4. Conclusion

The public outreach activities before the eclipse day were very effective in educating people about the solar eclipse, and avoid wrong understanding of Total Solar Eclipse. All of the eclipse observers were very amazed with the exotic moment of Total Solar Eclipse and expressed their admiration in their comments spontaneously. From the activities in Palu and Malang we found that most of the people have no experience with astronomical observation using telescope. The eclipse observation event became their first experience in using telescope to observe the Sun. Before the activity, most locals also did not know how to enjoy solar eclipse in a safe way, therefore most of them felt very happy to get new knowledge and experience through our team. The Total Solar Eclipse event has definitely increased the interest in astronomy among the people.

References

[1] Espenak F 2016 *Total Solar Eclipse of 2016 Mar 09* http://eclipse.gsfc.nasa.gov/SEgoogle/SEgoogle2001/SE2016Mar09Tgoogle.html (accessed on 2016-03-10)

[2] Ashford A 2016 *Get Ready for 9 March Total Solar Eclipse Astronomy*, https://astronomynow.com/2016.03/08/get-ready-for-the-9march-total-solar-eclipse/ (accessed on 2016-04-10)

[3] Rahardjo DPE 2016 *Dosa Pemerintah Tak Boleh Terulang Kompas* http://sains.kompas.com/read/2016/02/09/21593541/Sembunyi.di.Kolong.dari.Gerhana.Matahari.1983.Dosa.Pemerintah.Tak.Boleh.Terulang (accessed on 2016-04-20)

Journal of Physics: Conference Series **771** (2016) 012024 doi:10.1088/1742-6596/771/1/012024

Outreach activities in anticipation of the 2016 solar eclipse in Sorong

Endra Putra Raharja[1] and Yudhiakto Pramudya[2]

[1]Program Studi Pendidikan Fisika Universitas Ahmad Dahlan
[2]Pusat Studi Astronomi Universitas Ahmad Dahlan

E-mail: endra1310007006@webmail.uad.ac.id, yudhiakto.pramudya@pfis.uad.ac.id

Abstract. Sorong is located outside the narrow path of total solar eclipse on March 9th, 2016. The predicted obscuration of the sun was 94.2%. The public outreach to anticipate the solar eclipse was intended to educate students in junior and senior high school in Sorong Regency. Some of them are located in the remote area where the educational materials are difficult to find. The public outreach is unique, since it was run by the local person who is student of physics education. The student has both the ability to explain the solar eclipse phenomenon and able to adapt to knowledge level of students. The materials that were given to the schools are brochure and the eclipse glasses. Beside solar eclipse lectures in class, the pinhole workshop and observation practice were held. The limited materials and resources were faced during the public outreach. However, the enthusiasm was shown by the students and teachers. At least one of the schools held the solar eclipse observation on the day of the eclipse.

1. Introduction

1.1. Sorong Regency

Sorong is regency in the province of West Papua, Indonesia. The capital of the regency is located in Aimas. This regency is one of the main oil producers in Indonesia. Sorong regency has an area of 18.170 km^2. The population in Sorong Regency is 76669 people which consist of local Papuans and migrants from other regions in Indonesia. [1]

Regency location is far from the center of the national government. Hence, several educational facilities in the regency are still minimal. There are still many schools that are not connected to the internet. This condition generates problems of slow information propagation, including the solar eclipse. This should be a special concern of all parties in supporting the distribution of education in Indonesia. One that can be done is to do the public outreach. [1]

1.2. Partial Solar Eclipse in Sorong Regency

Sorong is located outside the narrow path of total solar eclipse on March 9th, 2016 as shown in figure 1. The predicted obscuration of the sun was 94.2%. The eclipse took place starting at 8:59:36 am local time and reached the maximum eclipse at 9:58:51 am local time. The eclipse finished at 11:28:35 am local time. [2]

Journal of Physics: Conference Series **771** (2016) 012024 doi:10.1088/1742-6596/771/1/012024

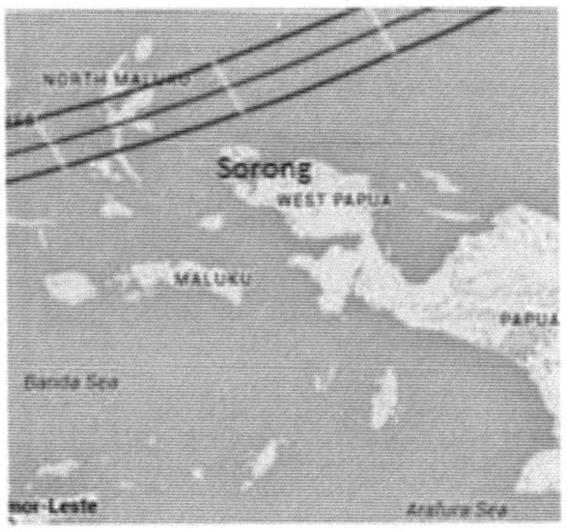

Figure 1. Path of total solar eclipse. Credits: NASA

1.3. Public Outreach
Education and public outreach by amateur astronomers in both formal and informal settings have been acknowledged as a critical focus for increasing the public's astronomical interest and scientific literacy [3]. Without amateur astronomers committing their time to education and public outreach efforts, much of the education and public outreach currently taking place would not exist. Storksdieck and Berendsen [4] estimate that 10,000 amateur astronomers engage in some level of education and public outreach. It is therefore critical to have an understanding of how amateur astronomer involvement in education and public outreach, as well as their level of involvement in education and public outreach, can be promoted and increased [4].

2. Methods
There were several approaches to perform public outreach to increase the public awareness in Sorong: mini lecture, workshop, and observation at school. These approaches ensured students and teachers get the hands-on experience learning astronomy especially on the solar eclipse. The physics education student in university level was the instructor for the public outreach.

2.1. Mini Lecture
The first approach in this public outreach is mini lecture in the class. The material consists of 3 sections which discussed the eclipse, the eclipse myths, and techniques of observation. Students were given the material on how the eclipses occur and how students should react to it. It also provided education about the myths so that the students will get the right concept of the eclipse.

2.2. Pinhole Workshop
The second approach in this public outreach is a pinhole workshop as shown figure 2. The students were given an explanation of the pinhole making as a safe tool for the observing the solar eclipse. The materials needed in the workshop were cardboard box, aluminum foil, needle, and glue.

Figure 2. Pinhole workshop at MTs Negeri Mariai.

2.3. Observation
The third approach in this public outreach is observation. Students were given the challenge of sun observations using a pinhole and eclipse glasses as shown in figure 3. Observation was held in the grounds of the school in the morning and afternoon. After the observation was completed, the school was given materials to educate students. The materials that were given to the schools are the brochure of solar eclipse and the eclipse glasses.

Figure 3. Observation at MTs Muhammadiyah Salawati.

3. Activities
The public outreach took place in five schools in Sorong. They are in the junior high school and senior high school. Some schools are public schools, the rest are private schools. Hence, the target schools are diverse in term of educational level and school managements. The schools were targeted since the students and teachers are potential agents to disseminate information and knowledge about the solar eclipse.

3.1. MTs Negeri Mariai
The school is located in the Mariai village, not far from the center of city.This school is one of the top schools in Sorong Regency who rely on religious education. It is a junior high school. Activities held on 1st of February 2016 with a sample of 70 students.

3.2. SMP IT Nurul Yaqin
The school is located in a boarding school that is quite far from the center of city. However, this junior high school has a good reputation in the Sorong Regency.There were 60 students joined the activities that were held on 2nd of February 2016.

3.3. MTs Muhammadiyah 1 Salawati
The school is located in the Makbalim village. It took one hour from the center of city to reach the school because its location is far. Activities held on 3rd of February 2016 with participation of 20 students.

3.4. SMA Negeri 7 Sorong
The school is located in the remote area where educational materials are difficult to find. Total participating students in this senior high school are 17 students mostly from local Papuans. Activities held on 4rd of February 2016.

3.5. SMA Negeri 2 Sorong
The school is located in the center of city and is the top schools in the Sorong Regency. There were 70 students joined the activities that held on 5th of February 2016.

4. Conclusion

The enthusiasm was shown by the students and teachers. Some of them have low awareness about the solar eclipse in Sorong Regency. The public outreach did not only increase the knowledge of the students and teacher but also gave the opportunity for the physics education student to start practicing his knowledge in teaching. MTs Negeri Mariai held the solar eclipse observation on the day of the eclipse.

Acknowledgment

This work is fully supported by Lembaga Penelitian dan Pengembangan Universitas Ahmad Dahlan

References

[1] Profil Kabupaten Sorong. 2016 http://sorongkab.go.id
[2] *Total Solar Eclipse of 2016 Mar 09. 2016* http://eclipse.gsfc.nasa.go
[3] Gada A., Stern A H, and Williams T R 2000 *Amateur—Professional Partnerships in Astronomy Astronomical Society of the Pacific Conference Series, ed. J. R. Percy and J. Wilson* 220 14
[4] Storksdieck M and Berendsen M 2007 *Science Educators Under the Stars: Amateur Astronomers Engaged in Education and Public Outreach, eds. M. G. Gibbs, M. Berendsen, and M. Storksdiek* (San Francisco: Astronomical Society of the Pacific) 30

Utilising Raspberry Pi as a cheap and easy do it yourself streaming device for astronomy

F Maulana, W Soegijoko and A Yamani
langitselatan, Indonesia

Abstract. Recent developments in personal computing platforms have been revolutionary. With the advent of the Raspberry Pi series and the Arduino series, sub USD 100 computing platforms have changed the playing field altogether. It used to be that you would need a PC or an FPGA platform costing thousands of USD to create a dedicated device for a a dedicated task. Combining a PiCam with the Raspberry Pi allows for smaller budgets to be able to stream live images to the internet and to the public in general. This paper traces our path in designing and adapting the PiCam to a common sized eyepiece and telescope in preparation for the TSE in Indonesia this past March.

1. Introduction

Astronomy education and outreach programs in this information era can benefit a lot from the advances in technology. One of the technologies that currently flourishes is live video streaming which has become a common thing for people who have access to Internet. Many people and organisations has already use this technology to broadcast astronomy related events.

To be able to capture astronomy event and broadcast it lives, the basic setup usually comprise of telescope, specifically design camera for telescope and computer or laptop. This setup is relatively expensive and not easy to acquire in Indonesia, especially for the camera.

A few months prior the 2016 Total Solar Eclipse (TSE), we had made plans to broadcast the event to the public around the observation area using a big screen and if possible to the Internet simultaneously. We were trying to build a cheaper and simpler setup for the live streaming event with hope it can be replicated easily so other people can also reap the benefits.

2. Astrophotography method using digital camera

There are three basic methods to capture an image from a telescope using a camera[1]:

(i) Prime Focus[2]: the camera is connected directly to the telescope, with no lens on the camera, and no eyepiece in the telescope so the telescope takes the place of the camera lens.

(ii) Eyepiece projection[3]: an eyepiece is used in the telescope but no lens is used on the camera. The eyepiece projects the image directly onto the camera sensor. This produces more magnification compared to prime focus.

(iii) Afocal[4]: the camera with the lens on it focused on infinity attached to the telescope with an eyepiece also focused at infinity. It's the best method for any camera with non-removable lens.

For eyepiece projection and afocal method, we have to calculate the distance between the eyepiece and the camera sensor or lens to get the focused image.

Journal of Physics: Conference Series **771** (2016) 012025 doi:10.1088/1742-6596/771/1/012025

3. Hardware design

We decided to use a Raspberry pi mini computer and pi camera module (PiCam) as the computer and camera sensor for the streaming hardware because it is cheap, simple, and relatively easy to be bought on online or offline electronic store. It's also open to be modified and customised to meet other requirement outside the video streaming purposes.

3.1. Raspberry Pi

The Raspberry Pi is a series of credit card-sized single-board computers developed in the United Kingdom by the Raspberry Pi Foundation with the intent to promote the teaching of basic computer science in schools and developing countries[5].

The Raspberry Pi 2 model B used in this paper has a Broadcom BCM2836 900 MHz 32-bit quad-core ARM Cortex-A7, 1GB onboard memory shared with Broadcom VideoCore IV @ 250 MHz 3D GPU with 1080p30 H.264/MPEG-4 AVC high-profile video decoder and encoder. It is powerful enough to be used as a video streaming device and it cost about USD 40 (IDR 550,000)

3.2. Pi Camera module (PiCam)

We are using the Raspberry pi camera module (PiCam) version 1.3 which can be connected directly to the Raspberry Pi board via CSI connector using a ribbon cable provided on the package. The camera module features a 5MP Omnivision 5647 CMOS sensor with 1.4 μm X 1.4 μm pixel size and fix focused lens able to deliver a 2592x1944 pixels resolution still image, or 1080p HD video recording at 30fps[6]. The PiCam module is cost about USD 30 (IDR 350,000)

3.3. Telescope attachment and accessories

To attach the Raspberry pi and PiCam to the telescope, we made an attachment out of PVC pipe. We used a $\frac{3}{4}$-inch pipe for the prime focus attachment and $1\frac{1}{2}$-inch pipe for the eyepiece projection and afocal, and other additional part from an acrylic sheet.

Figure 1. Prime focus adapter

Figure 2. Eyepiece projection and Afocal adapter

Other hardware we used was a USB Wifi dongle for network connection, a USB power 5V 2A adapter to run the system, and 3.5 inch LCD screen for raspberry pi to make the focusing easier. The total cost for the system is a little less than USD 100 and we can save more if we didn't use the LCD screen (cost about USD 17).

4. Software

Installing Rasbian Linux on Raspberry Pi is quite easy. After the installation is finished, run the post installation step and enable the camera feature in order to be able to use the PiCam.

Next we need to install ffmpeg software and get the YouTube stream URL before we can start to stream the video from PiCam. Testing the system can be done by running below command on the terminal[7]:

Journal of Physics: Conference Series **771** (2016) 012025 doi:10.1088/1742-6596/771/1/012025

Figure 3. Raspberry pi and PiCam mount using camera lens back cap

Figure 4. The complete setup attached to the telescope

```
raspivid -o - -t 0 -fps 25 -b 6000000 -f -ss 15000 | ffmpeg -re -ar 44100 -ac 2 -acodec pcm_s16le -f
    s16le -ac 2 -i /dev/zero -f h264 -i - -vcodec copy -acodec aac -ab 128k -g 50 -strict experimental
    -f flv [SERVER URL]/[STREAM NAME]
```

We should be able to see the video preview on the screen and on YouTube at the same time. Although the video on YouTube will have a few second delays depending on the internet speed.

All the detailed installation and setup process is available on the Internet and can be followed easily by any person with basic computer or technical skill.

5. Testing the system

We used a Celestron Travel Scope 70 400mm f/5.71 and Vixen Optics VMC110L 1035mm f/9.4 Maksutov-Cassegrain for the telescope. But for the 2016 TSE we used the Vixen telescope as it has a motorised mounting.

We tested the system on the moon as it relatively has the same apparent angular size as the sun. It's easier to find on the telescope and also safer for the camera sensor and our eyes in case we forget to attach the solar filter.

A recommended internet connection speed for a video broadcast is about 300kbps[8] (2G connection[9]). We used Telkomsel 3G connection on WIFI modem for the live streaming test as Telkomsel have the widest network coverage and 3G speed for a good video quality.

(i) For the prime focus method, we will need to remove the pi camera lens without damaging it by carefully removing the adhesive and rotating the lens counter clockwise.
The sensor size on the PiCam is small so the FOV will be narrow. Using the FOV calculator from `http://astronomy.tools` we saw that the resulting image of the moon will be cropped. Running the test command from before we saw a conforming result as the FOV calculator.

(ii) We used a 40mm eyepiece for the eyepiece projection and afocal method as the other eyepiece that we have has a narrow FOV. We couldn't get a focus image when we tested the eyepiece projection method due to problem on the adapter.

(iii) We were able to get a good quality and focused image of the sun using the afocal method. The FOV is also suitable for monitoring the solar eclipse.

6. Testing the system on 2016 TSE

During 2016 TSE, as part of the langitselatan program, we setup an outreach and educational event at SMKN 1 Maba, North Maluku. We used a 21" LCD TV screen for public viewing.

Figure 5. The prime focus image result using Celestron Travel Scope 70

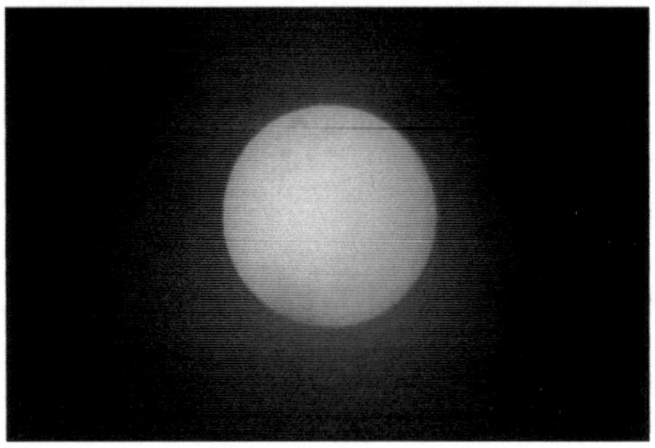

Figure 6. The afocal image result using Vixen Optics VMC110L and 40mm eyepiece

The sky during the event was cloudy and the sun was covered by clouds. A few minutes after the first contact it was raining so we had to stop everything, cover all the equipment, and move it to a dry and safer place. After the rain has stopped, the sky was still cloudy and we decided to cancel the public viewing and streaming for fear it would be rain again.

7. Result and conclusions

Outreach and education programs in astronomy can benefit from current technological advances such as live video streaming to spread the knowledge and message to all people without any physical boundaries.

The basic setup for providing a live video streaming event doesn't have to be expensive and can be achieved by utilising a sub USD 100 computing platforms such as Raspberry pi equipped with pi camera module.

Based on our test the image quality attained by the setup is good enough to be used for public viewing and with a little bit of adjustment and configuration, the system can be comparable with other dedicated astronomical camera system with the advantage of a simpler and more affordable setup.

References

[1] Aguirre E L 2006 *Astrophotography with a digital camera* http://www.skyandtelescope.com/astronomy-resources/astrophotography-tips/ astrophotography-with-a-digital-camera/ (accessed on 30 January 2016)
[2] *The prime-focus method* http://www.astronomyforbeginners.com/astrophotography/prime-focus.php (accessed on 30 January 2016)
[3] *Eyepiece Projection* http://www.astronomysource.com/2011/10/13/eyepiece-projection/ (accessed on 30 January 2016)
[4] *The Afocal Method* http://www.astronomyforbeginners.com/astrophotography/afocal.php (accessed on 30 January 2016)
[5] *Raspberry Pi* https://en.wikipedia.org/wiki/Raspberry_Pi (accessed on 9 April 2016)
[6] *Camera Module* https://www.raspberrypi.org/documentation/hardware/camera/README.md (accessed on 9 April 2016)
[7] Ryan K 2015 *Building a youtube live streaming camera with a rasp- berry pi* http://www.kianryan.co.uk/2015/10/buliding-a-youtube-live-streaming-camera-with-a-raspberry-pi/ (accessed on 28 January)
[8] 2012 *Hangouts system requirements on your computer* http://kenstechtips.com/index.php/download-speeds-2g-3g-and-4g-actual-meaning (accessed on 3 September 2016)
[9] 2012 *Download Speeds: What do 2G, 3G and 4G actually mean for you?* http://kenstechtips.com/index.php/download-speeds-2g-3g-and-4g-actual-meaning (accessed on 3 September 2016)

Measuring the level of public understanding of total solar eclipse from the mass media: Palembang as sample

F. G. Purwati, N. Ekawanti, Luthfiandari, P. W. Premadi

Astronomy Department, Institut Teknologi Bandung, Indonesia 40132

E-mail: fera.gustina@s.itb.ac.id

Abstract. The Total Solar Eclipse (TSE) on the 9th March 2016 received a huge attention from the mass media. Some of them intensively write articles about it even months before the TSE day. As we know media plays strategic role not only in raising public awareness but also interest. The aim of this project is to study the relation between the number of accesses to the media information and how well public learned the information delivered by the media. We prepared questionnaire consisting of seven semi-multiple choices on how public got information about TSE. We gave them choices of what they had heard to measure their basic understanding of TSE. Furthermore we add two "wrong" choices in the last questions to identify less serious respondents. We analyze 60 respondents of Palembang who visited Ampera bridge area. Our result shows no correlation between the number of information access and the level of understanding about TSE. We also found that local media did not provide the scientific content of TSE as well as the national media.

1. Introduction

As we know media plays strategic role in raising not only public awareness but also interest. Media can be a bridge to bring people to somewhere they never know before, rise up the imagination of knowledge, and a science communicator has more spirit to educate society too. Rosetta mission has successfully communicate the science of asteroid and how is important that mission for all, because so many feedback given from public such as 10 million live views on ESA livestream, 5.5 million page views on Rosetta blog, 92 kilo page likes on fan page FB Rosetta Mission and 287 kilo follower growth on twitter @ESA_Rosetta, data until November 2014 (Baldwin, 2016). In first quarter of 2016, Indonesia given the special astronomical event, which was total solar eclipse (TSE). TSE on the 9th of March 2016 received a huge attention from the mass media, especially in Indonesia as the only land passed by TSE, some of them intensively write article about it even months before TSE day. The amount of various media informed TSE had a correlation with public interest, such as Rosetta mission, but is that condition also correlated with increasingly public knowledge about that scientific event? We tried to find relation between the number of media accessed and their basic knowledge about TSE.

2. Method

We prepared questionnaire consisting of seven semi-multiple choices or using fixed response on how public got information about TSE. This questionnaire offered the respondent a closed set of responses from which to choose, we were not trying to capture new ideas or thoughts from the respondents. We gave them choices of what they had heard from all media information to measure their basic understanding of TSE, we designed list covers all possible alternatives choice. Before we distribute the questionnaire we conducted a small test to make sure the questionnaire clear enough to answer. The questionnaire is in Bahasa Indonesia, but we showed the simple questionnaire on the table 2 in English. We also collected the articles from mass media (newspaper) and compared between the information provided with public response. Furthermore we added two "wrong" choices in the last questions to identify less serious respondents, to later exclude their response in the analysis.

Table 1. Simply questionnaire in English version

1.	Reason for going to Ampera Bridge today
2.	Did you get any information TSE before these day?
3.	Since when you got information about TSE 9 March 2016?
4.	What kind of media did you get information about TSE today? (you can choose more than one media) A. Newspaper (tick or write the brand *optional) B. Broadcast media / TV/ Radio (tick or write the brand *optional) C. Social media (tick or mention the brand *optional) D. Website (tick or mention the brand *optional) E. Direct announcement/ talkshow (tick or mention the event *optional)
5.	Mention the frequency access those media (in average)
6.	Are the article content easily to understand?
7.	What kind of information that you got from media? ☐ The TSE process ☐ Sky condition will be like night during the TSE ☐ We cannot directly see the Sun ☐ The Sun will be bigger ☐ The chicken will be crowing during the TSE ☐ Duration of TSE only about 2 minutes ☐ The Sun will be oval ☐ Stars are visible in the sky during TSE ☐ We can see the sun's corona during TSE ☐ Explanation why TSE rarely happened

For the sample, we chosen respondent from one of the city which held big TSE event, there was Palembang. Ampera bridge became the centre of public for observing TSE in Palembang and then took random respondent out of the visitors. Respondent's age ranges from 13-50 years old.

3. Result and discussion

We got 60 respondents for this project. We checked the correlation between the response to question 7th "what are the information they knew from mass media" if they choose the wrong answer we compare it with they respond on question 3rd "when they accessed the information". We exclude 6 data because they are suspected as non-serious respondents and 2 respondents did not answer the questionnaire since they did know about TSE on that day. Just 86.67% samples were valid for the calculation. We divided age group based on Havighurst's developmental task theory and education stages in Indonesia. There were 3 groups data divided by their age, group 1 teenagers (13-18 years old) high school students, group 2 young adulthood (19-29 years old) university students, group 3 middle adulthood (30-60) parent and professional [2].

3.1. Information mass media versus public understanding level

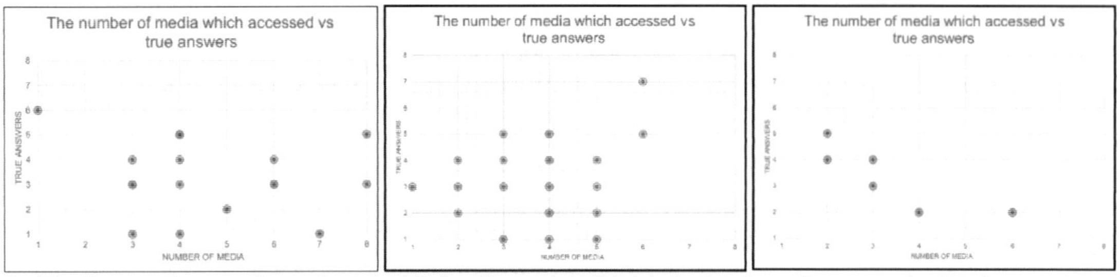

Figure 1. x = the number of accessed media, y = correct choice based on respondent's answer on questionnaire (number 7). From left to right, group 1, group 2, and group 3.

On the figure 1 maximum number of accessed media is 8 type of media, on y axis the highest true answers is 6. We can see that each group on figure 1 did not draw a relation pattern of the number of media public accessed and correct choice they choose in the questionnaire number 7 (basic knowledge). We cannot find that the more media they accessed the more basic knowledge of TSE they got. Even though we cut the extreme data, for example in group 1 who accessed 1 media but got 6 right choices and in group 3 who accessed 6 media but only got 1 right choice, we still could not find any relation.

3.2. Content of media that accessed versus public understanding level

We investigated scientific content that delivered by media about TSE. Three kinds of media with the widest access to our respondents are TV, Facebook, and Newspaper. In TV we knew that TSE was not only shared in news, but also spread on other programs such as infotainment. Unfortunately we could not traced TV programs to compare against responses, as no broadcast record is available for public. So we identified information from newspaper, we collected 55 national and local mass media a week before TSE day, from 2 – 8 March 2016. We marked some points in those articles based on our questionnaire and compared with the public response to guess their comprehension level.

Table 2. Number of article's content compared with questionnaire list. *Kompas* (national mass media), SP (*Sriwijaya Post*, local mass media), others (national mass media exclude *Kompas*), Public answers (data respondents from questionnaire)

Info	*Kompas*	others	SP	total	Public answers
The TSE process	6	1	0	7	50
Sky condition will be like night during the TSE	9	2	0	11	42
We cannot directly see the Sun	8	4	3	15	36
The Sun will be bigger	0	0	0	0	10
The chicken will be crowing during the TSE	0	0	0	0	1
Duration of TSE only about 2 minutes	3	1	2	6	38
The Sun will be oval	0	0	0	0	5
Stars are visible in the sky during TSE	1	0	0	1	5
We can see the Sun's corona during TSE	12	2	1	15	3
Explanation why TSE rarely happened	4	0	0	4	16
articles	20	9	5	34	

Data spread in wide range age and education stage, some questions were not sharp enough to get information needed from respondents. We realized that there was a double-barrelled question about idea so make it biased, such as in the question number 4, 5 and 7, which we asked about media and information written in there, but not asked specific type of media.

Large variations in number of sample of respondents for each age group. We took random respondents with a few sample made it hard to balanced age in each group that will be better if respondent are in the specific area and specific age.

From question number 4 we concluded that public in Palembang prefer to read local media (e.g. *Sriwijaya Post*) than national media (e.g. *Kompas*). However on table 2 shows that local media did not provide the scientific content of TSE as well as the national media, they mostly wrote about the festivities and announcement of *Sholat gerhana*, photo contest, etc before March 9th, 2016.

4. Conclusion

Our results shows no correlation between the number of information access and the level of understanding. The local media plays important role to rise public knowledge on scientific issue. Their big impact for public level of understanding would have been made better if local media also take the chance to increase the scientific content to educate public wisely. Inserting "wrong choices" is not enough to distinguish between their lack of seriousness in responding and the probability that they forgot about the whole of information and made guesses based on their feeling or experience.

5. Acknowledgement

We would like to thank to Mrs. Ely librarian of Bosscha Observatory who help to collect articles from mass media. This work also was supported by the Astronomy department which funds us in this symposium, and voucher scholarship from ITB.

References

[1] Emily Baldwin *et al.* 2016 *Communicating Astronomy with the Public.* **19** 30.
[2] Michael J M 1993 *A Comparison of Havighurt's Adult Social Roles and Developmental Events by Gender, Age, and Socioeconomic Status Levels* (Doctoral dissertation)

The use of astronomy questions as an instrument to detect student's misconceptions regarding physics concepts at high school level by using CRI (Certainty of Response Index) as identification methods

D N Utami and H R T Wulandari

Department of Astronomy, Faculty of Mathematics and Science, Institute Technology of Bandung

E-mail: dean.nurul@gmail.com

Abstract. The aim of this research is to detect misconceptions in the concept of physics at high school level by using astronomy questions as a testing instrument. Misconception is defined as a thought or an idea that is different from what has been agreed by experts who are reliable in the field, and it is believed to interfere with the acquisition of new understanding and integration of new knowledge or skills. While lack of concept or knowledge can be corrected with the next instruction and learning, students who have misconceptions have to "unlearn" their misconception before learning a correct one. Therefore, the ability to differentiate between these two things becomes crucial. CRI is one of the methods that can identify efficiently, between misconceptions and lack of knowledge that occur in the students. This research used quantitative-descriptive method with ex-post-facto research approach. An instrument used for the test is astronomy questions that require an understanding of physics concepts to solve the problem. By using astronomy questions, it is expected to raise a better understanding such that a concept can be viewed from various fields of science. Based on test results, misconceptions are found on several topics of physics. This test also revealed that student's ability to analyse a problem is still quite low.

1. Introduction

Based on the previous research, misconception has become one of the most possible reasons why students found some difficulties in learning physics. In the learning process, most of the students usually come to the class with some ideas or concepts that they have obtained from their previous learning and experience. Sometimes the ideas or concepts that they have is different from the ideas that have been agreed by the expert in the field at the current time, and this is what we call as misconception.

Some experts believe that misconceptions can interfere with the acquisition of new understanding and integration of new knowledge or skills. One of the reasons is because the student will try to relate a new concept that they get with a misconception that they have, which will lead to another misconception. That is the primary reason why misconceptions become something important that need to be solved.

While lack of knowledge can be corrected by the next instruction and learning, students that have misconception need to unlearn those misconceptions first before learning a correct concept. Therefore, the ability to differentiate these two things become crucial.

Several methods can be used to differentiate between students that have lack of knowledge and the students that have misconceptions. One of the methods is Certainty of Response Index (CRI). CRI method is a misconception detection method, which use a form of multiple choice questions where the respondents have to indicate, on a scale of 0-5, how certain they are that their one choice is correct [4]. The advantages of this method are in terms of ease, accuracy and speed the process for identifying and analyzing the data. Since one of the purpose of this research is to detect misconception, we found that this method is really compatible with our research.

2. Certainty of response index method

This method is mostly used in psychology test. In 1999, Hasan modified this method and used it for educational research. In this method, the respondents will do their own assessment regarding their certainty of their answer. The scale of certainty is in 6-point scale (0-5) where 0 indicates totally guessed answer (no knowledge) and 5 indicates a complete confident of an answer.

Regardless the answer was correct or wrong, the low CRI value (0-2) will indicate guessing which implies a lack of knowledge. While a high CRI (3-5) shows a high degree of confidence regarding a concept that they have. If the students with high CRI give a right answer, then we can conclude that they have a correct concept. But if the answer falls to an incorrect one, the high CRI will indicate that the students have a misconception.

3. Astronomy questions as a testing instrument

A concept in science should be understood holistically. When students learn about a concept, we want them to understand a concept not only from one field of science, but also to look at it in the various fields of science. Because the truth is, a concept actually can be applied in many fields of science. For example, the concept of energy, where energy cannot be created and destroyed, but can change from one form of energy into another form, is not only applicable in the domain of physics, but also applicable in biology, chemistry and astronomy. If students understand this, then these students will be able to answer questions related to the concept, although the question was asked from the different view of field in science.

On the other hand, astronomy has a special charm to attract student's interest. Based on experience, we all know that the students will be amazed when they see some amazing pictures that relate to astronomy. By this, astronomy could be a way to attract students so that they can get interested in various fields of science.

Based on this, we decide to use astronomy questions to detect misconception in physics at high school level. This is important to note that, there is no prior knowledge in astronomy needed to answer the question. If students understand the concept of physic (that has been taught), they can answer the question.

4. Method

In this research, we use a quantitative-descriptive method with ex-post- facto research approach where we will only detect the misconception without looking the reason why the student has these misconceptions. The numerical data obtained will be analyzed statistically, and then we will take a conclusion from it.

The research was conducted in one university at Makassar (first year students), 2 schools in Pekanbaru (accreditation A), and 4 schools in Bandung (accreditation A and B), with total subjects 784 students. The concept that we tested is the concept that has been taught in the second semester on 1st, 2nd and 3rd grade. The instrument (the questions) for this research has been validated by experts in the field, in this case experienced teachers and the lecturer of ITB.

5. Result

We realize that the research that we have done is not completely perfect, but we can use this data as a first step to find out the common topic where the students have a misconception. The data that we have

Journal of Physics: Conference Series **771** (2016) 012027 doi:10.1088/1742-6596/771/1/012027

can also be used to find out the common problem that the students have in learning, especially in physics. Based on the result of our research, we can conclude that the use of astronomy question as a testing instrument to detect misconception in physics was quite successful. Below is the result regarding the percentage of student's that have misconception in each grade:

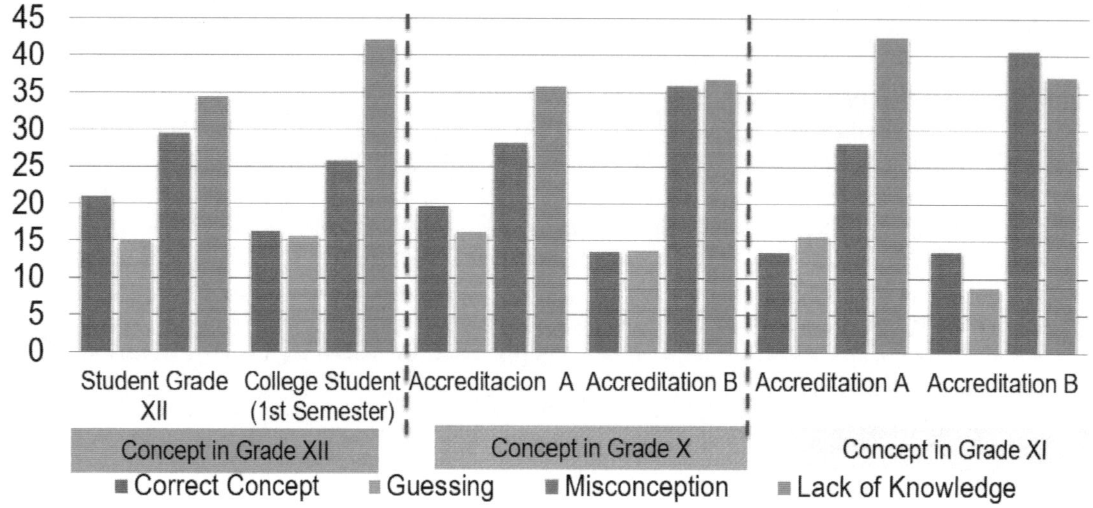

Figure 1. This graph shows a comparison between student in a school with accreditation A and school with accreditation B. The bar indicates a percentage of students that have a misconception as well as the students that have lack of concept and the students that have a correct concept

Based on the result, we can see that students in a school with accreditation B have a higher misconception than students in school with accreditation A. Several concepts that have a higher percentage of misconception listed as below:

1. The relation between light scattering and wavelength.
2. The relation between the energy of the radiation with a radius of the object.
3. The relation between the maximum intensity of radiation with a wavelength
4. The relation between the periods, the linear velocity and angular velocity in circular motion
5. Factors that cause objects to float and sink
6. The relationship between temperature, pressure, and volume
7. Reflection
8. Angular velocity of the rigid body rotation
9. The wave nature of light (scattering, flexing and polarisation).
10. Things that caused the greenhouse effect

In figure 1 above, a part of misconception percentage, we also notice that the percentage of students that have lack of knowledge is in a bigger number (we also include guessing as a lack of knowledge). Since the questions we have given are related with concepts of physics that they have learnt on the previous semester, we suspect that students are not having a full understanding of the concepts.

Although the main purpose of this research is to detect misconception, we also discover other things. In the test, we also put some detractors in the options to see how well is the student understand the concept. We also use 2 questions that have different form, for example; one is a calculation and the other one is comparison questions, but both of the questions have the same concept. By using these detractors and different form of a question for the same topics, we found out that students have another problem as below:

1. Students tend to memorize both concepts and formulas, without understanding them.
2. The student's ability to analyze a problem (questions) is still quite low.

3. Most students have some difficulties to read the relationship between the quantities in an equation/a formula.
4. Most students are not able to understand the relationship of the quantities in the form of graphs
5. Several students tend to memorize a unique number without understanding it
6. Students are not having a fully or complete understanding of a concept.

References

[1] Brown D E 1992 *Journal of Research in Science Teaching* **29(1)** 17-34
[2] Clement J 1987 Overcoming student's misconception in physics: the role of anchoring intuition and analogical validity, *The Second International Seminar on Misconceptions and Educational Strategies in Science and Mathematics Vol. III 84-9 Ithaca*, (NewYork: Cornell University)
[3] Comins N 1993 Sources of misconceptions in astronomy, *The Proceedings of Third International Seminar on Misconceptions and Educational Strategies in Science and Mathematics*
[4] Hasan S *et al.* 1999 Misconceptions and the certainty of response index (CRI) *Phys. Educ.* **34(5)** 294 - 299

Cluster compaction of two-dimension spherical particles binary mixture as model of forming process of an asteroid

S Viridi[1], B Dermawan[2]

[1]Physics Department, Faculty of Mathematics and Natural Sciences, Institut Teknologi Bandung, Jalan Ganesha 10, Bandung 40132, Indonesia
[2]Astronomy Department, Faculty of Mathematics and Natural Sciences, Institut Teknologi Bandung, Jalan Ganesha 10, Bandung 40132, Indonesia

Email: dudung@fi.itb.ac.id, budider@as.itb.ac.id

Abstract. Several two-dimension spherical particles are under influence of gravitational forces of each other and as they are colliding repulsion forces in form of linear spring-dashpot prevent them from collapsed into a single point. Gravitational constant larger than G is used for better and faster stabilization in forming cluster of the particles, accompanied with values of $k_N \approx 10^4$ N/m and $\gamma_N \approx 0.5$ N·s/m As initial conditions the particles are placed randomly with separation distances among each other. Molecular dynamics method implementing Euler algorithm is used to simulate the developing of particles cluster, which is intending to mimic the process of an asteroid forming. Time step of $\Delta t = 10^2$ s is chosen and results are reported every some period from 1 s to 1000 s, where after each period all particles velocity are forced to be zero. It is observed that not only physical parameters influencing the compaction of the asteroid but also simulation parameters.

1. Introduction

Asteroids are small bodies in our solar system whose surfaces have various features. There are high and low topography, large and small craters, boulders, small rocky materials, and regolith. From spectroscopy observations, taxonomy of asteroids provides several major surface types, i.e. carbonaceous, silicate and metallic. Solar wind and other space weathering process are believed to alter the spectral color of many types of asteroids. Asteroid rotates on its axis during its revolution orbiting the Sun. Some properties of internal structure of asteroid can be derived from its lightcurve. There are a barrier of rotation period of 2.2 hours for sizes of 1 km or larger, and remarkably, for sizes below 100 m the rotation periods can be 5 minutes [1]. This constraints its bulk density and level of (macro)porosity. Two examples of asteroids having distinct internal structure are 253 Mathilde (bulk density 1.3 g/cm^3, porosity 50%) and 216 Kleopatra (bulk density 6 g/cm^3, equilibrium figure of iron/metal) [1]. Results from some asteroid missions, e.g. to 433 Eros and 25143 Itokawa, have shown that asteroids are gravitational aggregate bodies (rubble-pile), not monolithic structure. However, for much smaller size of asteroids, whose rotation periods are very fast, it is believed that they are indeed monolithic bodies. There are many unexplained features and distribution of materials on asteroid surface. Smooth regolith and large rocky materials can be found side by side. Impact can shake seismically the asteroid body, so that surface and sub-surface materials can be changed and mixed [2].

Recent exploration of Hayabusa mission to Itokawa asteroid has successfully brought high resolution images showing how boulders, gravels, and sands distributed on its surface, where they have some preferred orientation at some places but not in other places [3]. Occurrences of larger particles, e.g. boulders, above smaller particles, e.g. gravel or sand, are addressed to the well-known phenomenon in granular world, i.e. the Brazil-nut effect (BNE) [4]. Simulation in 2-d, which is conducted under influenced only of gravitation of granular particles in the system, shows that there is a ring of larger particles formed near the edge of the granular aggregate [5]. Even further, collision of two such aggregates for much smaller particles has also been reported [6]. On earth surface, where gravitation is always toward center of the planet, decrease of porosity and increase of contactopy is observed during the process of BNE [7], which is interesting to observe it in the system under influence of gravitation of its own particles.

2. Simulation

Particles which later forming an asteroid are simplified into 2-d spherical particles, each with the same density ρ, while for each particle i the diameter D_i and mass m_i may be varied. Only two types of forces between particles are considered in this work. The former is gravitation force while the later is normal force. In order to get particles motion variables such as velocity and position, molecular dynamics (MD) method is used, which implementing Euler algorithm.

2.1. *Forces formulation*

Two types of force are considered, gravitation force F_G and and normal force F_N [8]

$$\vec{F}_{ij}^{G} = -k_G \frac{m_i m_j}{r_{ij}^2} \hat{r}_{ij} \quad (1)$$

$$\vec{F}_{ij}^{N} = \left(k_N \xi_{ij} - \gamma_N \frac{d\xi_{ij}}{dt} \right) \hat{r}_{ij} . \quad (2)$$

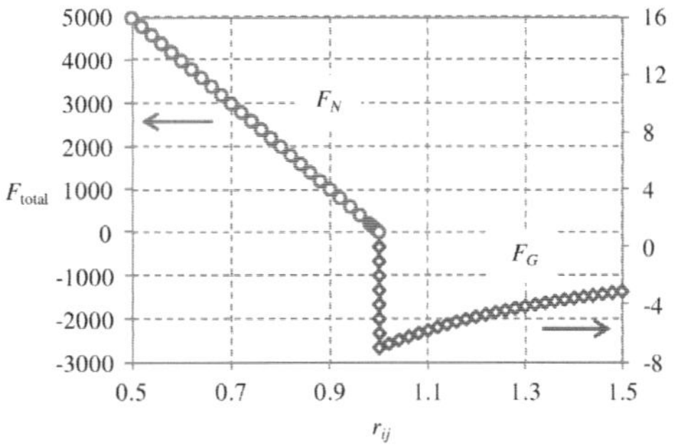

Figure 1. Total force $F_{\rm total}$ for $D_i + D_j = 2$, $k_N = 10^4$, and $G = 10^{-7}$, where F_N dominates for small range of r_{ij} and F_G dominates for large range of r_{ij}.

It can be seen in figure 1 that F_N dominates in short range while F_G in long range of distance between particles r_{ij}. Notice that values of F_N and F_G have different scale in order of magnitude. The distance about 0.999297 (can not be seen in figure 1) is equilibrium distance, where $F_N = F_G$.

2.2. *Molecular dynamics (MD) method*

The MD method has two steps: i. calculation of acceleration of all particle, and ii. calculation new velocity and new position of all particle. The first step is conducted using Newton second law of motion

$$\vec{a}_i(t) = \frac{1}{m_i} \sum_{j \neq i} \left[\vec{F}_{ij}^{G}(t) + \vec{F}_{ij}^{N}(t) \right] \quad (3)$$

and the second step can be performed using various numerical algorithm, but improved Euler algorithm is chosen in this work due to its simplicity

$$(\vec{v}_i, \vec{r}_i)(t + \Delta t) = (\vec{v}_i, \vec{r}_i)(t) + (\vec{a}_i, \vec{v}_i)(t)\Delta t . \quad (4)$$

Equations (4) are executed repeatedly until termination condition is reached.

2.3. Number of contact points

From contact energy U_c [9] formulation of number of contact points can be derived

$$N_c(t) = \frac{\partial U_c(t)}{\partial \gamma} \approx \sum_{i, j \neq i} \text{sign}[\xi_{ij}(t)]. \quad (5)$$

2.4. Microstates and macrostates

For $L \geq 2$ the macostates are defined slightly different since same ratio N_S/N could lead to different macrostates. For ratio $N_S/N = 0$ and $N_S/N = 1$ there are no other possible microstates. But it is getting more interesting for ratio $N_S/N = 1/7$ (or $N_S/N = 6/7$), where the same ratio N_S/N could lead to different final configuration since particle in center of the cluster will be different (it could be the smaller or the larger particle). It should have different number of contact points N_c and also different porosity ϕ. Possible value of N_S/N and related number of microstates for $L = 1, 2$ are given in table 1.

Tabel 1. Possible number of microstates for $L = 1, 2$.

***L* (*N*)**	***N*s / *N* (microstates)**
1 (3)	0 (1); 1/3 (3); 2/3 (3); 1 (1)
2 (7)	0 (1); 1/7 (1,6); 2/7 (6,6,6,3); 3/7 (6,6,3,6,6,6,2); 4/7 (2,6,6,6,3,6,6); 5/7 (3,6,6,6); 6/7 (6,1); 1 (1)

Number of microstates without smaller particles in two different layers (or in a layer, ial) can be calculated using

$$N_{\text{ial}} = C(n, k), \quad k > 1, \quad (6)$$

e.g. $N_S/N = 1/7$ only for first microstate $N_{\text{ial}} = 6! / (1! \cdot 5!) = 6 = (6)$, $N_S/N = 2/7$ after first microstate $N_{\text{ial}} = 6! / (2! \cdot 4!) = 15 = (6+6+3)$, $N_S/N = 3/7$ after third microstate $N_{\text{ial}} = 6! / (3! \cdot 3!) = 20 = (6+6+6+2)$, and the rests are similar to these. If there are particles in different layer, then possible combination should be multiplied by equation (26), e.g first row for $N_S/N = 3/7$ it would be obtained that $1 \cdot 6! / (2! \cdot 4!) = 15 = (6+6+3)$. It will be more complicated for $L \geq 2$.

3. Results and discussion

Number of contact points per particle is not monotonically increasing as number of layer increased. It fluctuates as shown in figure 2. Since forms of pores are similar for particle configuration with same diameter it is found that $\phi \approx 1.43\times10^{-4}$ independent to L. In previous work [10] the term of contactopy C is used instead of number of contact points per particles N_c / N, where $C \approx N_c$. From its initial configuration cluster of particles will evolve to more compact configuration as shown in figure 3. Similar results will be obtained for this work with different slopes. Final number of contact points per particle and porosity will be predicted since homogenous size of particle will again form hexagonal close packed configuration.

4. Summary

It can be summarized in this work that some programs have been developed to simulate compaction of two-dimension spherical particles as model of asteroid forming process. Number of contact points per particles fluctuates with number of layer for hexagonal close packed configuration. Porosity for three particles pairs is relatively constant.

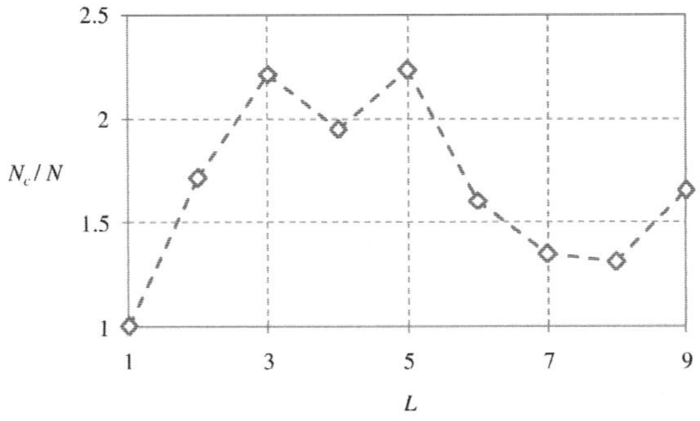

Figure 2. Average number of contact points per particle N_c / N as function of L.

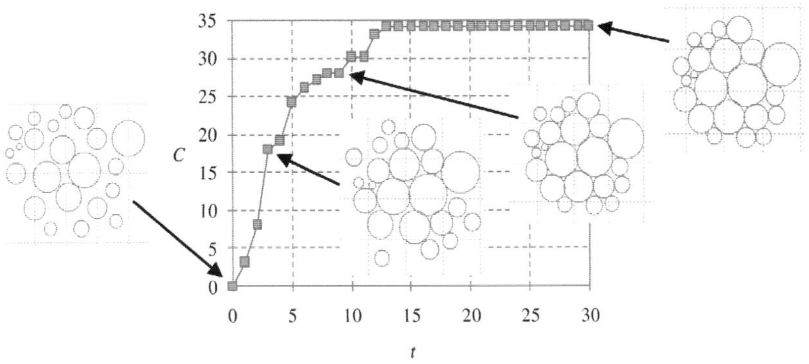

Figure 3. Evolution of contactopy C for non-homogenous particle size [10].

Acknowledgements

This work is supported by Institut Teknologi Bandung, and Ministry of Higher Education and Research, Indonesia, through the scheme Penelitian Unggulan Perguruan Tinggi – Riset Desentralisasi DIKTI in 2016 related to the vibration. Presentation of this work is supported by Committee of ISSEL 2016 and Faculty of Mathematics and Natural Sciences, Institut Teknologi Bandung.

References

[1] Asphaug E 2009 *Ann Annu. Rev. Earth Planet. Sci.* **37** 413
[2] Perera V, Jackson A P, Asphaug E and Ballouz R-L 2016 *Icarus* **278** 194
[3] Miyamoto H *et al.* 2007 *Science* **316** 1011
[4] Matsumura S, Richardson D C, Michel P, Schwartz S R and Ballouz R-L 2014 *MNRAS* **443** 3368
[5] Sánchez P, Scheeres D J and Swift M R 2010 *Proc.: 41st Lunar and Planetary Sci. Conf. (March 1-5, Woodlands)* eds. editors (Houston: Lunar and Planetary Institute) 2634
[6] Gunkelmann N, Ringl C and Urbassek H M *A & A* **589** A30
[7] Viridi S, Khotimah N, Widayani and Novitrian 2015 *AIP Conf. Proc.: 5th Int. Conf. on Math. and Natural Sci. (2-3 November, Bandung)* **1677** eds. Hidayat T *et al.* (Melville: AIP) p 070001
[8] Schäfer J, Dippel S and Wolf D E 1996 *J. Phys. I France* **6** 5
[9] Viridi S, Khotimah S N, Novitrian and Widayani 2015 *Adv. Mat. Res.* **1123** 12
[10] Viridi, S dan Dermawan B 2016 *Pros.: Sem. Nas. Fisika (28 Mei, Jakarta)*

On the nature of type 1 AGN: emission properties and correlations

Irham Taufik Andika[1], Mochamad Ikbal Arifyanto[1], Wolfram Kollatschny[2]

[1]Astronomy Study Program, Institut Teknologi Bandung, Jl. Ganesha No. 10, Bandung 40116, Indonesia
[2]Institut for Astrophysics, University of Goettingen, Friedrich-Hund-Platz 1, Goettingen 37077, Germany

E-mail: irham.andika@students.itb.ac.id

Abstract. We present a study of emission properties and correlations of 3191 type 1 AGN at $z < 0.35$, selected from Sloan Digital Sky Survey Data Release 12. We supplement the data with ultraviolet spectra from Hubble Space Telescope and International Ultraviolet Explorer along with X-ray properties from Chandra Source Catalog and XMM-Newton Serendipitous Source Catalog and radio measurements from FIRST. We find that the observed spectral diversity of type 1 AGN can be unified by Eddington ratio ($L/L_{\rm Edd}$). Objects with higher $L/L_{\rm Edd}$ tend to have narrower broad Hβ component, strong Fe II emission, systematic blueshift of C IV λ1549, soft X-ray excess, and lower ionization parameter compared to those of the lower $L/L_{\rm Edd}$.

1. Introduction

The term of active galactic nuclei (AGN) is related to the compact region at the center of galaxy that emits more radiation (up to several order of magnitude) than the rest of the host galaxy. The central engine of AGN are thought to be powered by accretion of gas (in the form of accretion disk) onto supermassive black hole (SMBH; [1]). This accretion disk generates energy which is then released in the form of radiation, winds, and/or jets which photoionizes the gas located near the center of AGN [2]. The region of the circumnuclear gas can be divided into two categories based on its physics and kinematics: broad line region (BLR) located closer (≤ 1 pc) to SMBH, which produces permitted emission lines with typical width thousands of km s^{-1}, and narrow line region (NLR), located further from SMBH, which produces permitted and forbidden emission lines with typical width hundreds of km s^{-1}.

In order to understand the physical mechanisms in BLR and NLR, many attempts have been made to find correlation among different observable properties of AGN. One of those interesting correlations is eigenvector 1 (EV1) correlation, originally found in the Palomar-Green (PG) quasar sample [3] and then studied further by several authors in context of 4DE1 (4 Dimensional Eigenvector 1; [4]). In term of EV1 correlation, objects that have strong Fe II emission tend to have narrow Hβ width and weak [O III] emission. Based on [1], the optical complex Fe II lines are believed to come from BLR while narrow Hβ and [O III] lines are emitted by NLR. The predominant factor that appears to be driving this correlation is Eddington ratio,

$L/L_{\rm Edd}$ [5]. The EV1 correlation is somewhat unexpected as BLR and NLR properties are physically different. NLR can be described as low-density and low-velocity gas extending from outer AGN torus up to thousands of parsecs along the general direction of ionization cones. This makes NLR emission not only driven by photoionization from accretion disk of AGN but also affected by the host galaxy. Therefore, it is important to study the connection BLR and NLR spectral properties while considering host galaxy contribution to AGN emission, especially for low luminosity AGN.

In this work, we present a new Sloan Digital Sky Survey Data Release 12 [6] based sample of selected type 1 AGN. Type 1 AGN is class of AGN which shows broad permitted and semi-forbidden emission lines and a bright, non-stellar, central point source visible at all wavelengths that are not contaminated by stellar light as the consequences of direct view to the nucleus based on AGN unification scheme [1]. Then, we supplement the data with spectral data based on Hubble Space Telescope (HST) and International Ultraviolet Explorer (IUE) archives and X-ray properties from Chandra Source Catalog [7] and XMM-Newton Serendipitous Source Catalog [8]. Using this data set, we explore the emission properties and correlation to study the nature of type 1 AGN.

Throughout this paper, we assume a Friedmann-Robertson-Walker cosmology with $\Omega = 0.3, \Lambda = 0.7, H_0 = 70$ km s^{-1} Mpc^{-1}.

2. Data

2.1. Main sample construction

We derived the data based on Sloan Digital Sky Survey (SDSS) Data Release (DR) 12 [6] in order to derive a well-defined sample of type 1 AGN at low redshift (z). We search the AGN data in SDSS DR 12 database using these criteria:

(i) The spectrum of the objects is classified as quasi-stellar object or "QSO" by SDSS pipeline.

(ii) $0.02 \leq z \leq 0.35$, to cover the Balmer series (especially Hβ and Hα), optical forbidden narrow lines [O II], [O III], [N II], [S II], and optical Fe II complex in the spectral lines.

(iii) S/N > 10 in the spectrum based on median S/N of whole pixels covered by the spectrum (available in the SDSS database), to improve the accuracy of NLR/BLR decomposition.

(iv) High confidence in redshift determination.

A total 9184 spectra pass our criteria. We name them as "main sample". The reduced one-dimensional (1D) spectral data are downloaded through the SDSS Skyserver (http://skyserver.sdss.org/dr12/en). To include soft X-ray photon index Γ_s measurement, we matched our AGN with the Chandra Source Catalog (CSC; [7]) and the XMM-Newton Serendipitous Source Catalog [8] with a matching radius of 1". The 1σ measurement uncertainties in Γ_s are typically $\sim$ 10% (relative uncertainty) for Chandra sources, and $\sim$ 20% (relative uncertainty) for XMM-Newton sources. To include radio properties, we match the main sample data with FIRST [9] catalog with a matching radius of 30". Then, we compute the integrated radio flux density at 20 cm and classify T1 AGN into radio loud or radio quiet AGN [10].

In order to derive a well-defined type 1 AGN sample, we utilize the diagnostic diagram to classify our sample [13]. Therefore, we exclude objects reside other than AGN regime in diagnostic diagram so that our sample now contains 3191 AGN. We named this sample as 'T1 AGN'. We obtained 31 UV spectral data from HST and IUE. The number of X-ray sources from CSC and XMM-Newton Serendipitous Source Catalog that matched with our T1 AGN is 38 and the number of radio source from FIRST that matched with our T1 AGN is 310. The T1 AGN reside at $42.4 < \log L_{5100} < 45.3$ with median $\log L_{5100} = 43.9$, where L_{5100} is averaged continuum luminosity at rest wavelength 5100 Å.

2.2. Spectral measurements

The data processing throughout this paper were done using custom software using *Python* (*Python* is an interpreted, interactive, object-oriented programming language). We use similar approach described by [11] with some modifications in continuum and emission lines fitting procedure. We separated the continuum windows into three parts: around [O II], Hβ, and Hα. For each region, we fit a local power-law continuum ($f_\lambda = A\lambda^{\alpha_\lambda}$), an iron template, and host galaxy spectra to the wavelength region around the line that is not contaminated by the emission line.

We compare our results with the $L_{[\mathrm{O\ III}]}$ measurements of SDSS AGN in [12], who model [O III] using multiple Gauss-Hermite (GH) functions. In the 2657 overlapping objects, the mean difference of our $L_{[\mathrm{O\ III}]}$ to Stern & Laor $L_{[\mathrm{O\ III}]}$ is 0.1 dex with dispersion of 0.08 dex. A total of 4293 objects with acceptable fits (reduced $\chi^2 < 2$) pass these criteria. We note that our rejection criteria may have been too restrictive, and potentially excluded objects with an unusual broad line profile.

3. Result and analysis

3.1. T1 AGN distribution in 4DE1 plane

The original optical plane of 4DE1 involved measures of FWHM bHβ and the $R_{\mathrm{Fe\ II}}$ [4]. The $R_{\mathrm{Fe\ II}}$ represents the ratio between equivalent width of Fe II within 4434 – 4684 Å to equivalent width of broad Hβ. The $R_{\mathrm{Fe\ II}}$ proposed by [4] only consider $\sim$ 25% Fe II strength in optical region. Here we modify $R_{\mathrm{Fe\ II}}$ into $L_{\mathrm{Fe\ II}}/L_{\mathrm{bH}\beta}$ because physically $L_{\mathrm{Fe\ II}}/L_{\mathrm{bH}\beta} \propto R_{\mathrm{Fe\ II}}$. We also account $L_{\mathrm{Fe\ II}}$ in the wider range, i.e. 4000 – 5500 Å, to account $\sim$ 100% Fe II strength in optical region. The $L_{\mathrm{Fe\ II}}/L_{\mathrm{bH}\beta}$ can be interpreted as physical condition of the emitting gas (density, ionization, and metallicity) while FWHM bHβ is widely used as a virial estimator. Higher $L_{\mathrm{Fe\ II}}/L_{\mathrm{bH}\beta}$ means higher ionization degree and metallicity in BLR gas while higher FWHM bHβ usually caused by higher center SMBH mass and vice versa [1].

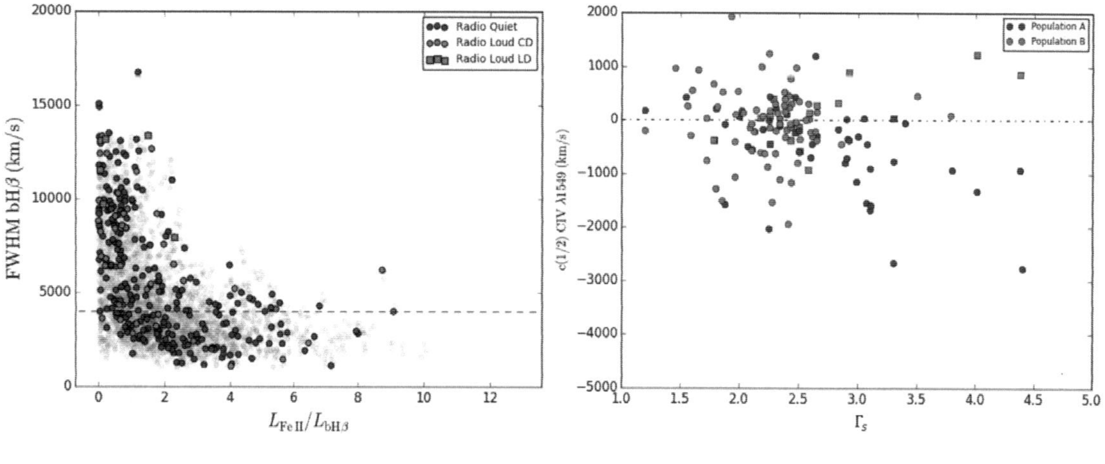

Figure 1. Left panel: T1 AGN in the 4DE1 optical plane. The blue dots represent the radio-quiet objects while the red dots are the radio-loud objects. Object without radio flux measurement available are marked with gray dots. Right panel: T1 AGN in the 4DE1 UV X-ray plane. The objects with dot symbol were taken from [15] while objects with square symbol were measured in this work.

The distribution of T1 AGN in 4DE1 optical plane shown in figure 1. We divide the T1 AGN into two populations with the boundary of separation at 4000 km s^{-1}. We adopt this as tentative boundary between population A and B as previously suggested by [4]. From the figure 1, it is clear that AGN with strong Fe II tend to have narrower Hβ. In terms of optical spectra, we can see there is no difference between radio loud and radio quiet objects because those two population distribution overlap each other in 4DE1 optical plane.

In the context of 4DE1, we can describe population A sources as showing narrower broad Hβ component, moderate or strong Fe II emission, a systematic blueshift or asymmetry of high ionization lines (in this case C IV λ1549) and a soft X-ray excess. On the other hand, population B shows broader broad Hβ component, weak Fe II emission, and almost absence of a C IV blueshift or soft X-ray excess. Figure 1 shows a correlation between soft X-ray excess (Γ_s) and the centroid shift of C IV (c(1/2)). A soft X-ray excess occurs in sources with the largest C IV blue shift which means population A often have larger soft X-ray excess than population B.

4. Conclusion

In this paper, we derive a well-selected sample of type 1 AGN at low redshift using large sample SDSS DR 12 data to get the optical spectra of the sample. Then, this data set combined with UV spectral properties from HST and IUE, along with X-ray properties from CSC and XMM-Newton SSC. Spectral reduction and analysis to optical and UV spectra carried out in order to derive emission lines parameters. We use the measured emission lines parameters to explore correlation and properties of type 1 AGN emission.

The collective evidence from T1 sample spectral properties in context of 4DE1 leads to simple interpretation. There is no significant difference in optical emission lines properties between radio loud and radio quiet AGN. By dividing T1 sample into two population based on FWHM bHβ, we found population A (FWHM bHβ < 4000 km s^{-1}) have narrower broad Hβ component, moderate or strong Fe II emission, a systematic blueshift or asymmetry of C IV λ1549 and soft X-ray excess. On the other hand, population B shows broader broad Hβ component, weak Fe II emission, and almost absence of C IV λ1549 blueshift or soft X-ray excess.

Acknowledgments

We thank Jelena Kovačević and Jonathan Stern for providing us the iron template and host galaxy eigenspectra. Also thanks to Jack Sulentic and Paola Marziani for help and useful discussion. ITA acknowledge the award of Erasmus Mundus Action 2, which supported the funding to stay at University of Goettingen, where much of this work was completed.

References

[1] Netzer H 2013 *The Physics and Evolution of Active Galactic Nuclei*
[2] Osterbrock D E, Ferland G J 2006 *Astrophysics of gaseous nebulae and active galactic nuclei*
[3] Boroson T A, Green R F 1992 *ApJS* **80** 109
[4] Sulentic J W, Zwitter T, Marziani P, Dultzin-Hacyan D 2000 *ApJ* **536** L5
[5] Boroson T A 2002 *ApJ* **565** 78
[6] Alam S *et al* 2015 *ApJS* **219** 12
[7] Evans I N *et al* 2010 *ApJS* **189** 37
[8] Watson M G *et al* 2009 *A&A* **493** 339
[9] White R L, Becker R H, Helfand D J, Gregg M D 1997 *ApJ* **475** 479
[10] Sulentic J W, Zamfir S, Marziani P, Dultzin D 2008 *in Revista Mexicana de Astronomia y Astrofisica Conference Series* 5158 (arXiv:0709.2499)
[11] Shen Y *et al* 2011 *ApJS* **194** 45
[12] Stern J and Laor A 2012 *MNRAS* **426** 2703
[13] Kewley L J, Dopita M A, Sutherland R S, Heisler C A, Trevena J 2001 *ApJ* **556** 121
[14] Shen Y, Ho L C 2014 **Nature 513** 210
[15] Sulentic J W, Bachev R, Marziani P, Negrete C A, Dultzin D 2007 *ApJ* **666** 757

Journal of Physics: Conference Series **771** (2016) 012030 doi:10.1088/1742-6596/771/1/012030

Spectroscopy and Photoionization Model of Planetary Nebulae : NGC 6543 and NGC 7662

Evaria Puspitaningrum[1], Hakim Lutfi Malasan[1], Hideyo Kawakita[2]
[1]Astronomy Study Program, Institut Teknologi Bandung, Jl. Ganesha No. 10, Bandung 40116, Indonesia
[2]Koyama Astronomical Observatory, Kyoto Sangyo University, Kyoto, Japan

E-mail: evaria.p@gmail.com

Abstract. Spectroscopic observations in the wavelength range of $\lambda\lambda4000 - 8000$ Å of the planetary nebulae (PN), NGC 6543 and NGC 7662, have been undertaken in 2014 using the compact spectrograph attached to the 28-cm Schmidt-Cassegrainian reflector at Koyama Astronomical Observatory. We derived physical parameters of NGC 6543 and NGC 7662 from the analyses of observed emission lines, i.e. electron temperature, density, and elemental abundances. To obtain more reliable physical quantities of NGC 6543 and NGC 7662, the use of photoionization model is needed. We use the computer code, CLOUDY C13.3 to build detailed photoionization models of NGC 6543 and NGC 7662. Model of NGC 6543 and 7662 can be constructed by providing sufficient information on physical parameters of the central stars and the nebula. In this work, input parameters to construct the photoionization model of NGC 6543 and NGC 7662 are distance, temperature and luminosity of central star and nebula parameters. The model results were compared with observational result taken in 2014. Observed and modeled temperatures were deduced from [O III] $\lambda\lambda(4959 + 5007)$ Å $/\lambda4636$ Å and the electron density from [S II] $\lambda\lambda$ 6716/6731 Å. There are some discrepancies between observed and modeled due to limitations inherent with the resolution and quality of spectroscopic data. Further spectroscopic observations with higher resolution of NGC 6543 and NGC 7662 is recommended.

1. Introduction

Planetary nebulae (PN) are isolated nebula, which are actually shells of gas that have been blown up in the fairly recent past by their central stars [8]. They emit strong emission-lines such as the recombination lines (hydrogen and helium) and collisionally excited lines of light elements that are observed in their optical spectra. Forbidden lines and two-photon emission were discovered from the study of PN [12],[8].

NGC 6543 is a helical nebula which lies near the north ecliptic pole. The spectral-line study of this PN is by [4]. NGC 7662 is a bright and spherical nebulae, the spectral-line study in the optical is by [3]. Harrington [2] did the photoionization modelling work on this PN.

The main target of this study is to construct a photoionization model of planetary nebulae by using information from the observation, especially low resolution spectroscopic data. This paper is organized as follows: Section 2 describe our new observational data. In Section 3, we describe an empirical analysis, our photoionization model and the derived result. Our final conclusion is in Section 4.

Table 1. Diagnostic for the electron temperature, T_e and the electron density, N_e. References: ^{a} – Hyung et al. (2000), ^{b} – Hyung & Aller (1997)

PN	Ion	Diagnostic	$T_e(K)$	Ref.	Ion	Diagnostic	$N_e(K)$	Ref.
NGC 6543	[O III]	$\frac{\lambda 4959+\lambda 5007}{\lambda 4363}$	8500	8300^a	[S II]	$\frac{\lambda 6717}{\lambda 6731}$	2800	6000^a
NGC 7662	[O III]	$\frac{\lambda 4959+\lambda 5007}{\lambda 4363}$	13700	13100^b	[S II]	$\frac{\lambda 6717}{\lambda 6731}$	3200	8000^b

2. Observation and data reduction

Spectra of NGC 6543 and NGC 7662 were obtained in 12 - 14 September 2014 with the 28-cm telescope in Koyama Astronomical Observatory and using a compact spectrograph (designed by Prof. Kawakita, one of our collaborator). The observations were done with a spectral resolution of $R \simeq 500$ in the 3800 - 8000 Å. The spectrograph has slit width $10''$ and slit length 7.5 mm.

The exposure time of 300 s for NGC 6543 observation yields a signal-to-noise ratio of $S/N \geq 10$ for the [O III] emission line. While the exposure time of 900 s yields a signal-to-noise ratio of $S/N \geq 10$ for the [O III] emission line in NGC 7662 observation. A spectroscopic standard star was observed for the flux calibration purpose, notably HR 7596. A series of flat-field frames and comparison lamp (F-Ne-Ar) by using HCT (Hollow Cathode Tube) were acquired for data reduction, flat-fielding and wavelength calibration.

Data reduction were carried out using `IRAF`. Data reductions involved two main processes of calibration, wavelength and flux calibration. All fluxes were corrected for reddening using $I(\lambda)_{corr} = I_{obs}(\lambda)10^{cf(\lambda)}$.The logarithmic $c(H\beta)$ value of the interstellar extinction for the case B recombination (T_e = 10000 K and N_e = 1000 cm^{-3}) has been obtained from the Hα and Hβ Balmer fluxes. We used the Galactic extinction law $f(\lambda)$ [5] for $R_V = A(V)/E(B-V) = 3.1$, and normalized such that $f(H\beta) = 0$. We obtained an extinction of $c(H\beta) = 0.1$ for the NGC 6543 and $c(H\beta) = 0.14$ for NGC 7662. Our derived nebular extinctions are in good agreement with the value derived by [14] for NGC 6543 and [7] for NGC 7662.

3. Result and analysis

3.1. Plasma diagnostic

The derived electron temperatures (T_e) and densities (N_e) are listed in Table 1. Due to low resolution spectrograph, not all emission lines appeared in the observation data. Therefore, (T_e) is only derived from the [O III] line ratio and (N_e) is derived from [S II] line ratio. We adopted the electron temperature and density from the observation for our CELs abundance analysis.

3.2. Ionic and total abundances from ORLs

Determining the ionic abundances, X^{i+}/H^+, by using the effective recombination coefficients [8] from the measured intensities of optical recombination lines (ORLs). Abundances of helium from ORLs are given in Table 2. The ionic and total helium abundances were derived from He I λ4471, λ5876 and λ6678 lines. We assumed the Case B recombination for the He I lines [11]. The total He/H abundance ratio is obtained by simply taking the sum of He^+/H^+ and He^{2+}/H^+.

3.3. Ionic and total abundances from CELs

We determined abundances for ionic species of N, O, Ne, S, and Ar from CELs (Collisional Excitation Lines). The ionic abundances, X^{i+}/H^+, can be derived from the line intensities of CELs. The equation given by [9] and [10] were used to derive the ionic abundances.

Table 2 shows total elemental abundances of nitrogen, oxygen, neon, sulphur and argon of CELs from the observation. Not all ionic abundances in each species can be determined due

Table 2. Input parameter for the CLOUDY photoionization model

PN	NGC 6543	NGC 7662		NGC 6543		NGC 7662	
Parameter				Model	Obs.	Model	Obs.
T_{eff} (kK)	63000	11300	He/H	11.04	11	10.81	10.97
$log(g_*/cms^{-2})$	7	6	C/H	8.75	-	8.9	-
$L_*(L_\odot)$	3510	2470	N/H	8.39	8.7	8.26	9.5
$N_H(\mathrm{cm}^{-3})$	2800	3200	O/H	8.98	8.78	8.77	8.75
D (kpc)	1	0.8	Ne/H	8.19	7.98	8.05	7.8
log r_{in}(cm)	16	16	S/H	7.27	-	7.49	7.57
log r_{out}(cm)	17.44	17.03	Ar/H	7.0	7.5	6.44	7.6

to low resolution. The *icf* (ionization correction factor) given by [6] was used to determine the total abundances.

3.4. Photoionization modelling

The photoionization code CLOUDY (version C13.3; [1]) was used to study the best-fitting model for NGC 6543 and NGC 7662. This code solve-consistently the radiative transfer of the stellar radiation field in a gaseous nebula with the defined total flux and spectral energy distribution of CS, density distributions and chemical abundances. We used the non-LTE model atmosphere from [13] as the ionizing source and used assumption for a spherical shell and homogeneous abundances. It generates the emission-lines spectrum, the thermal structure and the ionization structures of the nebula.

The best fitting model was obtained through CLOUDY's default optimization method of the "optimize lines" command over 100 iterations, involving the comparison flux intensities of some important lines relative to Hβ (such as He II λ4686 which highly depend on the temperature of CS, [O III] λ5007 and [N II] λ6584 describe ionization and thermal structure, He I λ5876 controls the He abundance, [Ne III] λ3869 controls Ne abundance, [S II] $\lambda\lambda(6716 + 6731)$ controls the sulfur abundances and [Ar III] λ7135 controls Ar abundance), with those measured from the observations. In NGC 6543, He II λ4686 was not detected so it is not used. χ^2 in CLOUDY is used to calculate the best fitting as follows :

$$\chi^2 = |\frac{F_i^m - F_i^o}{min(F_i^m, F_i^o)\sigma_i}| \quad (1)$$

The free parameters included distance and nebular parameters. We initially used the stellar luminosity and distance as listed in Table 2. Moreover, we adopted the nebular density and abundance derived from the empirical analysis in Section 3, but they have been gradually adjusted until the observed nebular emission-line spectrum was reproduced by the model.

3.5. Comparison of the emission fluxes

Figure 1 compares the 2-D spectra predicted by the best-fitting model and from the observations. Most emission-line fluxes presented are in reasonable agreement with the observations. χ^2 value for NGC 6543 model is 8.01, while NGC 7662 is 4.21. However, we notice there some discrepancies such as [Ar III], [Ar IV] in NGC 6543 and He II in NGC 7662. The discrepancy between our model and observed intensities of these lines can be due to the inhomogeneous condensations such as clumps and/or colder small-scale structures embedded in the global structure. It can also be due to the measurement errors of these weak and blended lines because of the low resolution instrument and blend with telluric line.

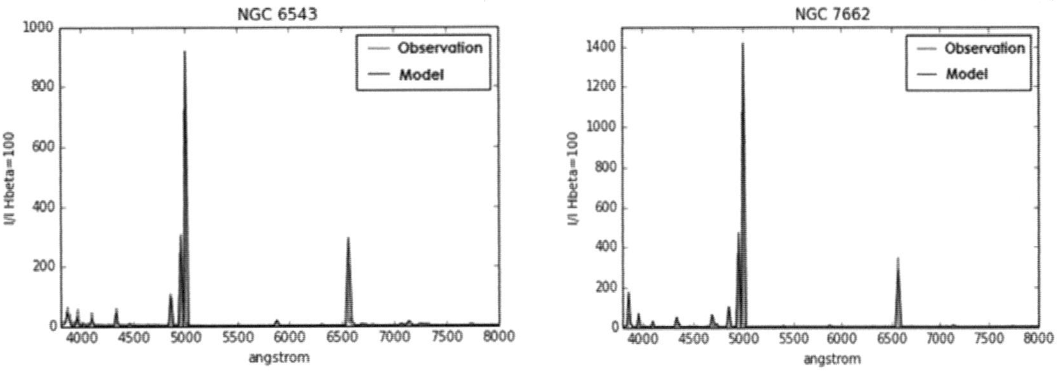

Figure 1. The 2-D spectra from the observation and model for NGC 6543 (left) and NGC 7662 (right). The red lines shows the observation result and the black lines shows the model results

4. Conclusion

Photoionization models based on CLOUDY can be build and were good applied for planetary nebula, NGC 6543 and NGC 7662. NGC 6543 has a good agreement with [4].

Photoionization model can be used to construct completed structure of PN with observational data as main constrain. A better observation data are required (i.e. higher resolution, etc) and more reliable model need to test another assumption, i.e. the nebula may not be spherically symmetric, an importance of temperature fluctuation, testing another different grain, etc.

Acknowledgments

This work would not happen without the instruments at our disposal. Therefore we thank Koyama Astronomical Observatory of Kyoto Sangyo University for providing all facilities, and the Sakura Science Program 2015 for the student exchange fund. We also thank FMIPA ITB.

References

[1] Ferland G J 2013 *PASP* **110** 761F
[2] Harrington J P, Seaton M J, Adams S, and Lutz J H 1982 *MNRAS* **199** 517
[3] Hyung S, Aller L H 1997 *MNRAS* **292** 71
[4] Hyung S *et al* 2000 *MNRAS* **318** 77
[5] Kaler J B and Jacoby G H 1989 *ApJ* **345** 871
[6] Kingsburgh R L and Barlow M J 1994 *MNRAS* **271** 257K
[7] Lame N J and Pogge R W 1996 *AJ* **111** 2320
[8] Osterbrock 2006 *Astrophysics of Gaseous Nebulae and AGN* (California: University Science Books)
[9] Pagel B E J, Simonson E A, Terlevich R J and Edmunds M G 1992 *MNRAS* **255** 325
[10] Pérez-Montero E and Diaz A I 2005 *MNRAS* **361** 1063 (PMD05)
[11] Porter R L, Ferland G J, Storey P J, Detisch M J 2012 *MNRAS* **425** L28
[12] Pottasch S R 1984 *Planetary Nebulae* (Dordrecht: Kluwer)
[13] Rauch T 2003 *A & A* **403** 709
[14] Wesson R, Liu X -W 2004 *MNRAS* **351** 1026

Membership determination of open cluster with parametric method: cross entropy

Itsna Khoirul Fitriana[1] and M. Ikbal Arifiyanto[1]

[1] Astronomy Program, Institut Teknologi Bandung, Indonesia

E-mail: itsnafitriana@gmail.com

Abstract. Stellar open clusters are useful for studying the structure and evolution of Galaxy. Determination of the open cluster members can be viewed from their kinematics: proper motion and radial velocity. It is because the star cluster members will have a motion in the same direction, towards a convergent point. Parametric method with a double elliptic bivariate gaussian function which contains 11 parameters is used as a model of the proper motion distribution. The value of these parameters are determined numerically by cross entropy method. This method is a method of data fitting using the likelihood function whose value is maximized to get the proper parameters. This method solves the problem of determining the initial value that is often difficult to determine in other numerical methods. There are four open clusters which we have analysed: NGC 2244, ASCC 100, NGC 5168, and NGC 2169. The resulting of proper motion parameters of this method matches the parameters in the literature.

1. Introduction

Study of structure and evolution of Galaxy can be complete by studying open clusters. Before performing further analysis of the open cluster, the membership determination of an open cluster is a process that should not be missed. This research aims to determine the membership of some open clusters with double elliptic bivariate gaussian function which contain of 11 parameters by [5]. The parameters in the equation are solved numerically using cross entropy (CE) method that contains of maximum likelihood function method. Actually there are many other methods to solve the equation, but the CE method is choosen because it has many advantages. The CE method uses the initial value in a range, it can answer difficult issues of other methods in determining the initial value.

2. Data

We used kinematic data (position, proper motion, and magnitude) of each cluster from different sources. We take data of NGC 2244 by [3]. This object is the benchmark model for this research and [5] which use same function. The other clusters, ASCC 100, NGC 5168, and NGC 2169 are taken from from UCAC4, the Fourth US Naval Observatory CCD Astrograph Catalog. We choose another object as the sample of an open cluster which has stars of field less than the stars belong to cluster (ASCC 100) and an open cluster which has field stars more than the stars belong to cluster (NGC 5168). The fourth object, NGC 2169, is the object whose data taken with twice the radius which is written in *Karchenko atlas page [2].* These objects had been determined by [1] using another gaussian model which solved numerically by CE method. The fourth cluster is used to check whether tour method can be used for a variety of data. Then, it is used on other clusters that haven't been checked before.

3. Method

The method used for the four clusters is similar. The method is CE method which has same steps with [4]. First, the determination of the initial value range using mean and dispersion value of proper motion data [5].Then, proper motion data is fitted with double elliptic bivariate gaussian function (equation 1) using CE method. Then, the best value of the parameters are used to calculate the membership probability of each stars with Bayessian model. We determine the stars belong to cluster which has probability more than equal to 0.5. Then the cluster members are plotted in a HR diagram to see the distribution of evolutionary track (figure 1). The last step is to compare the result with other literature. We compare the probability of the stars and the value of cluster's parameters in D value (equation 2). When the D value is close to 0, it means the comparison is fit.

4. Result and discussion

4.1 Results

The results of this research are:

- The model and method can be applied for the four open clusters: NGC 2244, ASCC 100, NGC 5168, and NGC 2169. The best parameter values are listed in the table 1

Table 1. Parameter values of open cluster

Parameter	NGC 2244	ASCC 100	NGC 5168	NGC 2169
N_f	0.50 ± 0.28	0,30 ± 0,16	0.52 ± 0.29	0.36 ± 0.22
σ_{xc}	0.09 ± 0.07	3.06 ± 1.77	2.65 ± 1.56	1.03 ± 0.58
σ_{yc}	0.09 ± 0.06	2.88 ± 1.62	2.53 ± 1.46	0.95 ± 0.54
ρ_f	0.02 ± 0.11	0.01 ± 0.14	-0.01 ± 0.16	0.00 ± 0.13
μ_{xf}	-0.04 ± 0.03	-2.58 ± 1.58	-3.44 ± 1.96	-2.63 ± 1.47
μ_{yf}	-0.06 ± 0.05	-2.14 ± 1.38	-3.41 ± 1.97	-2.75 ± 1.60
σ_{xf}	0.77 ± 0.46	20.83 ± 11.75	30.77 ± 17.91	22.68 ± 12.73
σ_{yf}	0.78 ± 0.45	20.59 ± 11.54	30.30 ± 17.13	22.93 ± 12.77
ρ_c	0.01 ± 0.12	0.04 ± 0.14	-0.03 ± 0.19	0.01 ± 0.16
μ_{xc}	0.01 ± 0.01	0.02 ± 0.07	-3.96 ± 2.24	-7.21 ± 4.74
μ_{yc}	0.02 ± 0.01	-2.41 ± 1.34	-4.05 ± 2.30	-1.48 ± 0.84

- D value between the research and first literature [5] for NGC 2244 and [1] for others) is more close to zero than D value of second comparison (literature by [2]). It can be seen in the table 2.

Table 2. D values for each cluster, D1 shows a comparison of the value of D to the data [1] except for NGC 2244 with the data [5] , while D2 shows the difference with [2]

	NGC 2244	ASCC 100	NGC 5168	NGC 2169
Nilai D1	1.13	0.43	0.02	0.24
Nilai D2	1.40	2.52	2.50	0.21

4.2 Discussion

The results show the successful performance of the model and method. So, it can be used to other open clusters which haven't been checked before. It will help understanding our Galaxy. But in practice there are still some steps that can be done in the future research. It aims to make research results getting better. Some suggestions put forward are as follows: (i) Radial velocity factor can be added in the membership determination. The study is conducted in three dimensions by adding a radial velocity component in addition to motion. (ii) Comparison between parametric and non-parametric method will be good to do to make sure the better method.

Appendices

$$\Phi(\mu_{xi},\mu_{yi}) = N_f\Phi_f(\mu_{xi},\mu_{yi}) + (1-N_f)\Phi_c(\mu_{xi},\mu_{yi})$$

$$= \frac{N_f}{2\pi\pi_{xf}\sigma_{yf}\sqrt{1-\rho_f^2}}\exp\left\{-\frac{1}{2(1-\rho_f^2)}\left[\left(\frac{\mu_{xi}-\mu_{xf}}{\sigma_{xf}}\right)^2+\left(\frac{\mu_{yi}-\mu_{yf}}{\sigma_{yf}}\right)^2 - 2\rho_f\left(\frac{\mu_{xi}-\mu_{xf}}{\sigma_{xf}}\right)\left(\frac{\mu_{yi}-\mu_{yf}}{\sigma_{yf}}\right)\right]\right\}$$
$$+\frac{(1-N_f)}{2\pi\pi_{xc}\sigma_{yc}\sqrt{1-\rho_c^2}}\exp\left\{-\frac{1}{2(1-\rho_c^2)}\left[\left(\frac{\mu_{xi}-\mu_{xc}}{\sigma_{xc}}\right)^2+\left(\frac{\mu_{yi}-\mu_{yi}}{\sigma_{yc}}\right)^2 - 2\rho_c\left(\frac{\mu_{xi}-\mu_{xc}}{\sigma_{xc}}\right)\left(\frac{\mu_{yi}-\mu_{yc}}{\sigma_{yc}}\right)\right]\right\}. \quad (1)$$

where N_f is a proportion of field to the number of stars in total, μ_{xi} and μ_{yi} are proper motion data in x and y directions, σxf and σyf are the standard deviation in the x and y directions for the field stars, μ_{xf} and μ_{yf} are proper motion parameter in x and y directions for the field stars, μ_{xc} and μ_{yc} are proper motion parameter in x and y directions for cluster stars, σxc and σyc are the standard deviation in the x and y directions for the cluster stars, ρ_f and ρ_c are the coefficient of rotation for the field and cluster stars.

$$D = \frac{(\mu_{xcCE}-\mu_{xcLIT})^2}{(\sigma_{xcCE})^2+(\sigma_{xcLIT})^2} + \frac{(\mu_{ycCE}-\mu_{ycLIT})^2}{(\sigma_{ycCE})^2+(\sigma_{ycLIT})^2} \quad (2)$$

Where CE declared value results of Cross Entropy method which we did and LIT stated value parameter of literature which we compare.

Journal of Physics: Conference Series **771** (2016) 012031 doi:10.1088/1742-6596/771/1/012031

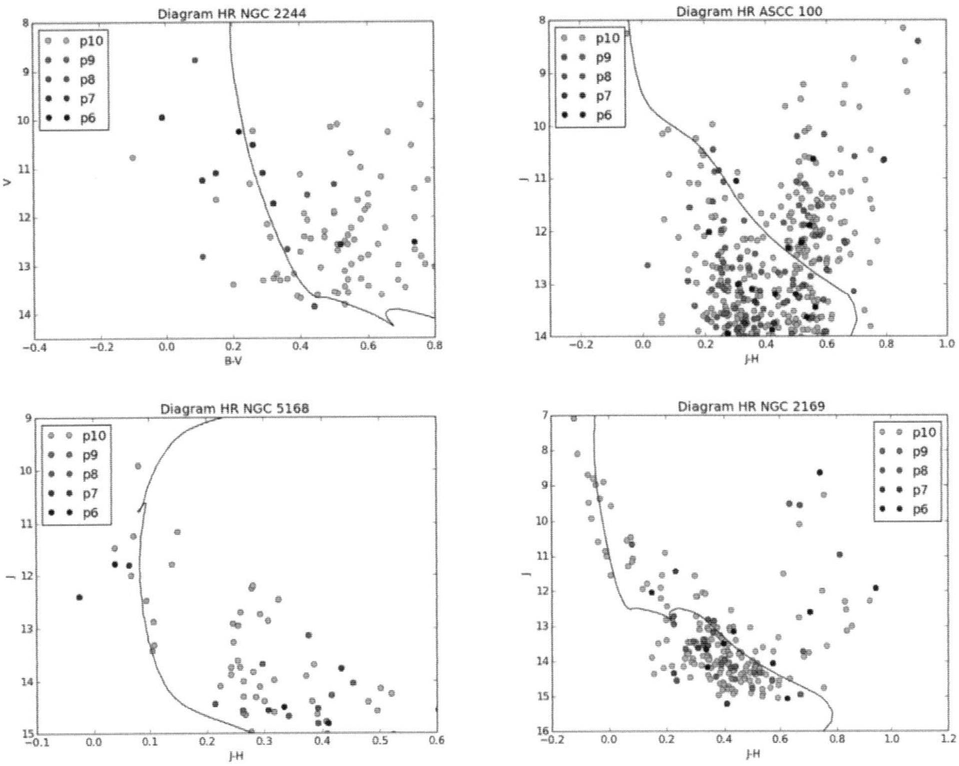

Figure 1. Color Magnitude Diagram (CMD) refer to diagram of Hertzsprung-Russell (HR) for the stars whose probability value ≥ 0.5. Color dots indicate the class of probability with 0.1 width of probability value: $0.5 \leq p6 \leq 0.6$; $0.6 < p7 \leq 0.7$; $0.7 < p8 \leq 0.8$; $0.8 < p9 \leq 0.9$; $0.9 < p10 \leq 1.0$

References

[1] Dias W S *et al.* 2014 *A&A* **564** A79
[2] Kharchenko N V, Piskunov A E, Röser S, Schilbach E, Scholz R D 2015 *A&A* **438** 1163K
[3] Marschall L, Van Altena W, and Chiu L 1982 *AJ* **87** 1497
[4] Monteiro H, Dias W S, Caetano T C 2010 *A&A* **516** A2
[5] Sabogal-Martnez B A, Garca-Varela J A G, Higuera M A, Uribe A, Brieva E 2001 Revista Mexicana de Astronoma y Astrofsica **37** 105

Search for streams in thick disk and halo of the Milky Way

Dian Puspita Triani and M Ikbal Arifyanto

Department of Astronomy, Institut Teknologi Bandung.

E-mail: Dptriani@gmail.com

Abstract. Study on the substructures of Milky Way is very important because it can be used as the stage indicator in the formation process of the Galaxy. Substructures are made by clumps of stars with similar velocity, called as stellar streams. We searched for the stellar streams in thick disk and stellar halo of the Milky Way using data from Geneva-Copenhagen survey. We built velocity curves from data and applied the wavelet transform to find dense area in the velocity distribution. For the significace test, peaks on the wavelet transform of the data are reduced using wavelet transform based on Monte Carlo simulation. We found five streams on thick disk and two in stellar halo. Based on the metallicity distribution, it is shown that both thick disk and halo streams have similar range of metallicity, thus we classified them as dynamical stream.

1. Introduction

Two most well-known theories about the formation of galaxies are bottom-up and top-down scenarios. In bottom-up scenario [1], galaxies are formed through merger of small clumps into larger structure. While in top-down scenario [2], galaxies are the result of a giant structure splitting up into smaller scale. The right scenario should be justified by how well it can explain the observed facts of galaxies, for instance, the components observed in Milky Way.

We could constrain each scenarios through the analysis of metallicity distribution and the stellar kinematics in the galaxy. Existence of structures and substructures observed in the Galactic components are very important to constrain the formation scenarios. These structure(s) appear as a group of stars moving with similar velocity, known as stellar streams. There are two kinds of stellar streams: tidal streams and dynamical streams. Tidal streams consist of stars with similar velocities suspected to have similar origin, for example from globular cluster or galactic satellite. Dynamical streams consists of stars from different origin but trapped in the certain phase space because of the dynamic resonance.

This paper focused on the method of stellar stream search on the thick disk and halo of the Milky Way from Geneva-Copenhagen survey. The thick disk member is classified based on elemental abundance trends [3] while the halo member is classified based on metallicity of the stars. We applied wavelet transform on the stellar velocity distribution to search for streams as explained by [4].

2. Data and method

To detect stellar streams as the substructures in the space configuration, there are two conditions should be fulfilled. First, the disruption process on the stream occured not so long ago, less than 10 Gyr [5], so that the structure are still not diluted and can be observed as clumps of data. Secondly, the survey should be conducted on a wide field of the sky. Therefore we used the data from Geneva-Copenhagen survey [6] via VizieR website. We got the parameters from 13.382 members of Milky Way in solar neighborhood. The important parameters that we used in this paper are metallicity[Fe/H] and velocity component in three directions: radial velocity (U), tangential velocity (V), and the velocity component to the north galactic pole (W).

2.1. *Thick disk members classification*

To find the possibility that a star is a member of a component of the Milky Way (we called it D, TD and H for the possibility of disk, thick disk and halo membership, respectively), we should constrain two relative possibility of the membership of thick disk compare to those in disk and halo:

$$\frac{TD}{D} = \frac{X_{TD}}{X_D}\frac{f_{TD}}{f_D}, \qquad \frac{TD}{H} = \frac{X_{TD}}{X_H}\frac{f_{TD}}{f_H} \tag{1}$$

X is fraction of the component member in the solar neighborhood and f is the Gaussian distribution function of stellar population velocity on each components. Stars are classified into thick disk if the fraction TD/D and TD/H is more than 4 [3]. There are 360 out of 13.382 data that fulfill these conditions and therefore we classified those stars as thick disk.

2.2. *Halo members classification*

Classification of halo is done using metallicity selection. Stars with log [Fe/H] less than -1 is classified as halo member. Only 64 stars in the Geneva-Copenhagen catalog that fulfill the criteria. Halo stars have high orbital inclination with respect to galactic plane.

2.3. *Search for streams in thick disk and halo*

Then we searched for the region with higher densities, then we called it clumps. To detect the stellar streams, we plot the eccentricity and angular momentum distribution of stars. In this plot, stellar streams will appear as clumps of datas. Unfortunately, we do not generally know the orbital parameters of stars in the solar neighborhood. Then we used Kepler approach to project the stellar stream members in plot consists of components analogue to the orbital parameter. Stellar stream in thick disk is detected by plotting data in diagram with V as the horizontal axis and $\sqrt{U^2 + 2V^2}$ as y-axis. V is the analogue of angular momentum and $\sqrt{U^2 + 2V^2}$ is the analogue of eccentricity. Velocity component in W direction is ignored because in thick disk, strong phase mixing will dilute the feature in W direction even in a short time. In halo, the Kepler approach angular momentum is analogued as V_{az} ($V_{az} = \sqrt{(V + V_{LSR})^2 + W^2}$) and eccentricity is analogued as $V_{\Delta E} = \sqrt{U^2 + 2(V_{az} - V_{LSR})^2}$. In [4], it is stated that tidal streams is concentrated in ($V_{az}, V_{\Delta E}$) around the progenitor orbital.

Then we applied the wavelet transform on the data to detect clumps clearly. Wavelet transform that we used is adopted from [4]. Data are divided into 10.000 bin. Wavelet transform then applied in each bin using Mexican-hat-shaped kernel. Parameter scale valued on 5 km/s.

We used Monte Carlo simulation using the same number of data with the sample (360 stars). Simulated data are generated using normal distribution function with velocity dispersion shown in table 1. U component peaked at 0, V component peaked at asymmetric drift velocity. Wavelet transform applied on the peak of simulation data based on Poisson noise, Monte Carlo simulation was done 5 times and then averaged. Bin number and parameter scale that we used in the simulation is similar with those we used in the data.

Wavelet transform from the data then is subtracted by the wavelet from Monte Carlo simulation. We only detect the positive value with significance bigger then 2σ. Wavelet transform after subtraction shown in figure 1 and 2.

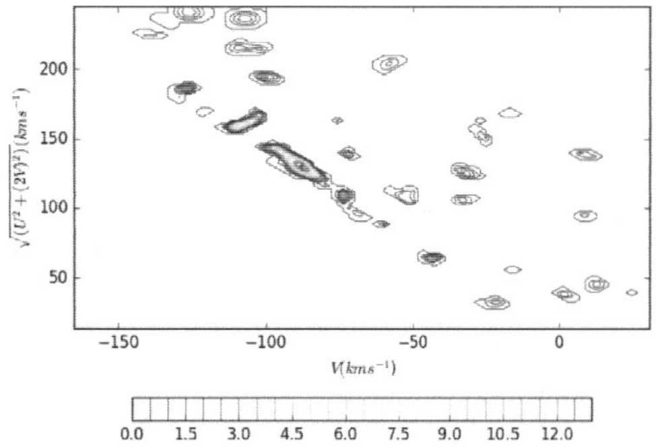

Figure 1. Stellar density in thick disk after subtraction by result from Monte Carlo simulation. The shown density have significance higher than 2σ. Colors signify the stellar density (seen in the colorbar)

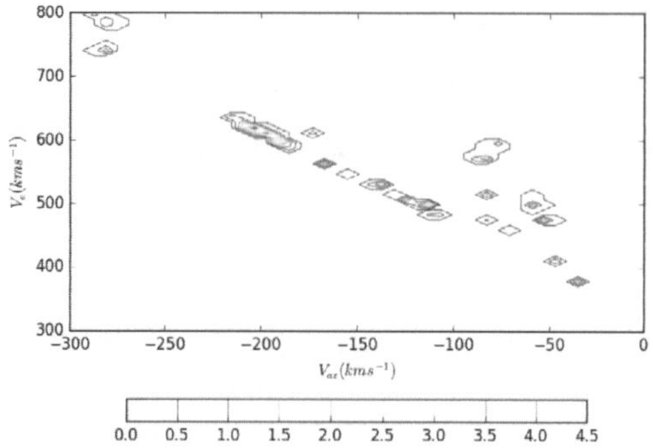

Figure 2. Stellar density in halo after subtraction by result from Monte Carlo simulation. Significance of the peak is larger than 2. Colors signify the stellar density (seen in the colorbar)

3. Result and analysis

3.1. Stellar streams in thick disk

Figure 1 shows several peak densities in velocity distribution with significance higher than 2σ. We compare these peaks with thick disk's stellar streams in [4]. The highest density peaked on V = -95 km s^{-1} and V = -120 km s^{-1}. The first peak is found by [7], namely AF06. The second peak is known as Arcturus stream. Both streams are correlated with the presence of resonance in the galactic bar.

Around V = -50 km s^{-1}, Hercules stream is detected. Hercules stream formed by the Outer Lindblad resonance wih the galactic bar. At V = -20 km s^{-1}, there is a weak indication of the presence of Hyades-Pleiades stream. We did not detect the Sirius stream that peaked at V = 4 km s^{-1}. At V= -110 km s^{-1}, we found a new peak with significance level around 7.5σ. This stream is not mentioned in [6]. We then count the number of stars, average and standard deviation of star's metallicity in each streams to analyse whether the streams are dynamical or tidal. Table 2 shows the data.

Table 2. Number of stars (N), average metallicity ([Fe/H]) and standard deviation ($\sigma_{[Fe/H]}$) of stars in each streams

Streams	N	[Fe/H]	$\sigma_{[Fe/H]}$
AF06	86	-0.38	0.32
Arcturus	26	-0.41	0.38
Hercules	5	-0.45	0.36
Hya-Plei	2	-0.24	0.17

From table 2, we can conclude that Geneva-Copenhagen survey shows significant detection on AF06 and Arcturus streams with number of data 86 and 26. Hercules and Hyades-Pleiades is detected weakly with only 5 and 2 data points. From this finding and previous literature, AF06, Arcturus and Hercules streams have large metallicity distribution, indicates that the stars in these streams have different origin. Therefore, these streams classified as dynamical stream. The new stream that we found at V = -110 km s^{-1} have 31 data points and large metallicity distribution. Therefore, this new significant stream is classified as dynamical stream.

3.2. Stellar streams in halo

Figure 2 shows the peak in wavelet transformed velocity distribution from Geneva-Copenhagen survey after subtraction by result from Monte Carlo simulations. Substructure number in stellar halo is much less than in thick disk sample. This is primarily because the number of sample in galactic halo from Geneva-Copenhagen survey is so small. Overall, we only detected two substructure in halo component.

The first stream ranged between V_{az} -140 km s^{-1} and -100 km s^{-1} and $V_{\Delta E}$ from 460 km s^{-1} and 530 km s^{-1}. We found 11 data points in this stream with average metallicity -1.5 and the standard deviation of metallicity 0.33. This stream is detected as S3 in [8] and as ωCen in [9] and [10].

[5] and references there in did not detect the second stream that we found. With V_{az} between -230 km s^{-1} and -180 km s^{-1} and $V_{\Delta E}$ between 550 km s^{-1} and 650 km s^{-1}, this stream have 14 data points with average metallicity -1.23 and standard deviation 0.17.

Metallicity of streams in galactic halo is far smaller than thick disk. Generally, this shows that thick disk is richer in metal (iron) than halo. Similar range of standard deviation of metallicity shows that stars in galactic halo could possibly come from different origins, therefore we classify them as dynamical stream.

4. Conclusion

Milky Way consists of several components:thin disk, thick disk, bulge and stellar halo. Presence of substructure in each component could be the tracer on the evolution and formation process in each component. Substructure can be observed as clumps in velocity distribution curve. Substructure is known as stellar stream since it consists of stars moving with similar velocity components.

There are two kind of stellar streams. Tidal stream consits of stars with similar velocity component and origin. Dynamical stream consists of stars from different origin but trapped on similar velocity distribution because of the dynamical resonance.

Here, we present the search for stellar stream in thick disk and halo using data from Geneva-Copenhagen survey. We detect 5 streams in thick disk, four of which had been detected by [4] : Hercules, Hyades-Pleiades, AF06 and Arcturus. We also found one stream that had not been detected in [4], this stream peaked on V = -110 km s^{-1}. Metallicity distribution in thick disk streams is quite large that we confidently conclude that stars in those streams have different origin, thus classify them as dynamical streams.

In galactic halo, we found two streams. The first is detected as S3 in [8] and the second is detected as ωCen in [9] and [10]. Metallicity distribution in both stream is quite large, indicates that origin in each stream is different, therefore we also classify them as dynamical stream.

References

[1] Nordstrom B *et al.* 2004 *A&A* **418** 989
[2] Eggen O J Lynden-Bell D Sandage A 1962 *ApJ* **136** 748
[3] Bensby T *et al.* 2003 *A&A* **410** 527
[4] Klement R 2010 *A&Arv* **18** 567
[5] Helmi A White S 1999 *MNRAS* **307** 495
[6] Klement R Fuchs B Rix H W 2008 *ApJ* **685** 261
[7] Arifyanto M I and Fuchs B 2006 *A&A* **449** 533
[8] Dettbarn B *et al.* 2007 *A&A* **474** 857
[9] Dinescu I 2002 *Astronomical Society of the Pacific Conference Series* **265** 365
[10] Brook B *et al.* 2004a. *Astronomical Society of the Pacific Conference Series* **327** 100

Sky brightness and twilight measurements at Jogyakarta city, Indonesia

Dhani Herdiwijaya

Astronomy Research Division and Bosscha Observatory, Institut Teknologi Bandung, Ganesha 10, Bandung, Indonesia 40132

dhani@as.itb.ac.id

Abstract. The sky brightness measurements were performed using a portable photometer. A pocket-sized and low-cost photometer has 20 degree area measurement, and spectral ranges between 320-720 nm with output directly in magnitudes per arc second square (mass) unit. The sky brightness with 3 seconds temporal resolutions was recorded at Jogyakarta city (110^0 25' E; 70^0 52' S; elevation 100 m) within 136 days in years from 2014 to 2016. The darkest night could reach 22.61 mpass only in several seconds, with mean value 18.8±0.7 mpass and temperature variation 23.1±1.2 C. The difference of mean sky brightness between before and after midnight was about -0.76 mpass or 2.0 times brighter. Moreover, the sky brightness and temperature fluctuations were more stable in after midnight than in before midnight. It is suggested that city light pollution affects those variations, and subsequently duration of twilight. By comparing twilight brightness for several places, we also suggest a 17^0 solar dip or about 66 minutes before sunrise for new time of Fajr prayer.

1. Introduction

Observing the starry sky on a clear, moonless night, from a location well away from population centers, can be unsettling moment. The enormous dome of the heavens may engender feelings of borderless and remoteness, and the ceiling of thousands of stars are visible to the unaided eye [1]. Moreover, the daily alternation between day and night, vice-versa is one process of nature whose regular repetition exerts a striking impact on the structure of organisms and the way of life of every inhabitant of our planet. The change does not occur instantaneously but over the rather protracted period known as twilight. Transportation, religious worship, the schedule of work in the fields, artificial lighting, in fact the whole routine of daily life must often be adapted to the varying illumination of twilight. The starry night and twilight belong to the few primeval natural phenomena that should be protected by all people, in about the same way as phenomena like nature, sunset or the beach. One should think that both directly affect our living conditions on so great a scale that science would long since have reached some depth of understanding as to the nature and behavior processes [2]. Night sky brightness means a cultural heritage. It refers to the condition in the night sky with moonless light. In the visible light regime, the sky brightness is affected by natural airglow, direct or scattered light, zodiacal light, and light pollution from man-made artificial light. Moreover, the city light pollution can extend many kilometers beyond the boundaries of the city. There are a number of different brightness units in use in the different fields of night sky brightness with their individual traditions and advantages. However, those have a conversion formula or tables [3], [4]. In this case, sky brightness was expressed in magnitude per arcsec square (mpass) unit [4].

Twilight is a neglected natural phenomenon with gradual and elusive sky brightening or darkening compare with more distinct light transition of sunrise and sunset. Twilight sky brightness is classified into three distinct phases according to solar elevation and the prevailing visibility conditions due to the illumination level: a) civil twilight (-6^0<solar dip<0^0), when terrestrial objects can still be distinguished by human observers, b) nautical twilight (-12^0<solar dip<-6^0), when only object outlines are visible, and c) astronomical twilight (-18^0<solar dip<-12^0), when the illumination level is low enough such that stars and other astronomical objects are available for observation [5]. In Islamic country, morning and evening twilights are related to obligatory prayers of Fajr (dawn) and Isya (night) prayer, respectively. Unfortunately, there is no world consensus for solar dip in both prayer times. Islamic countries use solar dip within interval 15 to 20 degrees. In Indonesia, the largest solar dip of 20 degree has been long accepted for starting Fajr prayer, without supporting some quantitative observational results.

All human outdoor lighting is functional for being the enhancement of the visibility and/or of the aesthetics (decorative lighting). Excessive outdoor lighting causes light pollution, which represents a loss as regards electric energy. Good lighting design ensures that the light comes where it is needed, and does not fall elsewhere. It may cause considerable economic and environmental losses [6]. Light pollution depends on two factors: light emission and scattering particle abundance. The latter is variable over the world. It depends of course on population density but even more so on affluence. Most countries do not have yet a legislature, which would permit the imposition of pollution control. Aerosols, particles floating in the atmosphere, can have many different sources. The most common are, apart from water droplets in haze and clouds, dust particles from industry, agriculture and deserts, salt crystals from the sea, soot from diesel engines, ash from volcanoes and forest fires [1]. The close relationship between light pollution and air pollution is then pointed out by [7] [8] [9]. Long term sky brightness monitoring has been done before, which was correlated with solar activity [10].

This paper will report single point measurement of sky brightness and twilight using simple, palm-size, and USB connected photometer. Unihedron Sky Quality Meter (thereafter SQM) was quantified the quality of the night sky with three seconds time resolution at Jogyakarta city, one of the famous cultural heritage cities in Indonesia, for 136 days on year 2014 to 2016. The effect of city light pollution and based on solar dip twilight were also described by comparing to other places with application to beginning of Fajr prayer which is important for Islamic community.

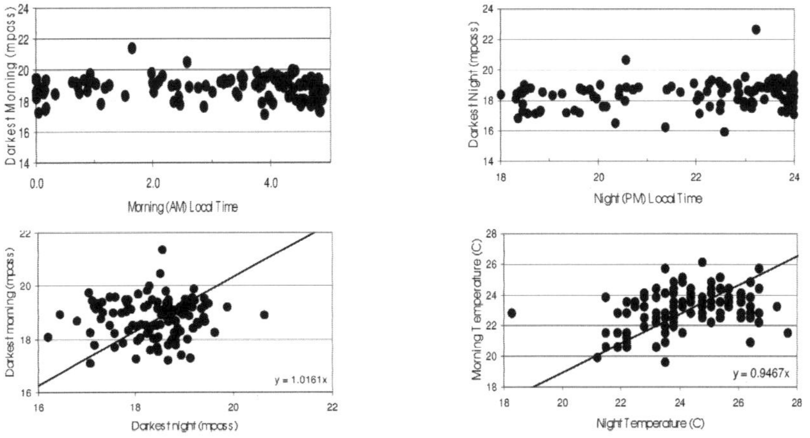

Figure 1. (from left to right) Distribution of the darkest sky brightness before sunrise, (a) and after sunset, (b). The relation between the darkest night and morning, (c) and their temperatures, (d).

2. Distribution of sky brightness and temperature

The darkest morning sky brightness (after midnight) could reach 21.34 mpass, with mean value of 18.81±0.70 mpass (see Figure 1.a). It is noted that the fluctuations can be obtained by calculating standard deviation of data. On the other hand, the mean value of night sky brightness (before midnight) gave 18.44±0.85 mpass with the darkest night reached 22.61 mpass (see Figure 1.b). It is seen that the darkest brightness at night and morning was directed toward midnight and before twilight in the morning, respectively. We also noted that sky brightness fluctuations in the morning were more stable (see Figure 1.c) and temperatures were cooler (23.25±1.15^{0}C) than before midnight (24.72±1.22^{0}C), as shown in Figure 1.d.

3. Twilight

The patterns of morning and evening twilight at Jogyakarta showed no significantly difference of about 4%, due to light pollution, as given in Figure 2.a. and 2.b. It means that the atmospheric layers are relatively the same structures in the east and west directions, respectively. Later, we selected only morning twilight. The sky brightness of evening twilight was shifted to the left at about 50 minutes after sunset, and showed darker sky brightness than morning twilight brightness, as seen in Figure 2.a.

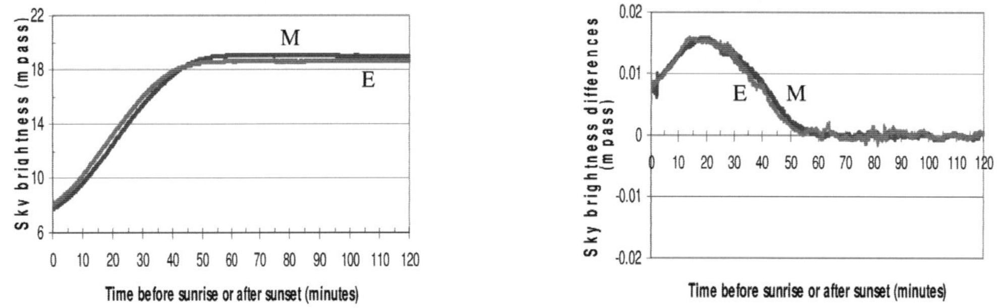

Figure 2. The morning with blue lines - M (a), and evening with red lines - E twilights (b)

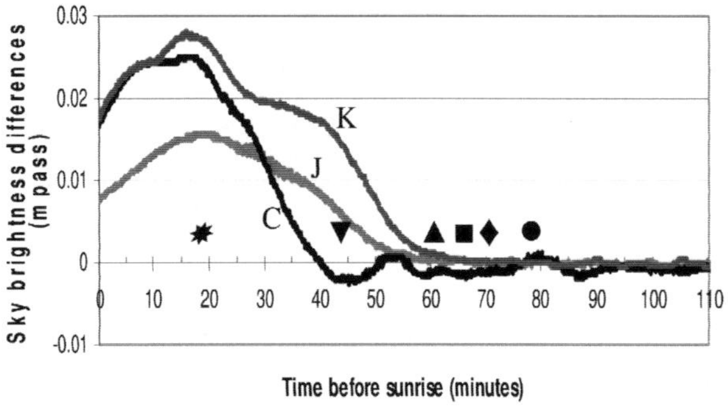

Figure 3. Morning twilight at Kupang (K), Jogyakarta (J), and Cimahi (C) cities with average solar dip 20^0 (circle), 18^0 (diamond), 17^0 (square), 16^0 (triangle), 12^0 (reverse triangle), and 6^0 (star)

Both light pollution and air pollution contribute to the sky brightness at Jogyakarta city, especially between 15 to 50 minutes after sunset or before sunrise. Aerosol particles in lower atmosphere will absorb the sunlight, so that they make darker sky. Then, they will scatter the city light, causing brighter sky for 50 minutes after sunset. The shifted sky brightness at evening twilight becomes one indicator for light pollution.

Figure 3 shows sky brightness differences of morning twilight with moving average curves for several cities, Cimahi (107^0 32'E; 6^0 53' S; elevation 700 m; Dec. 18, 2013), Kupang (124^0 0'E; 9^0 40' S; elevation 1300 m; May 10, 2013), and Jogyakarta (110^0 25'E; 7^0 52' S; elevation 100 m; May 18, 2015). It is clear that morning twilight at Cimahi had strongly affected by light pollution due to shifted curve, relative to intermediate level at Jogyakarta, and the darkest one at Kupang. Those cities have no any regulation about outdoor lighting. The brightening sky toward time of sunrise is clearly not linear, as showed in Figures 2 and 3. The different scale heights of atmospheric components allow each to be observed for different periods of interaction in twilight. However, smaller than 6^0 solar dip or civil twilight (about 19 minutes before sunrise), sky brightness differences gave the same pattern which implies the same lowest atmospheric height. The nautical twilight (12^0 solar dip and about 44 minutes before sunrise) was indicated by a hump at Kupang. The astronomical twilight was given by no sky brightness differences (about 70 minutes before sunrise). Figure 3 shows that sky brightness starts to increase at about 16^0 solar dip or about 62 minutes before sunrise. By using the above observational results, we argue that assuming 20^0 (about 79 minutes before sunrise) or larger solar dip for Fajr prayer needs a revision. We propose a 17^0 solar dip (about 66 minutes before sunrise), as a new reference angle for Fajr prayer.

4. Conclusions

We have used 136 days with high temporal resolution of sky brightness measurements at Jogyakarta city within year 2014 to 2016. We summarized as follow (1) the darkest of sky brightness fluctuations in the morning (after midnight) were more stable (18.81±0.70 mpass) and temperatures were cooler (23.25±1.15^{0}C) than before midnight 18.44±0.85 mpass and 24.72±1.22^{0}C, respectively, (2) light pollution contributed about 4% in the evening twilight at Jogyakarta, (3) twilight sky brightness is not a linear brightening, (4) we propose a 17^0 solar dip (about 66 minutes before sunrise), as a new reference angle for Fajr prayer.

References

[1] Mizon B 2002 *Light pollution; Responses and remedies.* Patric Moore's Practical Astronomy Series. (London: Springer)

[2] Rozenberg G V 1966 *Twilight: A Study in Atmospheric Optics* (New York: Plenum Press) p 160

[3] Leinert Ch et al. The 1997 reference of diffuse night sky brightness *Astron & Astrophys Suppl Ser* **127** 1

[4] Patat F 2003 UBVRI night sky brightness during sunspot minimum at ESO-Paranal *Astron & Astrophys* **400** 1183

[5] United States Naval Observatory Astronomical Applications Department 2005 *Multiyear interactive computer almanac, 1800–2050* (USA: Willmann-Bell)

[6] Schreuder D A 2000 Obtrusive light audits: A method to assess light pollution. The 3rd National Lighting Congress Special session on *Light Pollution* held on 23-24 November 2000 at Taskisla-Istanbul Technical University, Istanbul, Turkey. In: Anon, 2000

[7] Kosai H and Isobe S 1991 Organized observations of night sky brightness in Japan ed. Crawford p 35

[8] Garstang R H 1989 Night-sky brightness at observatories and sites *Astron. Soc. of the Pacific* **101** 306

[9] Schwarz H E ed. 2003 Light pollution: The global view. Proceedings of the International Conference on Light Pollution, La Serena, Chile. Held 5-7 March 2002. (*Astrophysics and Space Science Library* vol 284) (Dordrecht, Kluwer Academic Publishers)

[10] Matilla K, Vaisanen P, Appen-Schnur G F O 1996 Sky brightness at the ESO La Silla Observatory 1978-1988 *Astron & Astrophys Suppl Ser* **119** 153

Journal of Physics: Conference Series **771** (2016) 012034 doi:10.1088/1742-6596/771/1/012034

Coronal structure analysis based on the potential field source surface modeling and total solar eclipse observation

Johan Muhamad[1 2],Farahhati Mumtahana[1], Heri Sutastio[1], Irfan Imaduddin[3] and Gerhana P. Putri[3]

[1] Space Science Center, LAPAN, Jl. Djundjunan 133, Bandung, Indonesia
[2] Institute for Space-Earth Environment Research, Nagoya University, Nagoya, Japan
[3] Department of Astronomy, Institut Teknologi Bandung, Bandung, Indonesia

E-mail: johan.muhamad@lapan.go.id

Abstract. We constructed global coronal magnetic fields of the Sun during the Total Solar Eclipse (TSE) 9 March 2016 by using Potential Field Source Surface (PFSS) model. Synoptic photospheric magnetogram data from Helioseismic and Magnetic Imager (HMI) onboard Solar Dynamics Observatory (SDO) was used as a boundary condition to extrapolate the coronal magnetic fields of the Sun. This extrapolated structure was analyzed by comparing the alignment of the fields from the model with coronal structure from the observation. We also used observational data of coronal structure during the total solar eclipse to know how well the model agree with the observation. As a result, we could identify several coronal streamers which were produced by the large closed loops in the lower regime of the corona. This result verified that the PFSS extrapolation can be used as a tool to model the inner corona with several constraints. We also discussed how the coronal structure can be used to deduce the phase of the solar cycle.

1. Introduction

As the manifestation of the magnetic fields above the photosphere, solar coronal structure strongly depends on the magnetic fields generated from the interior of the Sun. These magnetic fields emerge to the photosphere, rise up to the corona and may extend to interplanetary space by some instabilities of its structure, which could result to the catastrophic impacts in the solar system. Therefore, one can infer solar corona as the extension of the magnetic fields dynamics on the solar surface. This should be valid for the assumption that the solar corona is force-free and plasma beta of the whole system is infinitesimal.

By using the vector magnetic field information on the photosphere, one can extrapolate magnetic fields in the corona on the base of assumption that there is no current flowing in the corona. This method, so called potential field, can give us general feature of the coronal structure and has been widely used to model the corona for variety of purposes. To generate global model of the coronal field, one boundary condition on top of the corona is needed where all magnetic fields in this spherical boundary are fully radial, which is called as source surface. This technique of modeling solar corona is known as Potential Field Source Surface (PFSS)[1].

In this paper, we generated model of the solar corona on 9 March 2016, which coincided with the total solar eclipse (TSE) happened in Indonesian archipelago. Since the structure of the corona can be observed during the total solar eclipse, observational and synthetic coronal structure from the model can be compared and studied extensively. In this preliminary results, we briefly compared and analyzed the structure of the corona generated by PFSS model and the observed corona from two different sites, which are Ternate and Penajam Paser Utara (PPU).

2. Data and methods

2.1. Vector magnetogram and PFSS model

The vector magnetic field data used in this work are derived from Helioseismic Magnetic Imager onboard Solar Dynamics Observatory (SDO). We used synoptic magnetogram data at 00:40:00 UTC in 9 March 2016 for one Carrington rotation taken from the database on the pfss_viewer module within SolarSoft SSW libraries.

PFSS method was generated by using radial component of the vector magnetogram data in the photosphere. Coronal fields were calculated as solution for the Laplace equation by using separation variable with some components derived from spherical harmonic components and associated Legendre polynomial [2]. The calculation was conducted by using PFSS module of SSW package in Interactive Data Language (IDL) environment developed by De Rosa and Schrijver with source-surface distance is 2.5 solar radius [3].

2.2. TSE observation from Ternate

The location site was in Pantai Falajawa, Ternate city, North Maluku, Indonesia, $0°47'$ N and $127°22'$ E. We used Extra Low Dispersion (ED) Apochromatic Refractor with aperture 103mm and focal ratio f/7.7 mounted in Sphinx SXD2, which were both manufactured by Vixen. The optical tube assembly was equipped with Seymour black polymer solar filter. Images were detected by 22.2 x 14.9 sized CMOS sensor with fast DIGIC 6 processor built in DSLR Canon EOS Rebel T6s, performing image size 6000x4000 and pixel size 3.7 μm. Unfortunately, the weather was cloudy that made the Sun covered with thin clouds during the peak time. Due to the initial target of the observation was to observe stellar background during the eclipse, most of the images were taken with low exposure which results to the dim corona in the images.

The chosen best images of the observation were first centered and then zoom (radially) blurred in GIMP. These blurred images were then subtracted from the original images before added each other to produce composite image. The composite image was then multiplied with the original images. Finally, these images were stacked and then some contrast and brightness adjustment were applied.

2.3. TSE observation from PPU

The observation site was Pantai Corong ($1°24'12.4''$ S, $116°39'14''$ E), 1.5 m above sea level in Kab. Penajam Paser Utara (PPU), East Kalimantan, Indonesia. The duration of totality was 1m 51s and the peak of totality was 00:34:02 UTC. The Sun was at 32.7 altitude when the peak was observed. By using William Optic Zenithstar 66SD (D=66mm, f=388mm) and Nikon D5100, we obtained 19 images. Some chosen images were processed by Adobe Photoshop CC 2015 and Adobe Lightroom.

The basic image processing steps are: 1) The coordinates of the center of the Moon were detected with high accuracy. 2) Applying radially blur filter from the center of the Moon. 3) Substracting centered images with its radial blurred images. 4) Multiplicating centered images with substracted images. 5) The aligned images were combined. The weights were proportional to the exposure time of each image. 6) Finally, in order to enhance the combined image, locally adaptive filters and noise reduction were applied.

Journal of Physics: Conference Series **771** (2016) 012034 doi:10.1088/1742-6596/771/1/012034

3. Results and discussions

Result of the PFSS model for 9 March 2016 solar corona, TSE observation from Ternate and PPU are shown in figure 1, 2, and 3, respectively. The orientation of the Sun in figure 2 and figure 3 are rotated in order to adjust the North-South orientation of the model.

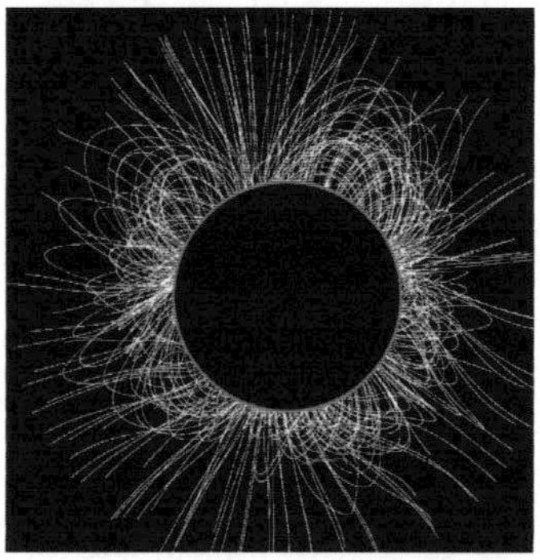

Figure 1. PFSS Model.

Figure 2. TSE observation from Ternate.

Figure 3. TSE observation from PPU.

TSE observation from Ternate shows only lower regime of the corona due to the combination of the cloudy sky and low exposure time of the observation. The detail structure of the corona is very difficult to be observed. However, we still can identify some small closed structures in the east limb of the Sun and at least three large loops in the western limb. Comparison for the observation image from Ternate and PPU shows that all the structures in figure 2 indeed correspond well with the larger structures of the corona shown in figure 3.

Figure 3 shows more detail structures with the extension of the corona beyond 2 solar radius from the photosphere. We can obviously identify several streamers originate from the corresponding loops of figure 2, particularly in the western side of the solar disc. More complex structures are observed in the eastern limb of the Sun with many loops overlaid each other from the line of sight of the observer.

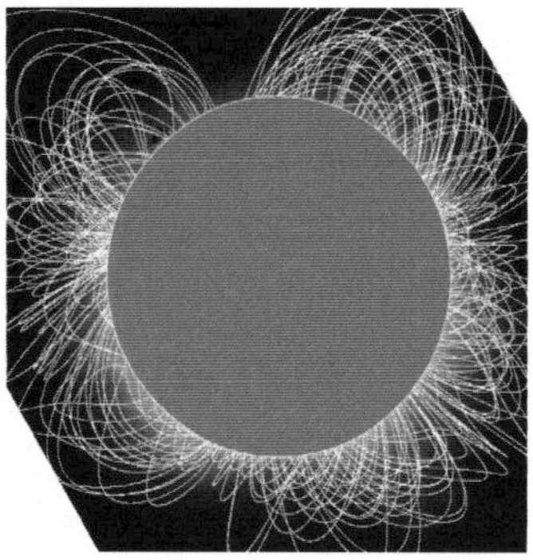

Figure 4. PFSS closed loop Model superimposed to the Ternate TSE observation.

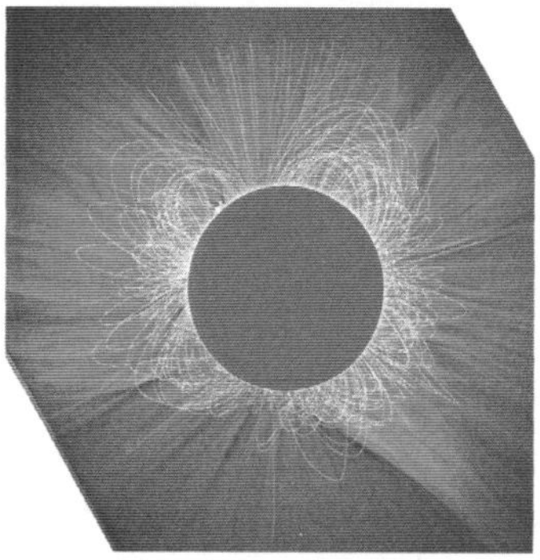

Figure 5. PFSS Model superimposed to the PPU TSE observation.

In order to emphasize the detail comparison between model and observations, we overlaid the PFSS model of the 9 March 2016 solar corona on the TSE observation images from Ternate and PPU. The results of the overlaid images for Ternate and PPU observations are shown in figure 4 and figure 5, respectively. Since coronal streamer always occur on top of the closed magnetic loop structure in the corona, we can compare the position of the coronal streamers from the observation with the closed loop structures in the model. The consistency between these two features can be used to qualitatively see how well the model can reconstruct real coronal structures.

Figure 4 shows the PFSS model of corona overlaid on Ternate TSE observation where only closed loops are shown. The observation from Ternate cannot provide large loop structure and it makes the comparison becomes difficult. However, in general we still can see the agreement between the observation of the TSE and the model from the consistency of the appearance of while light corona in the low altitude with some closed loops. In figure 5, more detail comparison between TSE observation in PPU and model can be possibly done. Here we see that coronal streamers in the TSE observation obviously related with the closed loops below it. PFSS model can reproduce those closed loops in the direction of the associated streamers although further analysis should be made in order to discriminate each source for different streamer. PFSS model can also predict the location of the open field with good agreement with the observation. It suggests that PFSS model is reliable as the model of corona, which can be used to model the source of the solar wind.

PFSS model and TSE observations show that the global coronal structure on 9 March 2016 are far from simple dipole structure. From the solar dynamo theory perspective, it means that the toroidal field still dominates the solar dynamo action in the interior of the Sun. This kind of property implies that the Sun is still active in producing many active regions in the low latitude regions [4]. Therefore, the Sun was in fact still descending from its maximum phase, but perhaps was still far to reach the lowest minimum phase of this solar cycle. Although here we still cannot yet exactly determine the phase by using this method, we suggest that some parameters of the symmetrical shape of the solar corona can be good representation to determine phase of solar activity for enhancing the determination of the solar cycle phase merely by measuring the flattening index.

4. Conclusion

In this paper, we show that PFSS method can be a good tool to model the general structure of the solar corona. We have briefly compared how the PFSS model can reconstruct the closed loops associated with the streamers appear in the TSE observations. We also show that the global structure of the corona is close to spherically symmetry shape, which suggest that the Sun is still far from the lowest phase of solar cycle. The more detail comparison and analysis of coronal structure as well as the method to represent symmetrical shape of the corona will be explained elsewhere in the future.

Acknowledgments

We acknowledge Muhammad Rayhan (Planetarium Jakarta and HAAJ) for his contribution in processing the image taken in PPU. JM, FM, and HS sincerely grateful to LAPAN for supporting this work. This work was partly carried out on the Solar Data Analysis System operated by the Astronomy Data Center of the National Astronomical Observatory of Japan.

References

[1] Mackay D H and Yeates A R 2012 *Living Rev. Solar Phys.* **9** 6
[2] Schatten K H, Wilcox J M and Ness N 1969 *Solar Phys.* **6** 442
[3] Schrijver C J and De Rosa M L 2003 *Solar Phys.* **212** 165
[4] Saito K, Poland A I and Munro R H 1977 *Solar Phys.* **55** 121

Analysis of ionospheric irregularities during total solar eclipse 2016 based on GNSS observation

A Husin, Jiyo, S Anggarani, S Ekawati and V Dear

Space Science Center, National Institute of Aeronautics and Space Jl. Dr Junjunan 133 Bandung Indonesia.

E-mail: asnawi@lapan.go.id

Abstract. A total solar eclipse occurred over Indonesia in the morning hours on 9 March 2016. Ionisations in the ionosphere which is associated with the solar radiation during the total eclipse provided a good opportunity to study the ionospheric irregularities. Using global navigation satellite system (GNSS) data taken from dual-frequency receivers in Manado, we investigated and analysed the total electron content (TEC) perturbations with a time resolution of 60 s to reveal ionospheric irregularities during total eclipse. Result showed that TEC conditions based on IPP were decreased during solar eclipse on March 9, comparing with the neighbour day. The maximum percentage deviation (DTEC) from the average value during eclipse period, 00:00 - 02:40 UT reach -41.5 %. The duration of maximum decrement in TEC occurs were around 2-30 minutes after the maximum obscuration.

1. Introduction

Solar eclipse is an interesting phenomena, since it is associated with ionospheric electron densities irregularities. Nevertheless ionospheric response to solar eclipse is quite complex. Studies of ionospheric responses to solar eclipse have been investigated for decades and many authors have been reported. Result in general showed that different regions behave differently during each solar eclipse event. The equipment used and the topic has been widely reported, such as electron density perturbation and wave dynamics in the height of the ionosphere layer during eclipse both in the middle latitudes and lower latitudes as well. At middle latitudes, studies the gravity wave vertical oscillations are raised during solar eclipse [1], response of the ionospheric total electron content (TEC) [2, 3], sporadic E layer (Es) generated during eclipse [4]. At low latitude, some author reported studies ionospheric irregularities during eclipse using ground-based equipment (radar, ionosonde) and satellite based observation (GPS), that provide data with better accuracy both in time and space [5, 6]. In this study, decrease of photo ionization during an eclipse is analysed in comparison with the neighbouring days. The studies done by analysing ionospheric TEC derived from GNSS observation.

2. Data and method of analysis.

The data used in this study are from two kind of GNSS receiver, one is the GISTM (*GPS Ionospheric Scintillation and TEC Monitoring)* GPStation-6 receiver Manado (MNDO) and the second is using IGS (International GNSS Service, https://igscb.jpl.nasa.gov/components/dcnav/sopac_rinex.html) Cibonong (BAKO) station which is obtained in the Receiver Independent Exchange (RINEX) format. Table 1 showed the geographic latitude and longitude of both two stations. The ionospheric irregularities behaviour during eclipse analysed by using the temporal evolution of total electron content (TEC) comparison with the neighbouring days and with particular attention to the satellite at the ionospheric pierce point (IPP) which is about 350 km altitude (figure 2). TEC (1TEC unit, TECU = 10^{16} electron m^2) measurements techniques from GISTM will not be discussed here but it can see in [7] and GNSS data from the IGS processed into TEC using software developed by [8].

The path of lunar umbral shadow moved from the South Sumatra through South Kalimantan, Sulawesi, and Maluku and ended in Pacific Ocean. As shown in figure 1, total eclipse time span start 00:20 UT at south Sumatra, Kalimantan at 00:30 UT, Sulawesi at 00:50 UT, Maluku at 01:00 UT and ended at Pacific Ocean about 01:30 UT. The dashed black line in figure 1, indicated the path of Northern and Southern limit of total eclipse, and the blue solid line is the center path of total eclipse above of the earth surface, http://eclipse.gsfc.nasa.gov/SEpath/SEpath2001/SE2016Mar09Tpath.html. While the red solid line is the path of total eclipse at ionospheric height which is converted based on mapping function using ionospheric single layer model (SLM)[9].

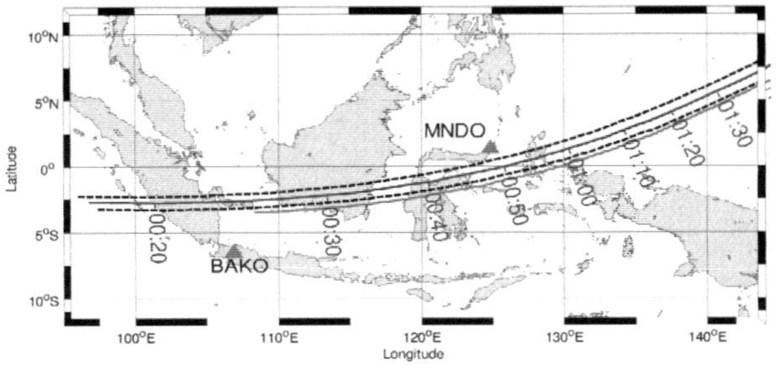

Figure 1. Map showing span time the path of maximum obscuration for total solar eclipse of March 9, 2016 (solid blue line) and the red solid line is the path of total eclipse at ionospheric height.

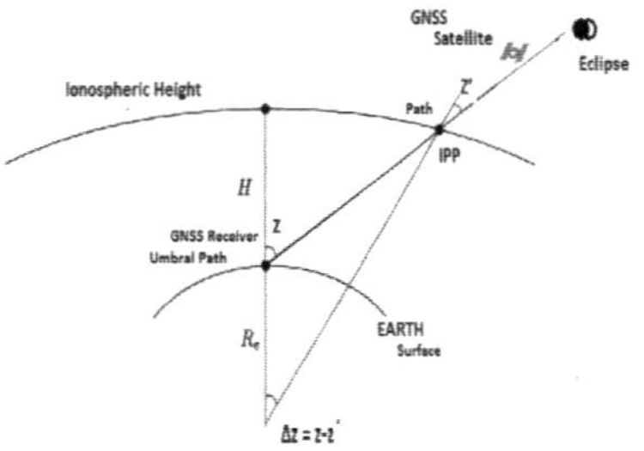

Figure 2. Geometry of the single layer model in the sun-fixed coordinate system

The basic geometry of the SLM in the sun-fixed coordinate system shown in figure 2. The signal transmitted from the GNSS satellite to the receiver and umbral shadow as well crosses the ionospheric shell in the so-called ionospheric pierce point (IPP). The zenith angle at the IPP is z' and the GNSS signal or umbral shadow arrives at the ground station with zenith angle z. In figure. 2 , $R_e \approx 6,370$ km is the mean Earth radius and H is the height of ionosphere layer (as single layer, in this study set off 350 km) and the mapping function F(z) is a triangular function [10] as:

$$F(z) = \frac{1}{\cos z'} = \frac{1}{\sqrt{1-\sin^2 z'}} \quad (1)$$

The line-of-sight TEC values derived from GNSS signals were converted to vertical TEC values (VTEC) using a mapping function F(z) in equation 1, and were associated to an ionospheric pierce point latitude and longitude. In similar way for umbral path at ground were associated to latitude and longitude at the peak ionospheric height of 350 km as illustrated in figure 2. The totality of eclipse over Indonesia lasted for a maximum period of about 1.5 minute to 3.5 minute. The eclipse time and percent obscuration at both stations are given in table 1.

Table 1. Local circumstances of total solar eclipse on 9 March 2016 at two locations.

Station Name	Station Code	Geographic Latitude	Geographic Longitude	Start of partial eclipse*	Maximum of partial eclipse*	Maximum obscuration*
Cibinong	BAKO	6,49° S	106,85° E	06:19 LT	07:21 LT	88.76%
Manado	MNDO	1.34° N	124.82° E	07:34 LT	08:49 LT	96.66%

*source: http://lapan.go.id/files_arsip/Info_GMT_2016.pdf

The irregularities periods of TEC during eclipse, determined by the detection percentage changes in TEC (% DTEC), which is obtained using the relation in equation 2 [11].

Journal of Physics: Conference Series **771** (2016) 012035 doi:10.1088/1742-6596/771/1/012035

The background TEC for the reference is obtained by averaging the TEC values of 8 and 10 March. These days are considered of a few days before and after the eclipse during the quiet period. The eclipse occurred during the local morning hours and continued for about one and half hour.

$$\%\text{DTEC}=\frac{\text{Eclipse day TEC - Background TEC}}{\text{Background TEC}} \times 100 \tag{2}$$

On the day of eclipse at the observation station, Manado and Cibinong, sky was slightly cloudy but mostly clear and no rainfall. However, two day before solar eclipse, the geomagnetic storm is reported on March 6, 2016. figure 3, show situation of geomagnetic storm before eclipse and scenario to selected data for analysis. As the solar eclipse occurred after the magnetic storm, we safely assume that the results presented in this paper are unaffected by the magnetic storm. Therefore, disturbance storm time index (Dst) data which is taken from world data center for geomagnetism WDC, http://wdc.kugi.kyoto-u.ac.jp/dst_realtime/201603/index.html, has been used to studies TEC variation during the magnetic storm in order to obtain description storm situation before solar eclipse.

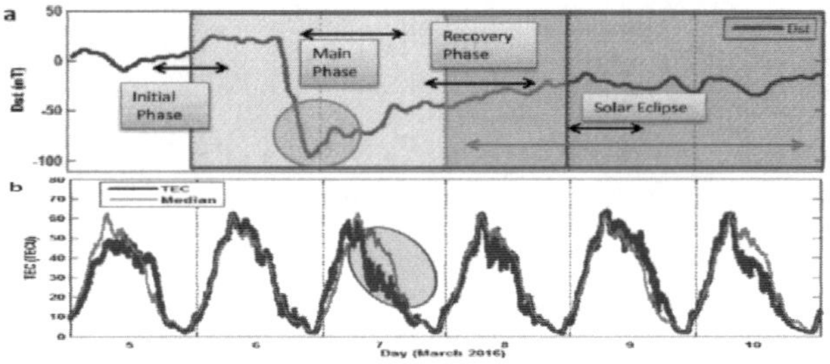

Figure 3. a) Dst index indicated moderate geomagnetic storm, before solar eclipse. b) Slight decrease in TEC (blue line) from quiet condition (red line) on March 7 as response of main phase and the next three day, 8, 9 and 10 March are unaffected of magnetic storm.

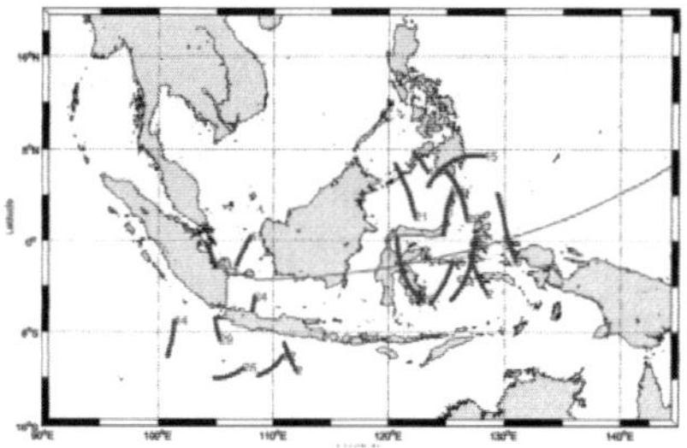

Figure 4. Path of all satellites over two stations (blue line) on IPP, along with the path of the maximum obscuration (red line) for period 00:00 UT – 01:00 UT on March 9

3. Results

As shown in figure 3, decrease of Dst index (H component) to about -96 nT (moderate storm) that was indicated the main phase of storm occurred on March 6 at 22:00 UT. The initial phase and recovery phase occurred on March 5 and March 7 respectively. TEC variation during geomagnetic storm could be seen in figure 3b. This impact has been studied by analysing TEC of 5, 6, 7, 8, 9 and 10 March from Cibinong (BAKO). The results (figure 3b, shows blue line) reveal a slight decrease in the TEC from its quiet condition (TEC median), in red line, on March 7 as response of main phase. Geomagnetic storm impact on TEC generally occurred only on March 7. However, data used to analysis of solar eclipse to the next three day, 8,9 and 10 March, when recovery phase geomagnetic storm progressing to the rest or quiet condition of the magnetic storm as shown in figure 3.b.

Figure 4, shown path of all satellites on IPP over BAKO and MNDO station, along with the path of the maximum obscuration (red line) for time period 00:00 UT – 01:00 UT, associated to period of solar eclipse event on March 9. The path of the satellite designated by PRN 29, 25, 24, 12, 5 and 2 are crossed the maximum path, while PRN 15, 19, 20 and 21 are slightly distant to the maximum path. These satellites report by MNDO station and selected to analysis of the occurrences of ionospheric irregularities associated with solar eclipse. For satellites report by BAKO station, are continuation of trajectory and mostly distant to the maximum path.

It can be seen in figure 5 that TEC conditions based on IPP were decreased during eclipse March 9, comparing with the neighbour day. The percentage deviation (DTEC) from the average value during eclipse period, 00:00 - 02:40 UT which was calculated by using equation 2, shown in figure 6, for PRN 12, with maximum decrease -40.8 % at 00:54 UT. As shown in figure 6, the vertical total electron content goes on decreasing with time, reach maximum at 00:40 UT (08:40 LT) and after 00:40 UT starts recovering. The percentage deviation TEC for others satellite which are crossed the maximum path i.e. PRN 29, PRN 25, PRN 24, PRN 5 and PRN 2 respectively -41.5% (00:45 UT), -37%, -29.6 % (01:00 UT), -34.3 % (00:59 UT) and -39.6% (00:46 UT). These results agree with author [3] and [11] by analysis of previous other eclipses.

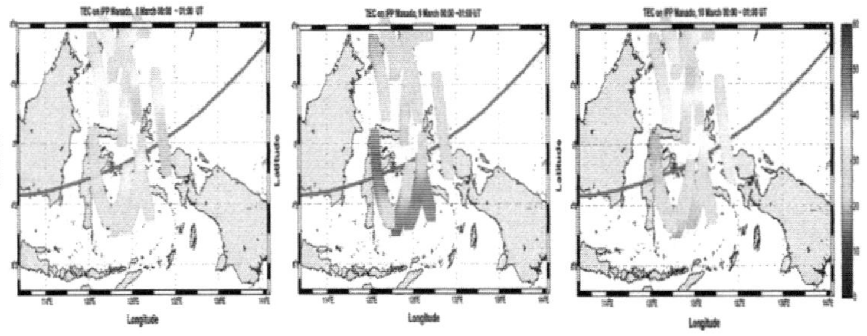

Figure 5. TEC conditions based on IPP were decreased during eclipse March 9, comparing with the neighbour day, before and after (March 8 and March 9) period, 00:00 - 01:00 UT

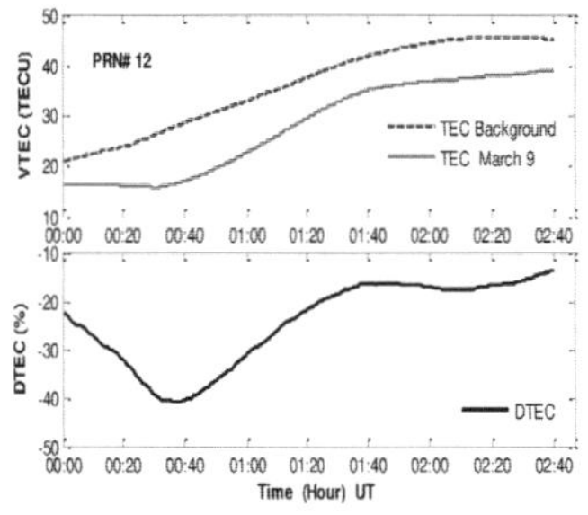

Figure 6. Percentage deviation (DTEC) and its evolution of PRN12

According to table 1, the eclipse starts at 07:34 LT and continues for 1 h and 15 min, with the maximum obscuration of 96 % at 08:49 LT (00:49 UT). Daily TEC as background has minimum at pre-dawn and tends to increase gradually with the time of day attaining maximum in the middle day as a result of photoionization due to EUV radiation. However all satellite (PRN) showed the reduction around maximum obscuration during TEC on-going increasing. It is indicated that ionospheric irregularities through TEC decreases during the present obscuration decreases, is directly related to the electron production by decreased photo-ionization process in the ionosphere. It is found that the maximum decrement in TEC occurs about time (2–30 min) after the maximum obscuration.

4. Conclusions

The ionospheric irregularity during the eclipse on 9 March 2016 has been investigated using TEC measurements with particular attention to the satellite at the ionospheric pierce point. The results showed the total ionization, as estimated by the TEC, reduces by approximately maximum -40%. This is generally in agreement with earlier findings for other eclipses. It is found the maximum reduction occurs a few minutes (2–30 min) after the maximum obscuration time. These delay related to timescale in photoionization and transport phenomena.

Acknowledgments
The authors would like to acknowledge IGS for GNSS data used and Sam Ratulangi University as collaboration partner.

References

[1] Altadill D, Sole J.G and Apostolov E.M 2001 *J Geophys. Res* **106** 21419-21428.
[2] Afraimovich E.L, Kosogorov E.A, Lesyuta, O.S 2002 *J Atmos Terr Phys* **64** 1933–1941.
[3] Jakowski N, Bettac H.D, Lazo B, Palacio L, Lois L 1983. *Physica Solariterrestris* **20** 110–116
[4] Pezzopane M, Pietrella M , Pignalberi A, Tozzi R 2015 *Adv. Space Res* **56** 2064-2072
[5] Bagiya,M. S, Iyer K N, Joshi H. P, Thampi S. V, Tsugawa T, Ravindran S, Sridharan R and Pathan B M 2011 *J. Geophys. Res.* **116** A01303
[6] Kumar S and Singh A K 2012 *Adv Space. Res.* **49** 75 -82,
[7] Asnawi 2012 *J Aerospace Sciences* **10** 13-22
[8] Rao P.V.S, Gopi K.S, Niranjan K, and Prasad SVVD 2006 Ann Geophys **24** 3279–3292
[9] Schaer S 1999 Ph.D dissertation Astronomical Institute University of Bern **59**
[10] Sardon E, Rius A, Zarraoa N 1994 *Radio Sci* **29** 3 577-586
[11] Kumar S and Singh A K 2012 *Adv Space. Res.* **49** 75 -82,

Effect of March 9, 2016 Total Solar Eclipse on geomagnetic field variation

Mamat Ruhimat[1], Anton Winarko[1], Fitri Nuraeni[1], Harry Bangkit[1], M. Andi Aris[1], Suwardi[2], Sulimin[3]

[1]Space Sience Center, Indonesian National Institute of Aeronautics and Space ,LAPAN
[2]Stasiun Geofisika Kelas III Ternate, BMKG
[3]Stasiun Meteorologi Kelas I Babullah Ternate, BMKG

E-mail: mamat.ruhimat@lapan.go.id

Abstract. During solar eclipse, solar radiation to the Earth is blocked by the Moon. Thus, the ionization process in the ionosphere is disrupted, as well as the variation of geomagnetic field. The disturbance of geomagnetic field is caused by electric current in the E layer of the ionosphere. At low latitude, the current which is dominant in quiet day is the Sq currents. The blocking of solar radiation cause decrement in electron density in the blocked region. The aim of the research is to find the effect of total solar eclipse to the geomagnetic field. The measurement of the geomagnetic field variation during total solar eclipse on March 9, 2016 was carried out at the Meteorological station of BMKG in Ternate (0° 49' 45.20 "N; 127° 22' 54.00" E). By eliminating the geomagnetic disturbance that occurred in a daily geomagnetic field variation, the pattern of quiet day which is usually in a shape of smooth curve became affected. During the total solar eclipse on March 9, 2016 from 00:30 until 02:00 UT, we found that the geomagnetic field variation of the quiet day decreased by -5 nT.

1. Introduction

Total solar eclipse (TSE) is a natural event that can also be observed indirectly by using fluxgate magnetometer. In the event of solar eclipse, the Sun, Moon and Earth are in straight line. So that incoming solar radiation to the Earth is blocked by the Moon. Total solar eclipse on March 9, 2016 has crossed across several areas in Indonesia. The condition of ionosphere in the area along the path of totality was more likely to change as a result of the eclipse. It is an excellent opportunity to investigate more deeply the effects on geophysics. Studying geomagnetic with the change in the radiation coming from the solar corona is important because of the action of ionization in the ionospheric dynamo changes [4]. The main effects that may occur in the geomagnetic field disturbance caused by electrical currents in the ionosphere E layer on a quiet day was dominated by the solar quiet day (Sq) current. Solar radiation blocked by Moon could cause a reduction in electron density in the ionosphere and thus significantly reduce the conductivity in its region [7].

To study its impact to geomagnetic field, we measured geomagnetic variation in totality zone. Indonesian National Institute of Aeronautics and Space (LAPAN), especially research group of geomagnetic and space magnetism, conducts research on geomagnetic disturbance originated from extraterrestrial. When sunlight heading towards earth is obstructed by the Moon, the radiation and

Journal of Physics: Conference Series **771** (2016) 012036 doi:10.1088/1742-6596/771/1/012036

ionization in ionosphere is decreased thus disturb ionosphere electricity. This event could affect geomagnetic daily variation. Geomagnetic variation describes the geomagnetic field fluctuations particularly quiet day variations, is heavily influenced by the process of heating and ionization layer of the ionosphere.

During the total solar eclipse, the ionization process will be disturbed. It is generally known that the daily variation quiet geomagnetic field caused mainly due to the current flowing in the ionosphere E layer. During the solar eclipse, part of the ionosphere is shielded from solar heating and ionizing radiation. It modifies the flow pattern in this area and affect the geomagnetic field observed at the surface of the earth. The ionosphere E layer change process and affect the geomagnetic daily variation which is commonly known as a Sq variation, originating mainly in the ionosphere. Therefore, the most significant changes appear in the ionosphere due to the close relation with the solar radiation. Obstruction of the Sun during solar eclipse potentially disturbs ionosphere.

The effects on the ionosphere during solar eclipse depends on several factors such as the level of solar activity and the level of geomagnetic disturbance [10, 11]. The purpose of this study is to determine the effect of the total solar eclipse to the geomagnetic variations in Ternate, North Maluku. Geomagnetic field measurements out from March 4 until 10, 2016.

2. Data and methods

Geomagnetic data used in this research was obtained from observations of geomagnetic daily variations in the meteorological observatory of class III Babullah BMKG Ternate with geographic coordinate (0° 49'45.20"N; 127° 22' 54.00" E) and geomagnetic coordinate (λ 08,01° S; β 156,31° W). We used fluxgate magnetometer with a resolution of 0.1 nT and sampling rate 1 Hz. The records within universal time (UT) and geomagnetic components were measured, they consist of components X (North-South), Y (East-West) and Z (Vertical). The horizontal component of the geomagnetic field (H) is the resultant of X component and Y components of the geomagnetic field. H component is used to reveal phenomena of extraterrestrial origin. The observation was conducted in Ternate from March 4 until 10, 2016, the period the geomagnetic condition on March 5, 2016 was quiet, while on March 9, 2016 there was geomagnetic storm.

To monitor the geomagnetic disturbance coming from solar activity, or ring currents, we used the geomagnetic activity data (Dst index) [12]. It was used to eliminate or separate the geomagnetic disturbance resulting from the ring currents in order to obtain a variation of the geomagnetic quiet. Daily variation of geomagnetic field was obtained from March 5, 2016 data, classified in one of the groups of geomagnetic quiet days. The geomagnetic quiet day pattern obtained by are eliminated daily geomagnetic variation from geomagnetic disturbances (Dst index) and the effects of the ring currents [5, 6]. The equation (1) and (2) are used to eliminate the effect of ring currents [7], and filtered using Fourier analysis.

$$X_D = X - Dst \; Cos\lambda \; Cos(-11^0) \tag{1}$$

$$Y_D = Y - Dst \; Cos\lambda \; Sin(-11^0) \tag{2}$$

X is a magnetic field with a north-south component, Y is the east-west magnetic field component, λ is the latitude geomagnetic observatory. The 11° angle is difference between the geographic and geomagnetic poles.

To obtain the equivalent current density in the ionosphere altitude of 110 km [7] we use equation 3.

$$J = \frac{2}{\mu_0}\sqrt{X_D^2 + Y_D^2} \tag{3}$$

Where μ_0 is the permeability in a vacuum.

3. Results and analysis

Horizontal (H) component of geomagnetic daily variation can be calculated by determining the geomagnetic resultant of the north-south (X) component and the east-west (Y) component. Based on the geomagnetic activity of the H component, we can select the quiet day geomagnetic variation.

The level of the geomagnetic disturbance index (Dst) [12] on 5 March 2016 was used to obtain a pattern of a geomagnetic quiet day in Ternate. By using equations (1) and (2), we calculated the variation of quiet horizontal component (H). The quiet day geomagnetic illustrates that there was a very weak geomagnetic disturbance ± 10 nT. It indicates that the geomagnetic disturbance caused by the ring currents or solar activity did not occur. Quiet day variation was filtered by using Fourier analysis and the result is known as the geomagnetic quiet day pattern in Ternate, where the condition geomagnetic on March 5, 2016, was included in the order of one day of quiet geomagnetic International [7].

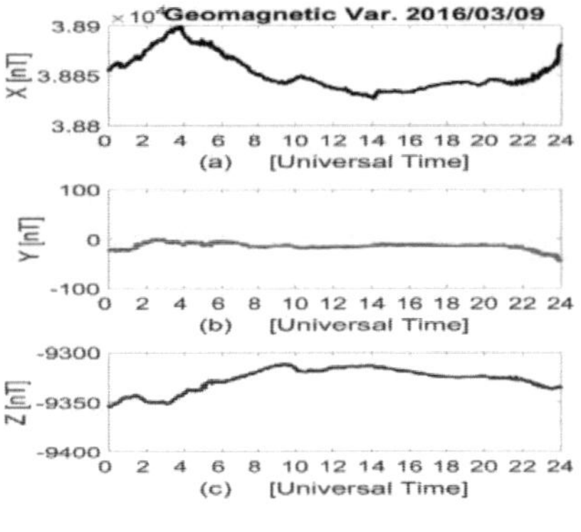

Figure 1. Daily variation of geomagnetic components X(north-south), Y(east-west) and Z(vertical) in Ternate on March 9, 2016. (X,Y,Z) is three orthogonal strength components of geomagnetic.

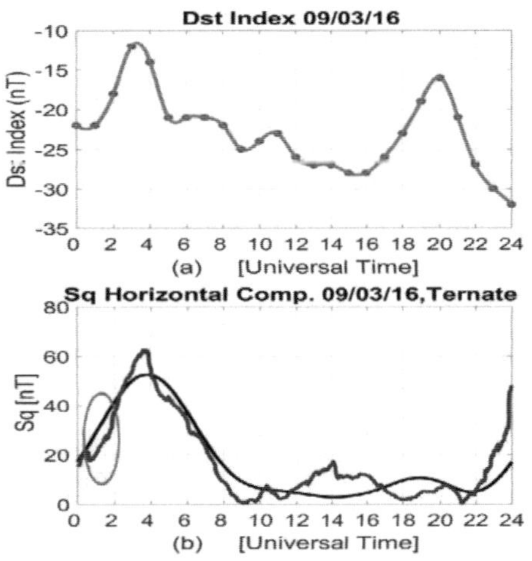

Figure 2. a) Dst index curve [12] and b) The pattern of quiet day variation in Ternate on March 9, 2016 (blue line) and March 5, 2016 (black line). The red ring is shown the period of total solar eclipse (TSE).

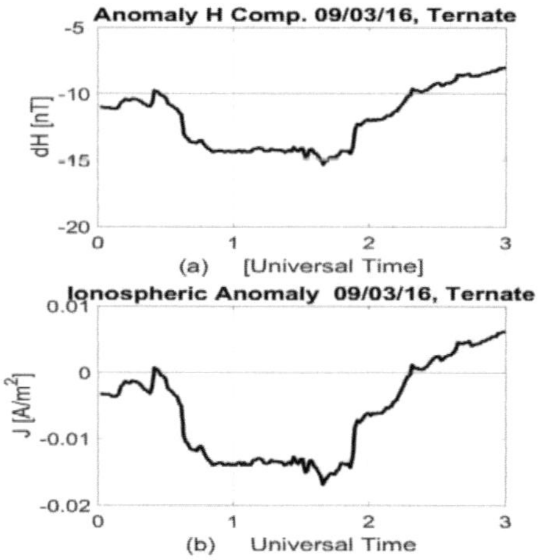

Figure 3. a) The curve of geomagnetic quiet day anomaly during the total solar eclipse on March 9, 2016, and b) The current density equivalent ionospheric anomaly curve. On period TSE, geomagnetic H component and currents density drops.

Figure 1 shows observation of the daily variations geomagnetic on March 9, 2016. The three components magnetic are measured. They are north-south (X), east-west (Y) and vertical (Z) components. On March 9, 2016, the magnetic field measurement data was affected by total solar eclipse event. During 00:30 UT until 02:00 UT there is a decrease of the daily variation of geomagnetic X component and an increase in the Y component. The magnetic field intensity during the total solar eclipse decreased. In the previous studies, several researchers have studied the geomagnetic field variations during the solar eclipse with most of the results showed positive anomalies in geomagnetic variations of the component east-west (Y) and decrease on the component north-south (X) [1, 2, 8, 9].

Variation of geomagnetic disturbance that occurred on March 9, 2016, is shown in the figure (2.a). Dst index curve ranges from -15 to -30 nT. Dst index variation is used to separate of disturbance in the quiet day from daily geomagnetic variation. In a similar way as data processing on March 5, 2016, variation in geomagnetic quiet day and the pattern of March 9, 2016 was obtained as shown a blue line in the figure (2.b). Variation in the quiet geomagnetic obtained is not as smooth as the pattern of the quiet day because there are other sources suspected that the effect of the total solar eclipse that lasts about 00:30 - 02:00 UT that caused an anomaly on quiet geomagnetic variation. The H component of quiet geomagnetic variation shows the decline from the quiet day pattern average -5 nT, shown in the figure (3.a). It is also known that the solar eclipse effects the field geophysics at Novosibirsk Russia as 5 nT decrease in the amplitude [3]. In figure (3.b) curve shows equivalent ionosphere currents density anomalies calculated from equation 3, where during a total solar eclipse March 9, 2016, equivalent ionosphere current density decreased 12 mA/m^2. The process of solar radiation to the earth disturbed by the presence of a total solar eclipse has led to the decrement of a current density of the ionosphere, so the magnetometer recorded the geomagnetic variations with the decline of the horizontal component (H).

4. Conclusion

Geomagnetic field measurement was conducted in Ternate from 4-10 March 2016. During the period, the geomagnetic quiet day was happened on March 5. The total solar eclipse occurred on March 9. From the recorded data on March 9, 2016, in the interval from 00:30 to 02:00 UT, we obtained that the X component of geomagnetic variation decreased while the Y component was increased during the eclipse. We obtained that there was an anomaly of variations of geomagnetic quiet day about -5 nT on H component which is equivalent with the ionospheric current density 12 mA/m^2. It is associated with lower ionization process on the layer of ionosphere caused by the total solar eclipse.

Acknowledgement
The authors would like to thank for Clara Yono Yatini on the advice in this study.

References

[1] Adam A, Vero J, Szendroi J 2005 *Annales Geophysicae* **23** 3487-3494.
[2] Ates A, Buyuksarac A, Beltas O 2011 *Turkish Journal of Earth Sciences* **20** 337-342.
[3] Babakhanov I V, Belenskaya A Y, Bizin M A, Grekhov O M, Khomutov S Y 2013 *Journal of Atmospheric and Solar Terrestrial Physics* **92** 1-6.
[4] Curto J J, Heilig B, Pinol M 2006 *Journal of Geophysics Reearch.* **111** A07312, 1-9.
[5] Daglis I A, Thome R M, Baumjohann W, Orsini S 1999 *Rev. of Geophys.* **37(4)** 407-438.
[6] Ganushkina N Y, Pulkkinen T I, Kubyshkina M V, Singer H J, Russel C T 2002 *Journal of Geophysics Research* **107** A7 1092 SMP 3 1-10.
[7] Korte M, Luhr H, Forster M, Haak V, Bencze P 2001 *Journal of Geophysics Research,* **106** A9 18563-18575.
[8] Ladynin AV, Smakov N N, Khomukov S Y 2011 *Russian Geology and Geophysics* **52** 343-352.
[9] Malin S R C, Ozcan O, Tank S B, Tuncer M K, Yazici-Cakin O 2000 *Geophys. J. Int* **140** F13-F16.
[10] Onovughe E 2013 *International Journal of Astronomy*, **2(4)**, 51-55.
[11] Tomas AT, Luhr H, Rotlur M 2009 *J Annales Geophysicae*, **27** 4449-4461.
[12] http://wdc.kugi.kyoto-u.ac.jp/dst_realtime/201603/index.html (acccessed on 14 March 2016)

Changes of NmF2 and hmF2 over Biak (1°S, 136°E) during total solar eclipse on March 9, 2016

Sefria Anggarani, Jiyo, Asnawi, Varuliantor Dear and Sri Ekawati

Indonesian National Institute of Aeronautics and Space

E-mail: sefria.anggarani@lapan.go.id

Abstract. Ionospheric responses to solar eclipse is interesting to learn. In this paper, we presented ionspheric F2 responses to total solar eclipse on March 9, 2016 over Indonesia from Biak station (-1 °; 136.0 °) which is within the obscuration 86.55%. The total solar eclipse over Biak occurred at 00:58 UT (09:58 local time) in F2 layer height of ionosphere (h ≈ 300 km). Observation of the ionospheric F2 over Biak during the total solar eclipse investigated using ionosonde with 15 minutes resolution. Ionization and recombination processes that occur in the ionosphere proceed by solar radiation. At the time of the ionosphere receives solar radiation, ionization process dominated in the F2 layer. Diurnal ionosphere parameters changed following the solar radiation. During the solar eclipse, the process of ionization in the F2 layer disturbed. Total solar eclipse on March 9, 2016 affected almost simultaneous decrease in maximum electron density (NmF2) ~52% and increase hmF2 ~24% at the elapsed time of totality observed from Biak Station.

1. Introduction

Total solar eclipse on March 9, 2016 that occurred over Indonesia crosses over the archipelago with its path of totality passing over 12 provinces and partial solar eclipse with obscuration more than 60% to the rest. The eclipse occurred during declining phase of solar activity cycle-24 (http://www.swpc.noaa.gov/). This event provides unique and rare opportunities to study the ionosphere effects associated with an accurately estimated variation of solar radiation. During this event, the amount of solar radiation that reaching earth is decrease which result in reduction of plasma. It is the opportunity to do ionospheric study especially on electron density of F2 layer (NmF2) and the height of F2 layer.

Study on ionospheric response to solar eclipse have been conducted since the 1940s [1]. One of the results of the study was an ultraviolet effecting the ionoszation in F2 region of ionosphere. The next study by [2], has shown that the F2 layer has unequally behavior, with the result that was not surprising that a difference of phenomena occur during eclipses. There are some things that control the upper ionosphere, which are transport processes, photoionization and loss rate. [2, 3, 4]. The F2 region have different behavior, during eclipse events. So the objective of this study is to obtain information on ionospheric electron density of foF2 (nmF2) and the height of F2 (hmF2) during total solar eclipse.

2. Data and method

The data used in this study are maximum electron density of F2 layer (NmF2) derived from critical frequency of F2 layer (foF2) and maximum height of F2 layer (hmF2) observed by ionosonde from Biak Station. The value of NmF2 and hmF2 compared with normal day (moving average 5 quiet days data based on quiet geomagnetic storm). The data also calculated to the relative value using a modified form analysis of Adekoya [5] (equation (1) and (2)). At Biak station, on Earth's surface (h ≈ 0 km), eclipse started at 08:46 LT (UT+9) and ended at 11:40 LT. Assuming that Sun's altitude was ~ 60° at maximum eclipse, we can approximate the time of eclipse at F2 layer (h ≈ 300 km). At this height, eclipse occurred ~5 minutes earlier. It is explained from the study before that there is a shifted zone of Eclipse based on geometry of Earth [6]. The shifted zone of Eclipse is showed in figure 1. On F2 ionospheric height at Biak Station, eclipse started at 09:58 LT and ended at 10:14 LT. The maximum obscuration 86% occurred at 10:09 LT.

$$\text{Re}lative\, NmF2 = \frac{NmF2(obs) - NmF2(average)}{NmF2(average)} \tag{1}$$

$$\text{Re}lative\, NmF2 = \frac{hmF2(obs) - hmF2(average)}{hmF2(average)} \tag{2}$$

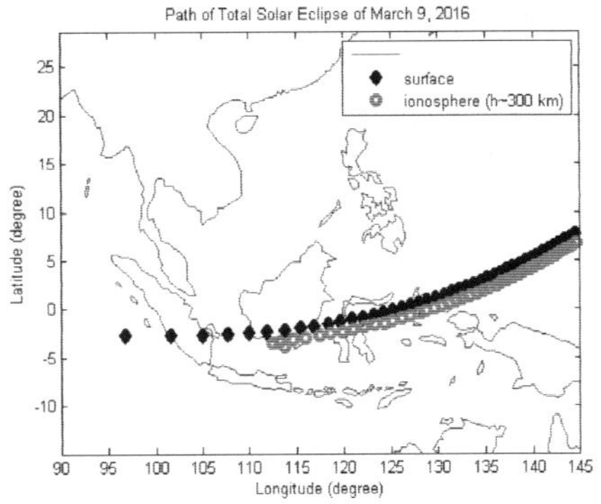

Figure 1. Path of total solar eclipse on March 9, 2016. Black diamond mark observation on surface. Red diamond mark observation on ionosphere (h ≈ 300 km).

3. Result and discussion

3.1. Decrement of maximum ionospheric density (NmF2)

Figure 2 shows the value of maximum electron density on F2 layer compared by normal day. The blue shade marks the time of eclipse. The maximum electron density started to decrease and drop almost simultaneously with the first contact and gradually increase at the end of eclipse. The relative value of NmF2 over Biak is shown on figure 3. The NmF2 relative plot shows negative value in most time of eclipse. The negative value during eclipse almost equal to negative relative value at night and morning. At night until just before sunrise, the electron density of the ionosphere is at the minimum value. If the 5 days average data is considered as the normal condition, then the density decreased maximum ~52% during eclipse.

The decrement of NmF2 during eclipse can be caused by reduction in the ionosphere ionization due to the reduction in the intensity of the sun because of the closure of the moon's shadow. The decrement in solar radiation causes change in the photochemical of the atmosphere with night time chemistry dominating [7]. However, the F-2 behavior may be quite different as it is governed by photochemical processes as well as by electro dynamical and neutral forcing. When the F-2 layer is normal in state of quasi-equilibrium, the obscured changes in electron density during eclipse are almost entire due to loss by recombination and production by photo-ionization [8]. Noteworthy that the ionospheric behavior is governed by photochemical processes, so the decrease in solar radiation during solar eclipse leads to the decrease in electron production [9].

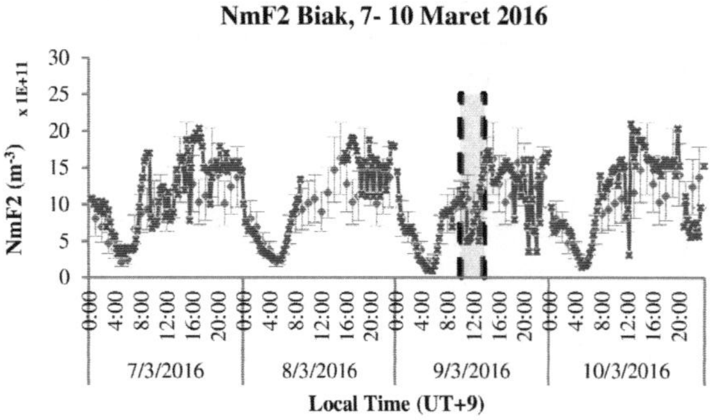

Figure 2. Maximum electron density at F2 layer (NmF2) over Biak Station (red) superimposed with moving average 20 days NmF2 (blue) with 30% deviation. Grey shade mark the time of total solar eclipse.

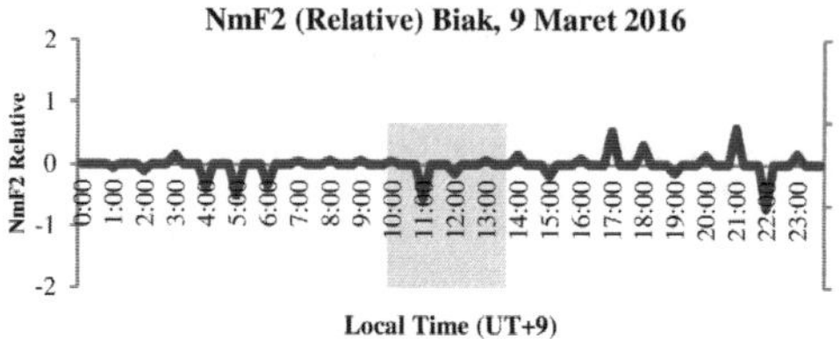

Figure 3. Relative Maximum electron density at F2 layer (NmF2) over Biak Station (blue) Grey shade mark the time of total solar eclipse.

3.2. Increment of F2 region height (hmF2)

Figure 4 shows the height of F2 layer ionosphere (hmF2) compared with normal day. The gray shade marks the time of eclipse. During eclipse, the value of hmF2 tend to increase. The relative hmF2 showed on figure 5. The relative hmF2 plot shows the vary value. Positive value occurred at beginning of eclipse and negative value occurred at end of eclipse. During the solar eclipse, maximum height of ionosphere (hmF2) increase ~24% from the normal condition.

The increment of hmF2 is most likely due to the influence of gravity in the event of a solar eclipse is greater than the force of gravity. The increment can also caused by electro dynamic processes. The vertically upward E×B drift pushes the plasma upward in altitude. In the meantime, the electrons tend to diffuse along the magnetic field lines to higher latitudes creating the fountain effect [10]. Biak station that located at ~10° affected by that effect. Le et at also showed that total solar eclipse on March 9, 2016 that peaked at 10:20 LT is more unaffected by morning [11].

Generation of Gravity waves due to the transit of locally cooled region of the atmosphere moving at supersonic speed and a reduction in the plasma density as the source of ionization is switched. The Earth's atmosphere may produce wave motions due to localized time dependent heating or cooling action. A solar eclipse, by interfering with the heat balance in the shadowed portion of the atmosphere, is expected to generate atmospheric gravity waves (AGW) [5].

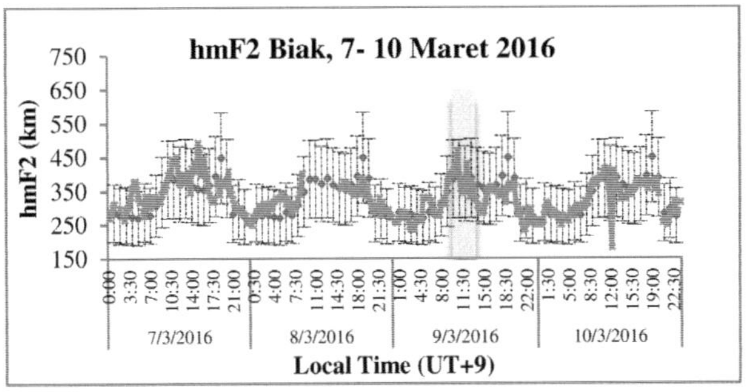

Figure 4. Maximum height of F2 layer (hmF2) over Biak Station (blue) superimposed with moving average 20 days NmF2 (green) with 30% deviation. Grey shade mark the time of total solar eclipse.

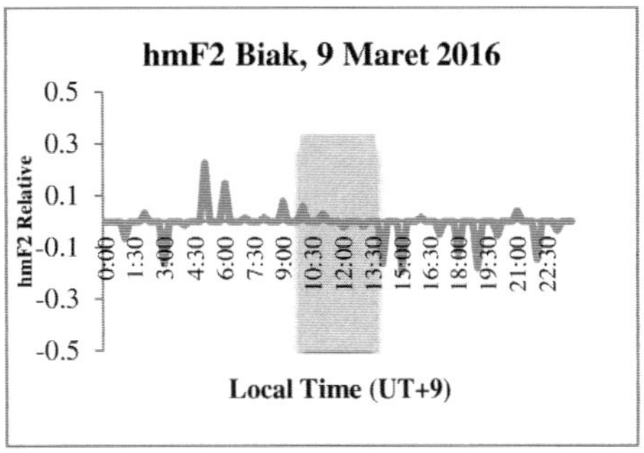

Figure 5. Relative height of F2 layer (hmF2) over Biak Station (green). Grey shade mark the time of total solar eclipse.

4. Conclusion

Total solar eclipse on March 9, 2016 that passes through some areas in Indonesia led to a state of the ionosphere was different from the conditions on a normal day. The change was shown from observations of values NmF2 and hmF2 over Biak. The eclipse affected almost simultaneous decrease in maximum electron density (NmF2) ~52% and increase hmF2 ~24% at the elapsed time of totality observed from Biak Station.The decrement of maximum electron density comparable with the part of the result by Nayak et al. [10]. Ionospheric parameters change was not too large because of the solar eclipse occurred in the morning.

Acknowledgement

The authors of this paper thank to ionogram scaling team that provide scaled data from ionogram.Also to the researchers in Space Science Center – LAPAN for discussing this topic.

References

[1] Higgs A F 1942 *Royal Astronomical Society* **162** 451
[2] Rishbeth H 1968 Solar eclipses and ionospheric theory *Space Science Reviews* **8** 543
[3] Adeniyi J O *et al* 2007 *J. Geophys. Research* **112** A06314
[4] Adeniyi *et al* 2009 *J. Geophys. Research* **114** 11303
[5] Adekoya B J and Chukwuma 2012 *Indian Journal of Radio & Space Physics* **41** 594-605
[6] Jiyo et al 2015 Totality zone determination of total solar eclipse on March 9th 2016 at the level of ionosphere *Prosiding Worshop Cuaca Antariksa dan Peluang Pemanfaatannya*
[7] Gerasopoulos E *et al* 2008 The total solar eclipse of March 2006: overview *Atmos Chem Phys Discuss* **8** 5205-5220
[8] Skinner N J 1967 Eclipse effect in the equatorial F-region *J Atmos Terr Phy* **29** 287-295
[9] Le H *et al* 2008 The ionospheric responses to the 11 August 1999 solar eclipse: observation and modeling *Ann. Geophys* **26** 107-116
[10] Nayak C K *et al* 2012 The equatorial ionospheric response over Tirunelveli to the 15 January 2010 annular solar elipse: Observation. *Ann Geophys* **30** 1371-1377
[11] Le H *et al* 2008 *Journal of Geophysical Research* **113** A013012

Stellar background observation during Total Solar Eclipse March 9th 2016

Farahhati Mumtahana[1], Anton Timur Jaelani[2], Johan Muhamad[1,3] and Heri Sutastio[1]

[1]Space Science Center, LAPAN, Jl. Dr. Djundjunan 133, Bandung, Indonesia
[2]Astronomical Institute, Tohoku University, Japan
[3]Institute for Space-Earth Environment Research, Nagoya University, Nagoya, Japan

E-mail: farahhati.mumtahana@lapan.go.id

Abstract. We report observation and an early analysis of stellar background from total solar eclipse in Ternate, Indonesia. The eclipse phenomena which occurred on March, 9th 2016 was observed with certain portable instruments in order to obtain the stars behind the Sun in particular field of view and resolution. From our observation site in Ternate city, solar eclipse occurred in the late morning when the weather was unfortunately cloudy. However, during the darkness of totality, we obtained several point source objects between the gaps of the moving clouds and we suspected them as very faint stars due to their appearance in several frames. Those so called stars have been identified and measured with respect to their positions toward the center of the Sun. The main purpose of this research is to revisit strong lensing calculation of the Sun during total solar eclipse by measuring the deflection angle of the background stars as it had been calculated by Einstein and proved by Eddington at a total solar eclipse in 1919. To accomplish this aim, we need to conduct another observation to measure position of the same stars in the next period when those stars appear in the night sky.

1. Introduction

Our team proposed a research to revisit strong lensing calculation by measuring the displacement of the background stars during Total Solar Eclipse (TSE) March, 9th 2016. The deflection angle was calculated by Einstein with General Relativity Theory ($\alpha = 4GM/c^2 R$), which leads to an apparent displacement at the solar limb amounting to 1"75 [1]. It was successfully proved by Eddington and Dyson in TSE May, 29th 1919 and also been repeated several times by astronomers around decades [2] [3]. However, among 70 TSEs occurred between 1916 until 2015, not all observation were successful due to some challenges such as weather, location, instruments, very short TSE moment, politics, and so on. Therefore, we consider TSE March, 9th 2016, as a notable rare phenomena, is important to observe particularly to revisit lensing calculation. Nevertheless, to obtain the displacement, we still have to observe the stars in the night when there is no Sun's gravitational effect as comparison field taken at night five months later [2]. Accordingly, in this paper, we only report about TSE observation aiming for early analysis of stellar background from TSE March 9th 2016 by identifying and measuring several detected point sources.

This noteworthy eclipse event was observed at Eastern part of Indonesia, Ternate, which lies at longitude 0°47N, latitude 127°22E, and altitude of about 5 m. In Ternate, the occurrences of TSE began at 00:51:43.3 UT with the maximum at 00:53:0 UT, and lasting until 00:54:19.1 UT (2'46"). Considering the distant expedition, we were indeed planning to use only portable telescope and simple detector. However, in order to obtain the desired resolution and field of view, several configurations of telescope and detector were calculated and also tested. In special case of portable telescope, small telescope can perform better resolution than telescope with wider diameter and longer focal-length in visual wavelength [4]. It means low cost simple telescope can also give good performance with the suitable detector. Finally, we used Vixen ED Apochromatic Refractor 103mm f/7.7 mounted in German Equatorial SXD2 and 22.2 x 14.9 sized CMOS sensor built in DSLR Canon EOS Rebel T6s, performing image size 6000x4000 and pixel size 3.7µm. This detector features a new high resolution of 24.2 Mega Pixels coupled with 7560 pixels RGB, and has maximum ISO at 25.600. This configuration gives desired performance of resolution 0.96", area 1.71sq, and limiting magnitude around 13.

2. Method

2.1 Simulation of displacement

Displacement of the background stars due to gravitational lens during TSE was simulated in figure 1. The first zone which is illustrated in yellow bar indicates solar radius (R⊙), therefore no star can be seen at this region. Second zone, 1< R⊙<2, is region with the biggest deflection angle of the sun, however the stars are hardly seen due to coronal intensity. Meanwhile, region R⊙>2 as the radius further away from corona, the deflection angle is become smaller.

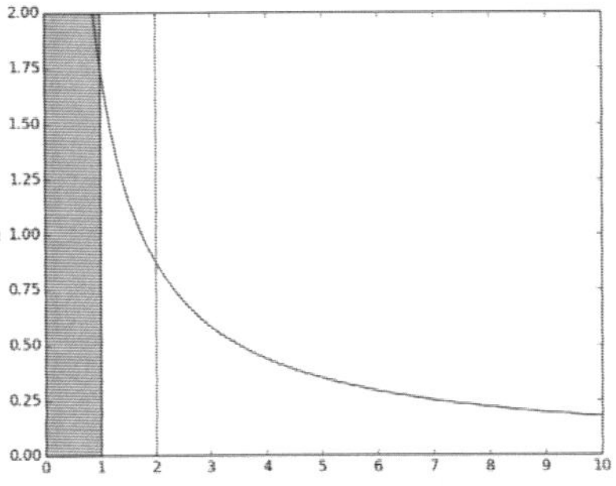

Figure 1. Hyperbolic plot derived from Einstein equation which shows deflection angle estimation vs R⊙.

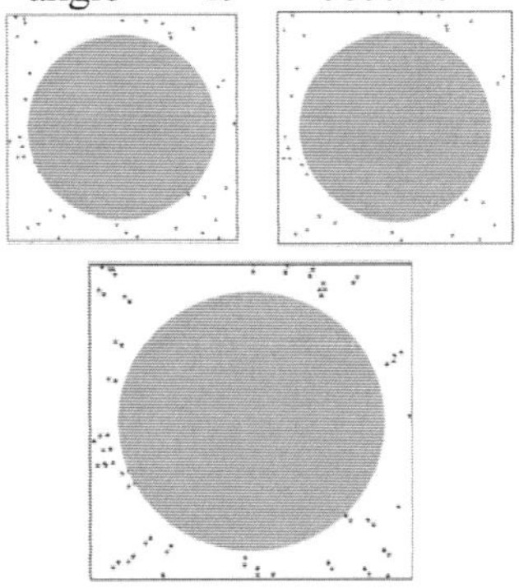

Figure 2. Simulation of displacement of the background stars due to Sun's gravitational lens during TSE.

Simulation of figure 2 shows the displacement between random blue dot stars and red dot stars. Red dots correlate with the random stars after going through gravitational lens effect of the sun. Both simulation figures are important as consideration in choosing appropriate instrument configuration and for the preparation.

2.1 Observation method and processing

We began to observe and take image of the sun since 07.26 LT, about an hour before first contact. Ternate was chosen due to its good weather prediction for the eclipse day. However, this long term meteorological prediction in particular area was incorrect as the sky around the sun was covered by clouds since the first to the fourth contact of 9 March 2016 solar eclipse. Finally 29 frames of TSE were acquired with different ISO and exposure adjusted to the clouds, even though all of them were not as good as we expected and even some of them were very noisy.

Although no ideal image were taken, corona composite from raw data was perform quite well for cloudy data following the steps from Espenak [5] and Druckmuller, et all [6]. However, no star appears at the final composite still. Therefore we considered another steps for quick look and preliminary analyst. First, with the naked eye we searched appearance of dots in each frames. Among 29 frames, there was no visible star obviously until we zoomed them in to see details and finally found some point sources that we suspect as stars. Same sources appear in several frames, while the others only appear once. Then using MaxIm DL, images with same position of point sources were stacked for higher SNR [4]. The real measured profiles are close to two dimensional Gaussian profile [7]. Accordingly, we fitted Gaussian profile to collect information (FWHM & SNR). After that we measured the position towards the center of the Sun with AstroImageJ. We determine position of the sun and point sources using true direction seen from crescent solar partial eclipse.

3. Result and discussion

4.1 Source identification

We obtained 4 point sources from 3 group frames as it seen on figure 3 and listed on Table 1. This table gives information of name list of the objects suspected as stars which is shown that SNR source 2_1 is very low to be categorized as star.

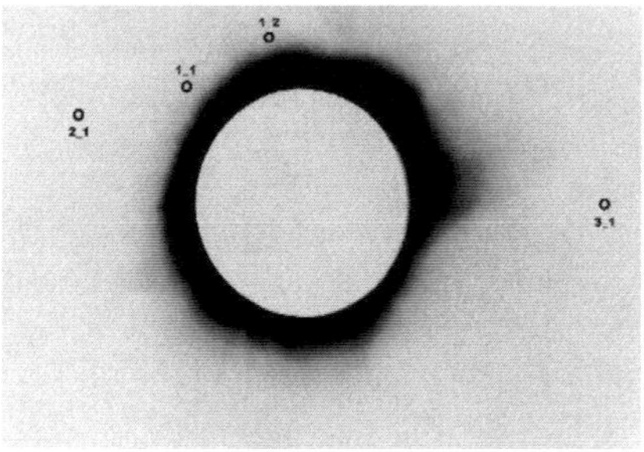

Figure 3. Identification of point sources

Table 1. Point sources information

name	frame	SNR	FWHM	Intensity
source 1_1	71,72,74	6.3	8	1598
source 1_2	71,72,74	1.52	17.052	484
source 2_1	84,85	0.142	3	87.98
source 3_1	90,91,92	23	1.5	1339

Figure 4 shows each images with their Gaussian plot and area plot. The stars image with the Gaussian profile is usually spread over several pixels [7]. Therefore all of them are suspected as faint stars in TSE according to intensity and Gaussian plot.

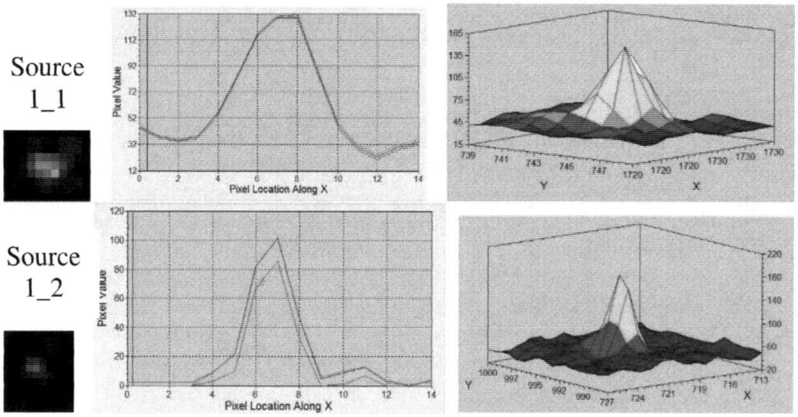

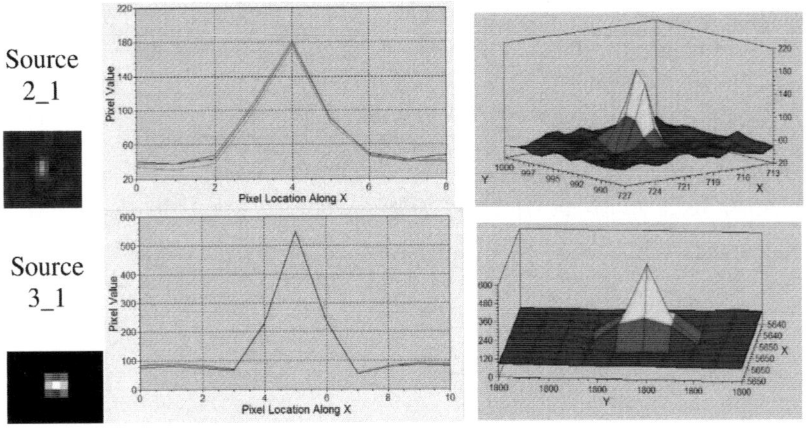

Figure 4. Sources, Gaussian profile, and area plot.

4.2 Source measurement

The apparent position toward the center of the sun can be seen in figure 3, while the measurement is shown is table 2. As gravitational lensing from the Sun has more effect to the sources in 1< R⊙<2, source 1_1 and source 1_2 will result better in displacement measurement later on.

Table 2. Measurement of Each Sources

Name	Length (pxl)	Length (km)	ratio to R⊙	X	Y
r sun	1463.452	695.7	1	2812	1779
source1_1	2212.81	1052.341	1.513	1725	744
source1_2	2181.955	1036.629	1.490	2490	287
source2_1	3261.912	1550.68	2.229	720	994
source3_1	4191.998	1994.734	2.867	5646	1799

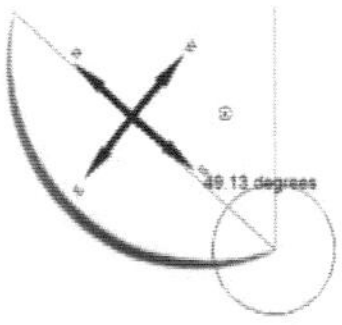

Figure 5. Sun's direction obtained from partial eclipse phase

In figure 5, true direction is measured using partial eclipse image taken right before TSE. From our field of view, the sun's image tilts around 49.13^0 toward the real north position of the sun. This true direction can be used for measuring the point sources coordinate and further identification.

4. Conclusion

It is rather difficult to recognize very faint sources around the sun during TSE with thick clouds. However, four points sources were identified and measured toward the center of the sun even though one of them (source 2) might failed to be categorized as a star due to its low SNR. As it seen on Gaussian fitting, all of them still have a possibility to be the stars. Simple measurement shows that two sources (source 1) would be a good example of lensing effect. After all, it needs better measurement and coordinate transformation in right ascension and declination to obtain the exact position and the real name of the sources to be compared with the next observation.

Acknowledgements

We are thankful to some lecturers and graduate students in ITB, and astronomers in LAPAN for the discussion. And we also thank Space Science Center - LAPAN for supporting our expedition in Ternate and joining ISSEL. And we also thank the committee for giving us the opportunity to publish our research.

Journal of Physics: Conference Series **771** (2016) 012038 doi:10.1088/1742-6596/771/1/012038

References

[1] Dyson F W and Eddington A S 1919 *Astrophysical Journal* **XLVI** 249
[2] Jones B F 1976 *The Astronomical Journal* **81** 6 455-463
[3] Matakuma T *et al.* 1941 *Japanese Journal of Astronomy and Geophysics* **18** 51
[4] Herdiwijaya D 2015 *Applied Mathematics and Materials* **771** 129-132
[5] Espenak E 2000 *ASP Conf. Series* **205**
[6] Druckmuller M, Rusin V, Minarovjexh M 2006 *Astron. Obs. Skalnate Pleso* **36** 131-148
[7] Hroch F 2000 *Experimental Astronomy* **9** 4, pp 251-25

The 2016-2100 total solar eclipse prediction by using Meeus Algorithm implemented on MATLAB

A Melati, S Hodijah

Physics Department, Faculty of Science and Technology, UIN Sunan Kalijaga Yogyakarta, Jl. Maksda Adisucipto No. 1, Yogyakarta, Indonesia

E-mail: asih.melati@gmail.com

Abstract. The phenomenon of solar and lunar eclipses can be predicted where and when it will happen. The Total Solar Eclipse (TSE) phenomenon on March 09th, 2016 became revival astronomy science in Indonesia and provided public astronomy education. This research aims to predict the total solar eclipse phenomenon from 2016 until 2100. We Used Besselian calculations and Meeus algorithms implemented in MATLAB R2012b software. This methods combine with VSOP087 and ELP2000-82 algorithm. As an example of simulation, TSE prediction on April 20th, 2042 has 0.2 seconds distinction of duration compared with NASA prediction. For the whole data TSE from year of 2016 until 2100 we found 0.04-0.21 seconds differences compared with NASA prediction.

1. Introduction

Total solar eclipse (TSE) is not a sign of human mortality and natality. As a matter of fact, it has a predetermined time as the appearance of the hilal as said by Islamic phylosopher Shaykh al-Islam Ibn Taimiyah. The eclipse on March 9th, 2016 was clearly visible in many parts in Indonesia, including Central Sulawesi and Ternate. It was becomes special momentum and astronomy euphoria. The high interest of public indicates that most of Indonesian people eager to learn astronomy. This euphoria was different from TSE phenomenon in 1983 which enormous people were anxious. This research aims to predict next TSE using Meeus Algorithm. Meeus Algoritm could predict maximum duration of TSE [1][2]. We simulated TSE from year of 2016 until 2100 using MATLAB R2012b software.

2. TSE calculation based on Meeus Algorithm

To calculate or predict solar eclipse, we need to know Besselian Elements according to the date of the solar eclipse. The Besselian elements of unification algorithms are VSOP87 (for the Sun) and ELP2000-82 (for the Moon). The method was developed by Friedrich Wilhelm Bessel in 1842 and was repeatedly refined since then. The basic idea of the method is that the Besselian elements describe the motion of the lunar shadow on a suitably chosen, so called *fundamental plane*. The fundamental plane crosses the centre of the Earth and is perpendicular to the axis of the shadow cone. The Besselian elements shown in figure 1 below[3][5]. Where observer plane is an observer on the Earth's surface. L_1 is a radius of the penumbra cone in the fundamental plane, L_2 is the radius of umbra cone in the fundamental plane, L_1' is the radius of penumbra cone in observer plane on the Earth's surface, L_2' is the radius of umbra cone in observer plane on the Earth's surface, f_1 is angle between the penumbra cone and the shadow axis of Moon's, f_2 is angle between the umbra cone and the shadow axis of Moon's.

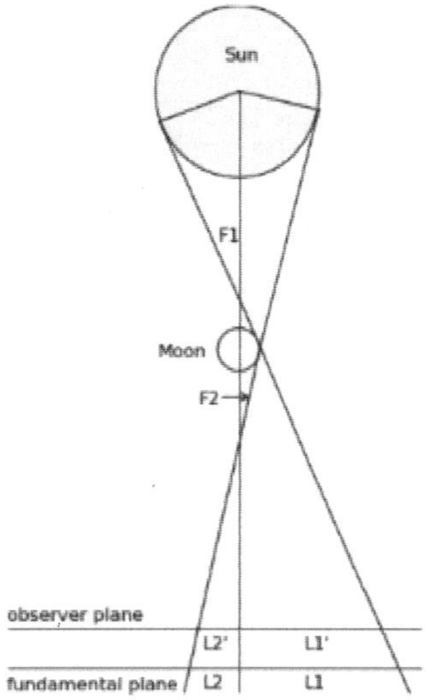

Figure 1. Description Besselian element in the from of image (http://www.gautschy.ch/~rita/archast/solec/solec.html).

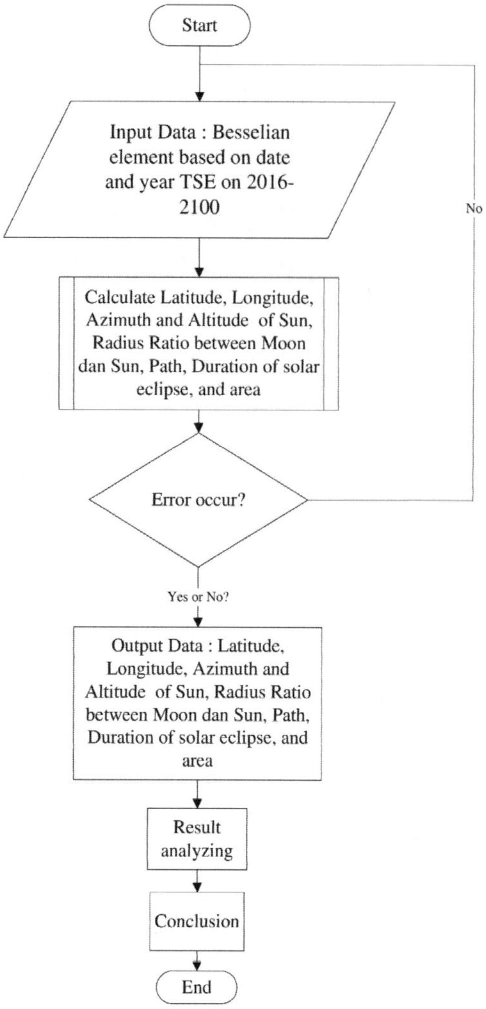

Figure 2.Flow chart of TSE predictions.

In the case of a central eclipse, the type of the eclipse can be determined by the following rules: if $u < 0$, the eclipse is total; if $u > +0.0047$, it is annular; if u is between 0 and +0.0047, the eclipse is either annular or annular-total. u represents radius of the Moon's umbral cone in the fundamental plane. In the latter case, the ambiguity is removed as follow

$$\omega = 0,00464\sqrt{1-\gamma^2} > 0 \tag{1}$$

If $u < \omega$, the eclipse is annular-total; otherwise it is the annular one. In the case of a partial solar eclipse, the greatest magnitude is attained at the point of the surface of the Earth which comes closest to the axis of shadow. The magnitude of the eclipse at that point is

$$\frac{1,5433 + u - |\gamma|}{0,5461 + 2u} \tag{2}$$

γ represents the least distance from the axis of the Moon's shadow to the center of the Earth, in units of the equatorial radius of the Earth [4].

3. Methods

Prediction steps of a total solar eclipse are presented in the form of a flow chart shown figure 2.

4. Result and discussion

Based on our TSE predictions there are ± 68 total solar eclipse that will occur from year of 2016 to 2100 (see figure 3, prediction of the total solar eclipse on April 20th, 2042) using Besselian element illustrated on Table.1

Table 1. Besselian Element for TSE on April 20th, 2042[5]

Elemen Bessel	X	Y	d	M	L1	L2	tan f
0	-0,282156	0,190376	11,51744	210,25866	0,534676	-0,011391	
1	0,520432	0,2561881	0,01367	15,003454	0,0000561	0,0000558	-0,0000127
2	0,0000418	-0,0000819	-0,000003		-0,0000128	-0,0000127	0,0046311
3	-0,00000861	-0,00000444					

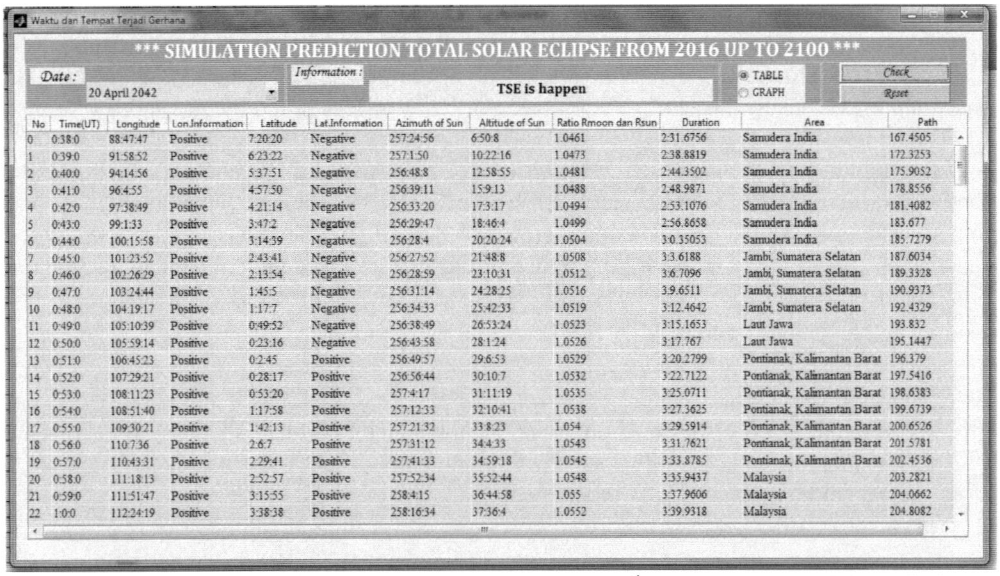

No	Time(UT)	Longitude	Lon.Information	Latitude	Lat.Information	Azimuth of Sun	Altitude of Sun	Ratio Rmoon dan Rsun	Duration	Area	Path
0	0:38:0	88:47:47	Positive	7:20:20	Negative	257:24:56	6:50:8	1.0461	2:31.6756	Samudera India	167.4505
1	0:39:0	91:58:52	Positive	6:23:22	Negative	257:1:50	10:22:16	1.0473	2:38.8819	Samudera India	172.3253
2	0:40:0	94:14:56	Positive	5:37:51	Negative	256:48:8	12:58:55	1.0481	2:44.3502	Samudera India	175.9052
3	0:41:0	96:4:55	Positive	4:57:50	Negative	256:39:11	15:9:13	1.0488	2:48.9871	Samudera India	178.8556
4	0:42:0	97:38:49	Positive	4:21:14	Negative	256:33:20	17:3:17	1.0494	2:53.1076	Samudera India	181.4082
5	0:43:0	99:1:33	Positive	3:47:2	Negative	256:29:47	18:46:4	1.0499	2:56.8658	Samudera India	183.677
6	0:44:0	100:15:58	Positive	3:14:39	Negative	256:28:4	20:20:24	1.0504	3:0.35053	Samudera India	185.7279
7	0:45:0	101:23:52	Positive	2:43:41	Negative	256:27:52	21:48:8	1.0508	3:3.6188	Jambi, Sumatera Selatan	187.6034
8	0:46:0	102:26:29	Positive	2:13:54	Negative	256:28:59	23:10:31	1.0512	3:6.7096	Jambi, Sumatera Selatan	189.3328
9	0:47:0	103:24:44	Positive	1:45:5	Negative	256:31:14	24:28:25	1.0516	3:9.6511	Jambi, Sumatera Selatan	190.9373
10	0:48:0	104:19:17	Positive	1:17:7	Negative	256:34:33	25:42:33	1.0519	3:12.4642	Jambi, Sumatera Selatan	192.4329
11	0:49:0	105:10:39	Positive	0:49:52	Negative	256:38:49	26:53:24	1.0523	3:15.1653	Laut Jawa	193.832
12	0:50:0	105:59:14	Positive	0:23:16	Negative	256:43:58	28:1:24	1.0526	3:17.767	Laut Jawa	195.1447
13	0:51:0	106:45:23	Positive	0:2:45	Positive	256:49:57	29:6:53	1.0529	3:20.2799	Pontianak, Kalimantan Barat	196.379
14	0:52:0	107:29:21	Positive	0:28:17	Positive	256:56:44	30:10:7	1.0532	3:22.7122	Pontianak, Kalimantan Barat	197.5416
15	0:53:0	108:11:23	Positive	0:53:20	Positive	257:4:17	31:11:19	1.0535	3:25.0711	Pontianak, Kalimantan Barat	198.6383
16	0:54:0	108:51:40	Positive	1:17:58	Positive	257:12:33	32:10:41	1.0538	3:27.3625	Pontianak, Kalimantan Barat	199.6739
17	0:55:0	109:30:21	Positive	1:42:13	Positive	257:21:32	33:8:23	1.054	3:29.5914	Pontianak, Kalimantan Barat	200.6526
18	0:56:0	110:7:36	Positive	2:6:7	Positive	257:31:12	34:4:33	1.0543	3:31.7621	Pontianak, Kalimantan Barat	201.5781
19	0:57:0	110:43:31	Positive	2:29:41	Positive	257:41:33	34:59:18	1.0545	3:33.8785	Pontianak, Kalimantan Barat	202.4536
20	0:58:0	111:18:13	Positive	2:52:57	Positive	257:52:34	35:52:44	1.0548	3:35.9437	Malaysia	203.2821
21	0:59:0	111:51:47	Positive	3:15:55	Positive	258:4:15	36:44:58	1.055	3:37.9606	Malaysia	204.0662
22	1:0:0	112:24:19	Positive	3:38:38	Positive	258:16:34	37:36:4	1.0552	3:39.9318	Malaysia	204.8082

Figure 3. Simulation TSE Predictions Display on April 20th, 2042 Using MATLAB Software

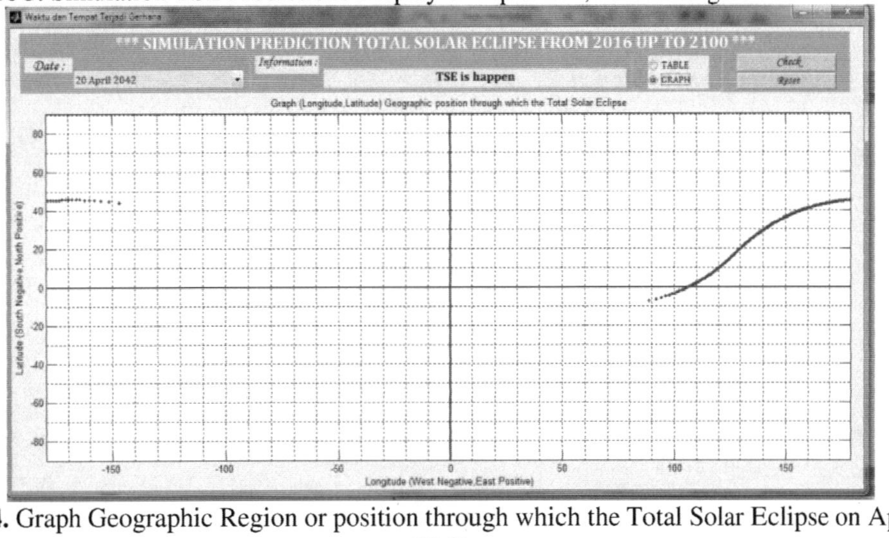

Figure 4. Graph Geographic Region or position through which the Total Solar Eclipse on April 20th, 2042.

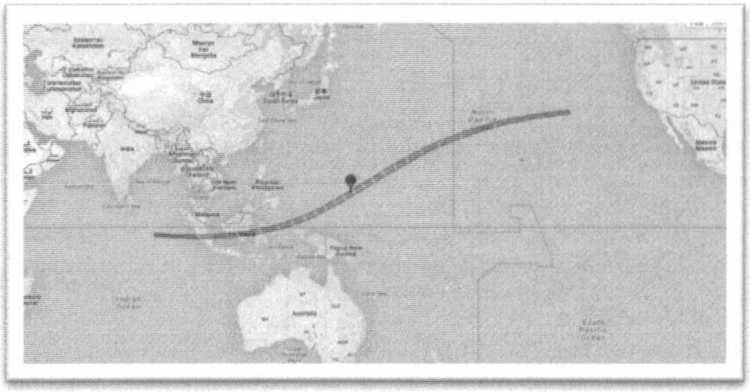

Figure 5. Geographic Region or position through which the Total Solar Eclipse on April 20th, 2042.

Figure 4 and 5 shows TSE coverage on 20 April 2042. The TSE will be observed in Indian Ocean, then across Jambi town and southern Sumatra, Java Sea, Bangka and Belitung Island, Java Sea, the city of Pontianak, West Kalimantan, then across Malaysia, South China Sea, Brunei, Sabah-Malaysia, the Philippines and ended up in the North Pacific Ocean. The longest duration of totality based on the algorithm is 4 minutes 51 seconds.

All of the obtained TSE are different compared with NASA prediction. NASA prediction will have 0.2 seconds longer duration than our research. Based on the results that have been obtained, simulation TSE predictions with Besselian method combined with VSOP87 (for the Sun) and ELP2000-82 (for the Moon) algorithms on MATLAB software could be used as reference for the next total solar eclipse predictions until 2100[4]. Our results show 0.04-0.21 seconds longer duration of totality compared with the results from NASA [6]

Acknowlegment
We would like to thank Astronic Study Club Sunan Kalijaga Islamic State University for support this research, Dr Rinto Anugraha lecturer Physics Department Gajah Mada University and Anggara Dwi for helping operate several label programming.

References
[1] Lewis Isabel M 1931 *American Astronomical Society* **6** 265 – 266
[2] Meeus J 2003 *J. British Astronomical Association* **113** (6): 343-348
[3] Gautschy R 2012 *Canon of Solar Eclipse from 2501 BC to 1000 AD* (Swiss: Swiss National Science Foundation)
[4] Meeus J 1998 *Astronomical Algorthms second edition* (Virginia, USA: Willmann Bell) p 379-388
[5] Meeus J 1989 *Elements of Solar Eclipses 1951-2200* (Virginia, USA: Willmann Bell) p 150
[6] Espenak F and Meeus J 2009 *Five Millenium Catalog of Solar Eclipses: -1999 to +3000 (2000 BCE to 3000 CE)*-Revised NASA/TP-2009-214174 p. A-164

The determination of area and time comparison of the partial solar eclipse at space science center, LAPAN

S Filawati[1], Gammamerdianti[2], E E Hidayat[1], Y Suryana[1], R Kesumaningrum[1]

[1] Space Science Center, LAPAN, Indonesia
[2] Atmospheric Science and Technology Center, LAPAN, Indonesia

E-mail: siska.filawati@lapan.go.id

Abstract. The observation of the partial solar eclipse in Bandung on March, 9th 2016 was done to measure the surface coverage area and to compare to NASA calculation. We have calculated solar disk coverage area based on image data at the maximum contact until final contact of the eclipse. We obtained that maximum eclipse was at 7:21 a.m. (UT+7) with 84.81 % area of solar disk covered by Moon and the end of eclipse was at 8:31 a.m. (UT+7).

1. Introduction

The partial solar eclipse (PSE) occurred in Bandung is in conjunction with the solar eclipse at some regions in Indonesia. Some of Indonesia regions had total solar eclipse occurrence i.e. Bengkulu, Bangka Belitung, Palembang, Palangkaraya, Balikpapan, Palu, Poso, Ternate, and Halmahera. While other regions including Bandung only had a partial solar eclipse.

Solar eclipse occurs when the Moon is between the Sun and the Earth. There are three types of solar eclipse: total, annular, and partial. The types of the solar eclipse depend on the distance and position of the Sun, Moon, and Earth. The total solar eclipse will occur if the angular diameter of the Moon is equal to the angular diameter of the Sun with the position of the Sun, Moon, and observers on the Earth in a line. Partial solar eclipse occurs when the position of the Sun, Moon, and observers on Earth is not perfectly in a line. Whereas, annular solar eclipse occurs when the Moon is too far from the Earth, so the angular size of Moon will be smaller.

The solar eclipse that occurred in Bandung is a partial solar eclipse, which occurred at 6:19 a.m. and ended at 8:31 a.m. based on NASA's prediction. Observation of partial solar eclipse in Bandung conducted by Pussainsa Eclipse Observation Team. The observations were made on the rooftop Pussainsa. There were also Network Team, Security Team, and Public Relations Team who participated in this event. The preparations of Eclipse Observation Team began at 05:30 a.m. and ended at 9:00 a.m.. Solar eclipse observation has been done by using a telescope system with image recording interval time between 30 to 60 seconds, depend on eclipse's phases. This paper aims to determine the area of Moon that covered the Sun and time for the maximum contact and final contact of the eclipse. The results of these calculations will be compared to NASA calculation or prediction.

2. Method

Image data of the partial solar eclipse was observed and documented at 6:53 p.m.. Although partial solar eclipse began at 6:19 a.m.. This was caused by cloudy conditions. The eclipse observation was done by using BabyQ Takahashi telescope, mounting VIXEN AXD ATLUX DELUX 2115, VIXEN Star Book Ten, Nikon D5500, and VIXEN tripod. Image data were taken every 60 seconds, but approaching the maximum of partial solar eclipse, the data were taken every 30 seconds. The size of images is 4000×6000 pixels. We used MATLAB and Photoshop software to calculate the surface area of the Sun that are not covered by the Moon. However, we had intermittent cloudy conditions. Therefore, the eclipse data were converted into a binary image so that they can be easily proceed. The differences between binary image processing by MATLAB and Photoshop will be explained in the discussion. Calculation of the area was conducted on the binary image, which means only the images that are worth 1 or white will be calculated. So that the output is the surface area of the Sun not covered by the Moon to determine time of maximum and end of solar eclipse.

3. Discussion

3.1. Binary image processing

Comprehensive calculation method used in this paper is by turning the data into a binary image (black and white). To change the image into a binary, MATLAB was used. However, because the cloudy conditions during observation, the results of binary image by MATLAB were poor. See an example of the conversion process for image at 6:59 a.m. in figure 1 and figure 2.

Figure 1. Partial solar eclipse at 06.59 a.m.

Figure 2. Binary image of partial solar eclipse at 06.59 a.m.

Based on the results of Photoshop processing as in figure 3, we determine area of eclipse using MATLAB.

Figure 3. Results of photoshop processing on partial solar eclipse at 06.59 a.m. (UT+7).

3.2. Determination of the area

Calculation has been performed from the surface area of the Sun that has not been covered Moon. The results would be compared to the whole surface area of the Sun when the Moon does not cover the Sun. Data were taken, during the observation, every 60 seconds, but approaching the maximum of the eclipse at 7:19 a.m. until 7:31 a.m, the data were retrieved every 30 seconds. Images were recorded till the end of the solar eclipse at 8:31 a.m..

Table 1. Results of the Sun area that was not covered by moon.

Time (UT+7)	Area (Pixel Unit)
06:53:18	4.97×10^5
07:19:16	1.43×10^5
07:20:00	1.42×10^5
07:20:16	1.41×10^5
07:21:00	1.41×10^5
07:21:16	1.42×10^5
07:22:00	1.43×10^5
08:31:16	9.27×10^5

Table 1 shows the result of Sun area not covered by the Moon. To find out what percentage of the surface area of the Sun being covered by the Moon can used a simple equation.

$$A_E(\%) = 100\% - \left(\frac{A_i}{A_S} \times 100\%\right) \tag{1}$$

A_E is *area of eclipse* that the Sun covered by the Moon, A_i is area of *i* data, and A_S is area of the Sun. Covered area at 6:53 a.m. reached 46.34%. We found differences of area of the Moon covering the Sun between the data converted to black and white by Photoshop and MATLAB as seen in table 2.

Table 2. The results of area of the Moon covered the Sun

Time (UT+7)	Area of The Moon Covered The Sun % (Photoshop)	Area of The Moon Covered The Sun % (MATLAB)
07:19:16	84.60 %	87.34 %
07:20:00	84.64 %	87.36 %
07:20:16	84.78 %	87.43 %
07:21:00	84.81 %	87.49 %
07:21:16	84.68 %	87.59 %
07:22:00	84.60 %	87.50 %

The difference of percentage area in table 2 is due to differences in the edge detection method of the image area. Figure 4 shows the results of area calculations that processed by Photoshop (a) and MATLAB (b). It appears that the images edges processed by Photoshop are smoother than by MATLAB.

(a) (b)

Figure 4. Calculation of Solar Eclipse by (a) Photoshop (b) MATLAB

Table 2 shows that the maximum area reached 84.81% at 7:21 a.m.. NASA predicted that the maximum area of the Moon covered the Sun was 85.86% at 00: 21: 44.8 UT or 07: 21: 44.8 a.m. (UT+7). A significant differences are in area calculation and time occurrence of maximum phase of partial solar eclipse. The differences of area and time occurrence between the calculation and the reference are 1.05% and 44.8 seconds. Respectively, differences may be caused by less subtle edge of the picture. Additionally, the timing method is important, because we recorded time sequence by the camera setting, instead of using GPS timing and images were stored in JPEG format.

4. Conclusion

Solar eclipse image obtained at 6:53 a.m. has percentage of the covered area of 46.34%. Data were taken every 60 seconds and approaching the maximum phase of the eclipse (at 7:19 a.m. until 7:31 a.m) the data were taken every 30 seconds. The time for the maximum contact by calculation was at 07:21:00 a.m. with percentage of 84.81%, while NASA predicted the time for maximum contact was at 07:21:44.8 a.m., with the percentage of 85.86% eclipse. The difference value is caused by the lack of precision of location and time during telescope setting because we did not use GPS to set right variables. Moreover, this difference is also due to the edge determination.

References

[1] http://eclipse.gsfc.nasa.gov/SEgoogle/SEgoogle2001/SE2016Mar09Tgoogle.html (accessed on March 2016)
[2] http://space-facts.com/wp-content/uploads/eclipse-types.png (accessed on March 2016)
[3] Utami R Z, I Made B S, Bulkis K 2015 Menentukan Luas Objek Citra dengan Teknik Deteksi Tepi *Dielektrika* **2** 1:11-17

Prototype of sun projector device

Ihsan[1] and B Dermawan[2]

[1] Undergraduate Program in Astronomy, Faculty of Mathematics and Natural Sciences, Institut Teknologi Bandung, Jalan Ganesha 10, Bandung 40132, Indonesia
[2] Astronomy Research Division, Faculty of Mathematics and Natural Sciences, Institut Teknologi Bandung, Jalan Ganesha 10, Bandung 40132, Indonesia

E-mail: budider@as.itb.ac.id

Abstract. One way to introduce astronomy to public, including students, can be handled by solar observation. The widely held device for this purpose is coelostat and heliostat. Besides using filter attached to a device such as telescope, it is safest to use indirect way for observing the Sun. The main principle of the indirect way is deflecting the sun light and projecting image of the sun on a screen. We design and build a simple and low-cost astronomical device, serving as a supplement to increase public service, especially for solar observation. Without using any digital and intricate supporting equipment, people can watch and relish image of the Sun in comfortable condition, i.e. in a sheltered or shady place. Here we describe a design and features of our prototype of the device, which still, of course, has some limitations. In the future, this prototype can be improved for more efficient and useful applications.

1. Introduction

Describing the Sun is one among various ways to give introduction and education on astronomy to public, including students. Before doing that and the related activities, we should know properly some hazards of direct viewing the Sun. Some points are needed to take care of, e.g. not to use unfiltered optical device when viewing the Sun directly. In general, there are at least two safe view methods. First, using aperture filters attached to the device to reduce intense solar radiation; and second, the technique of projection. The second method belongs to indirect way, so that it is still the safest of all the ways in observing the Sun [1]. We do not have to worry about the filter coming off the telescope.

There are many devices that has been built for indirectly observing the Sun. Some classic and popular devices are known as coelostat and heliostat, which are manifestation of the remarkable ideas of the projection's techniques. In those devices some mirrors and lenses are used [2], incorporated with particular gears to locate and track smoothly the daily motion of the Sun.

By improving the scheme of light deflection and image projection, there are a number of advantages when describing the Sun to a group of people. The screen can be placed in a sheltered area or indoor, so that people simultaneously can watch the sun image in a shady place without basking under the sun. If needed, not only device related descriptions can be explained by a tutor, some folks and astronomy related stories can also be comprised additionally in a suitable place.

Here we describe our prototype of such a device of sun projector. The device is a compact and light, easy to transport and set up, uses minimum optical elements and self-power-supplied solar-tracking. The device is designed to be reproducible using low-cost and easily get hold of materials.

However, the device prototype has some limitations that are subjects for better improvements in the future. The benefits of development of this device can be widely understood.

2. Prototype design

Before designing the device some aspects of its capabilities are considered. Capabilities of the device include: constructing clear image of the sun disk in a sheltered place, tracking daily motion of the Sun, not requiring electricity network, being easily set up, being compact, light and high portability, being low-cost and simply reproducible.

Based on its functionality, the device has three schemes, comprising optical, electronic (with motor), and mounting systems. On the other hand, in order to construct image in a sheltered place, we separate the device into two parts, i.e. Part A that is located outside, basking under the Sun, and Part B which is located in a sheltered place or indoor. Tracking motion of the Sun and deflecting its light are main purposes of Part A, and then projection of the sun disk image seen on the screen is a role of Part B.

2.1. Optical system

Diameter of sun image on screen is an important consideration. Although in general, sun image can be projected in any size, but the light intensity decreases very rapidly with increasing scale of the projection [3]. A sun image diameter of 15 cm on screen is adequate [1,3], so that lens magnification and the distance to the screen should be examined.

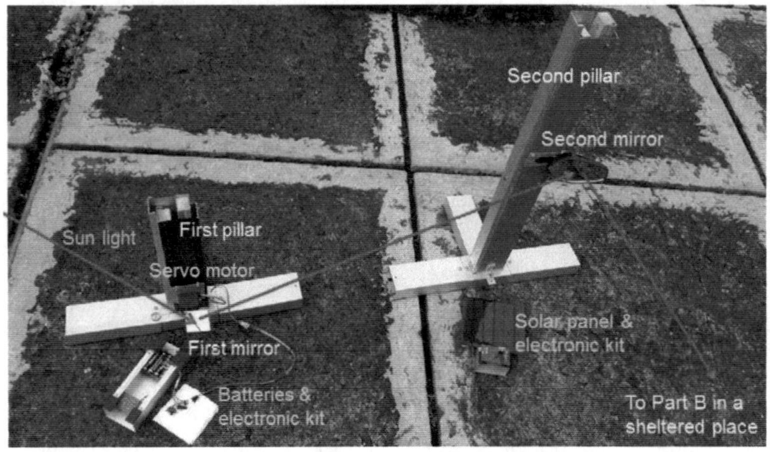

Figure 1. Part A includes optical (mirror), electronic (and motor), and mounting systems.

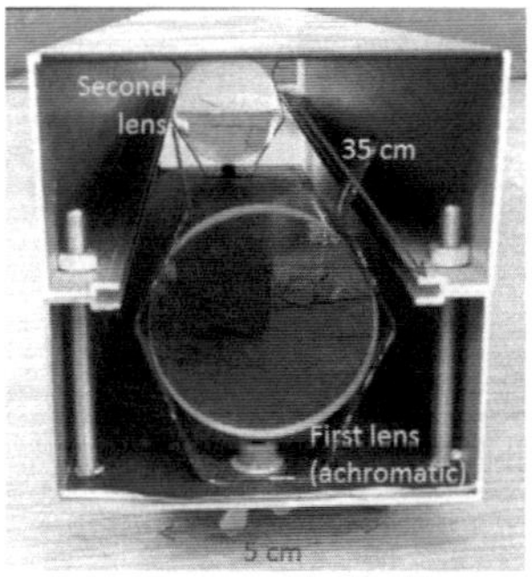

Figure 2. Part B for sun disk projection on screen

Basic concepts of optics, such as principles of mirror reflection and lens refraction, are available in references, e.g. [2]. After surveying materials available in the market, we set that optical system consists of two flat mirrors in Part A (figure 1) and two convex lenses in Part B (figure 2). A servo motor is attached on the first mirror in Part A in order to track motion of the Sun. The second mirror is fixed, but manually adjustable, to deflect the sun light coming from the first mirror to Part B located in a sheltered place (see figure 1). Part B has mainly an optical system (figure 2), comprising a configuration of an achromatic lens (first lens with focal length 30 cm) and a lens (second lens with focal length 5 cm). The second lens can be taken off if necessary.

2.2. Electronic system

A set of four rechargeable batteries (type: AA) are arranged as power supply (figure 1) for the servo motor. The motor is mounted on the first pillar of Part A with adjustable elevation for tracking motion of the Sun. An electronic (timer) kit/circuit is developed for this purpose (figure 3). In line with this, a solar panel (9 volt) with a corresponding electronic kit (figure 1) are also developed for the purpose of charging the other set of batteries (figure 3). The two sets of batteries can be mutually changeable when one set of them is low in power. The electronic kits are equipped by switches and put underneath the pads (figure 1).

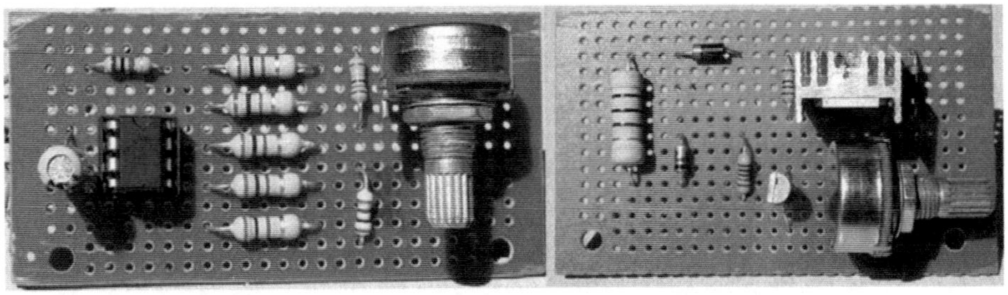

Figure 3. Electronic circuits for controlling speed of the servo motor (left) and recharging a set of batteries using solar panel (right). Length of a kit board is 7 cm.

2.3. Mounting system

Mounting system is a dominant component for Part A. It is made of aluminum because of its light, strong, and low-cost. The system consists of two separate mountings (group of base-pads-pillars, figure 4), but if needed, a pad can be mutually used by both mountings. Pillars are developed in segmentations for easy arrangement in packing. A segment of pillar can be augmented with other segment when installing motor or mirror at higher elevation.

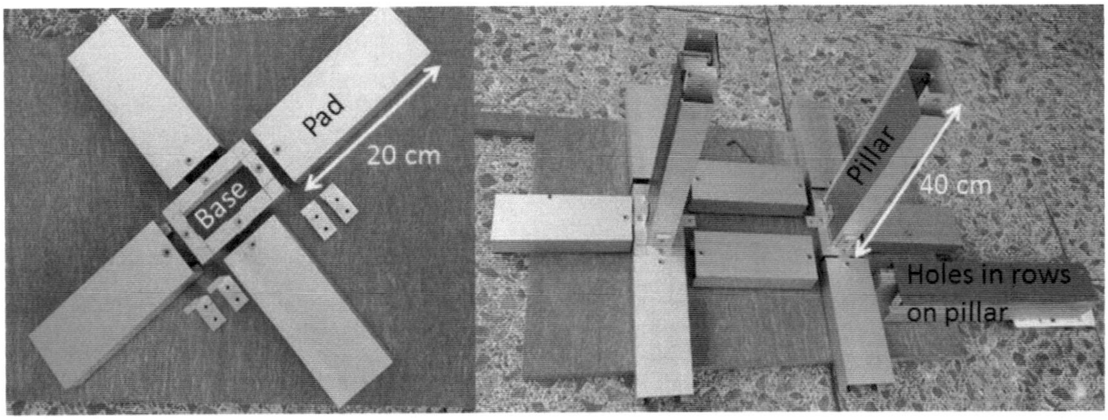

Figure 4. Left: A base for supporting and foundation of pillar, and pads for maintaining stability and its underneath for storage of the electronic kits. Right: Pillar has holes in rows for adjusting elevation of the attached motor and the deflection mirror. The pillar segment can be augmented with other segment if necessary.

3. Discussion

We have performed many experiments related to optical, electronic, and mounting systems. Optics provided in the market are rather limited but some are usable. We use ordinary mirrors in Part A since it is rather difficult and expensive to get high quality mirror. However, we use fine lenses in Part B, and eventually, the sun image can be suitably projected on screen although it is not very sharp. We believe that the precision and alignment of the optical system contributes essentially to quality of the image.

Because of lack of software programming capability, we finally chose a servo motor instead of using stepper motor. Motion of the servo motor has a time-lag, not smoothly as that of programmable stepper motor. We have set an electronic timer circuit so that the constructed image stays on the screen. Using stepper motor is better for smoothly tracking the sun motion in the sky. This will be one subject among the future improvements.

Our surveys reveal that aluminum is the well-chosen material for developing the mounting system. By two people, it takes about 30 minutes to set up the device. We have tested also the stability of a mounting exposing it to strong wind from a fan having high speed propeller, and it works well.

For reason of portability and easy to transport, we use a tool box (figure 5) that can store the entire components of the device. Total weight is less than four kilograms. Fortunately, the bottom of the box inside can also be used as a screen since it is covered completely by the box's sides. The sun disk image can be seen satisfactorily.

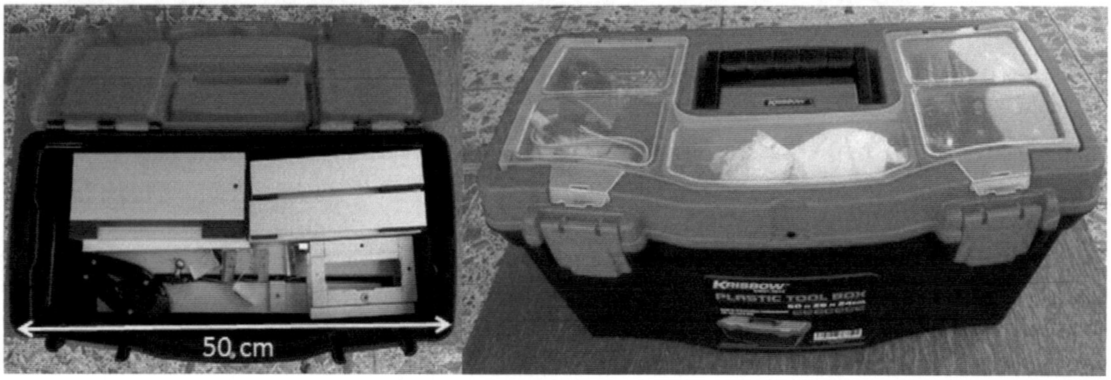

Figure 5. A box for packing the whole components of the sun projector device.

The device is also provided with appropriate instructions, which are simple and contain decent lesson plans. For the whole components of the device, we spent less than one million rupiah, including a box for packaging system. We believe that our device belongs to low-cost instrument and it can also be reproducible easily.

4. Summary

A prototype of sun projector device has been developed. People can watch the sun disk in more comfortable way in a sheltered place. The device is portable and self-support (solar panel) power supply using rechargeable batteries. With this support, it is capable to track the daily motion of the Sun. In the future the prototype needs more improvements.

Acknowledgments

The authors are thankful Mr. Satria for discussions and lending many tools and utensils when building the device. Faculty of Mathematics and Natural Sciences, Institut Teknologi Bandung is warmly acknowledged for financial support.

References

[1] Taylor P O 2008 *Observing the sun* (Cambridge: Cambridge Univ. Press)
[2] Hecht E 2014 *Optics* (New York: Addison–Wesley)
[3] Nicklas H 1994 *Compendium of practical astronomy* vol 1 ed G D Roth (Heidelberg: Springer-Verlag) pp 124

Chasing the shadows, a trip to spice island

A Yamani[1], W Soegijoko[1], A A Baskoro[1,2], R Satyaningsih[1,3], F M Simatupang[1], F Maulana[1], J Suherli[1], R Syamara[4], L Canas[5,6], T Stevenson[7], F Oktariani[1], I Santosa[1], F Ariadi[1], N Carvalho[1] and K Soegijoko[1]

[1] **langit**selatan, Bandung, Indonesia
[2] sekolahalam Minangkabau, Padang, Indonesia
[3] Research and Development Center of Agency for Meteorology Climatology and Geophysics (BMKG), DKI Jakarta, Indonesia
[4] Planetarium Jakarta, DKI Jakarta, Indonesia
[5] IAU Office for Astronomy Outreach, Mitaka, Tokyo, Jepang
[6] National Astronomical Observatory of Japan (NAOJ), Mitaka, Tokyo, Jepang
[7] Sydney Living Museums, Sydney, Australia

E-mail: avivah@langitselatan.com

Abstract. The 2016 Total Solar Eclipse provided us an opportunity to introduce astronomy to a much wider audience. The path of totality crossed the Indonesia from Sumatra to the Maluku Islands and ended its journey in the Pacific Ocean. Its path crossed over 4 major islands, 12 provinces and many cities. Most of the cities have minimum exposure to astronomy. **langit**selatan travelled to observe the eclipse and to do astronomy outreach at the eastern most island under the eclipse path. We chose Maba, a small village in East Halmahera, North Maluku as our site to observe the eclipse as well as conduct a workshop for teachers and students. The aim of the workshop is to introduce astronomy taking advantage of the eclipse as well as raise awareness and curiosity among students. In this paper, we will share a short report regarding the whole trip and event in Maba.

1. Introduction

Astronomy is a unique and unusual scientific topic for a high school in Indonesia. It may be interesting to many people but it is uncommon as a scientific discipline at school. In most high schools in Indonesia, astronomy is not a standalone subject in curricula but a small part of Geography and Physics.

Indonesia has some astronomy clubs for students and the general public in several big cities but the overall interest in Astronomy is still low. Interest in astronomy in general has grown notably since the International Year of Astronomy 2009. Most of the astronomy enthusiasts in Indonesia are interested in "sky shows" such as meteor showers, lunar and solar eclipses, comets or asteroid passing by and even lunar perigee / apogee.

March 9, 2016 was a very special day for Indonesia public as the Moon crossed the face of the Sun and a good portion of the country experienced total darkness for a few minutes. On average, total solar eclipse is visible every 18 months at any particular region on Earth [1]. Previous total solar eclipse passed over Indonesia on October 24, 1995, hence make the one on March 9, 2016 very special, especially those who live along the path of totality. The Total Solar Eclipse event also provided a good opportunity for science education especially in remote regions where science resources are often lacking.

We planned the trip a year in advance and choose North Maluku as the observation and public outreach location. North Maluku or Spice Island has had a rich history due to the presence of the valued spices indigenous to the area. We conducted a survey in 2015 to several cities in North Maluku to look for the optimal location for observation as well as for education and public outreach (EPO).

2. The plan

The Total Solar Eclipse (TSE) March 9, 2016 project targets primarily high schools students and teachers as our main subject for the EPO. Our main target are high school students since they have highest probability to sharing information with parents, friends and siblings and have the highest amount of science background than other age groups in the area. The purpose for doing outreach is to introduce solar eclipse and safe observation techniques with tools from available materials.

As for the locations, after considerations for seeing, weather probabilities and audience size we choose Maba, East Halmahera, North Maluku as our observation site. Maba is the eastern most city in TSE central line to pass through Indonesia before the Sun heads to the Pacific Ocean affording Maba has the longest totality time on land. Maba was also convenient for observational purposes as the Sun is well above the horizons (> 30°) at the time of first contact. At the time of the survey and the subsequent decision making time, North Maluku had the best predicted weather and cloud cover at time of eclipse.

Education and Public Outreach in Maba was done through TSE Introduction workshop prior the eclipse and TSE observation on the eclipse day. We collaborated with SMKN 1 Maba (a vocational school) to be our main contact and venue for the EPO session. The workshop itself will cover the introduction of the TSE, safe observation techniques, and hands-on activities to build safe observation tools from available material. The latter is important as it will be difficult for people in Maba to access solar filter and astronomical instruments.

3. The tools

At TSE Observation, **langit**selatan also welcomed the public to join and experience the eclipse. Telescopes were not the only tools that we took with us, we also provide several other tools that can be used for solar observation. All our telescopes are equipped with solar filters but we also prepared tools with simple projection techniques for the public. The purposes of the tools are eye safety. The easiest tool we provided for TSE observation was the Pinhole Box, which is easy to build using available material. This pinhole box was also part of our Do It Yourself workshop for teachers and students. The goal of this project is to introduce simple optics to the students.

Thousands of international and domestic tourists visited many places in Indonesia, including Maba to view this eclipse. This was the reason for us to provide an Eclipse Viewer called the *KacaMatahari*. *KacaMatahari* is not only for eye safety during TSE but also as a way to introduce Indonesia's unique culture to the tourist and designed as a Barong Mask equipped with solar filter to view the Sun. Barong is a lion-like creature and character in the mythology of Bali, Indonesia. **langit**selatan provided 1000 *KacaMatahari* which was supported by ITB'85, NAOJ and IAU OAO. We also received 300 eclipse glasses from Cosmoquest and another 200 from Sydney Observatory. All eclipse viewer were distributed at the observation site and several other region in Indonesia.

"You are Galileo!" telescopes for children, were also provided during this event. This is a small telescope which is the same size Galileo Galilei used for his observations. The idea of this telescope is so that children can have fun looking at celestial objects. Even though this telescope is not designed for solar observation, we added solar filters for safe viewing. We used this telescope as a tools to introduce basic principles of telescopes and lenses for students.

Another tool for solar observation was the Solar Scope. It has larger field of view and is based on the principles idea of a pinhole and a small telescope, hence it is safe to use for solar observation. The Public will never have to look to the sun directly, but instead at its projection.

Last but not least, **langit**selatan had also planned to broadcast the eclipse from telescope onto the screen. This will cut the queue/ line to telescope and the public can see the whole eclipse from a larger screen. To do so, we built a cheap streaming tool using raspberry pi and the pi-cam. The main driver for this project is to show that 'broadcasting' is not expensive and anyone can build the streaming tools easily. The development of this tool is reported in separate paper in this issue.

Figure 1. TSE Workshop at SMKN 1 Maba. Photo Credit: Fikry Maulana / Wicak Soegijoko / Ferry M. Simatupang / langitselatan

4. The execution

langitselatan conducted the TSE 2016 workshop in Maba, in collaboration with SMKN 1 Maba. We invited 8 schools from Maba and nearby towns. Each school sent their physics / science teachers along with several students. Out of 8, we had 6 nearby schools who attended the workshop.

Through the workshop we aimed to (i) introduce Solar Eclipse and its observation tools; (ii) build astronomy awareness among teachers and students in Maba and its vicinity; (iii) build a network with schools in Maba; and (iv) prepare teachers and students as ambassadors to share the knowledge to the rest of the school as well as their families and friends. In order to achieve our targets, we divided the workshop into following sessions:

- Solar Eclipse introduction: to understands eclipse and sun earth interaction.
- Safe Observation: to introduce how to do safe solar observation and why it is dangerous to observe the sun directly without any protection.
- DIY Pinhole Projectors: to learn simple optics and participants can build their own tools at home without any special equipment.
- You Are Galileo telescope: to learn basic telescope and how to build a telescope.
- Testing the tools: participants use the tools including the small telescope to see the sun. All "You Are Galileo" telescope has its own solar filter.

During the workshop, **langit**selatan also provided several telescopes for students to observe the Sun. Teachers and students whose not attended the session choose to have on location discussion and observation with astronomer from **langit**selatan.

On March 9, 2016, we observed Total Solar Eclipse from SMKN 1 Maba. Many students and public also joined the session even though it was cloudy. We did not broadcast (streaming) the observation due to weather problems but the public were able to observe the solar eclipse using Solar Scopes, *KacaMatahari,* pinhole projectors and the telescopes we provided in the location. As for the eclipse itself, the sun played hide and seek with us but we were able to capture most of the moments leading up to and including totality.

5. Lesson learned

Post TSE Workshop and Observation, **langit**selatan had several notes as lessons learned that we need to improve in the future especially if the venue is in remote area. The first one was an education gap between what we observed and what we are accustomed to. It is understandable that such gaps exist between the city and remote sites, especially in science, math and language arts. We need the ability to be able to customize the material quickly as we better understand the audience and its culture. A common core with the capability to be customized, standard teaching material for teachers and students is required to overcome this gap. Simplification of concepts and less technical sounding terms are needed. Longer workshop time is needed to cover the whole material including theory, tools, do it yourself sessions and observation.

Distance constraints for covering schools in a larger areas with longer travel time has effect on schedule and logistics. This applies for **langit**selatan as well as for the teachers and students who traveled some distance to see us. **langit**selatan needs to be mindful of these issues in the future, some of the attendees travelled more than 2 hours to partake in the workshop.

We have observed that we should start the day with a combination of observation & shorter indoor sessions. This combination will help participants have a better understanding of the topic as well as maintaining their focus on the workshop. If time constraints is not the issue, then it is better to divide the workshop over a few days to cover the whole material and having more time for hands on and discussion (Q&A) session.

As much as possible send workshop material (including basic science) ahead of time and have them available on the **langit**selatan website so students and teachers can download and absorb the material at their own pace and have questions ready.

We have found that Internet access is a necessity, however at remote sites this requirement might become difficult logistically or financially. Internet access will help us not only to broadcast and share our activities live to the global community, but it also can help students and teachers to have access to distant learning in the future.

6. Conclusion

The workshop session was successful. Many students and teachers have became even more curious about astronomy. We do not have the data to support this, but we can see from their curiosity and from the variety of questions they ask during the sessions and at TSE observation day. Despite rain on TSE-day, a large number of students and teachers showed up and took part in the observations. Even though we choose Maba as our EPO and Observation site, we also supervised similar workshop and observation in another city. TSE workshop were conducted in Ambon and Padang in collaboration with Amboina Astronomy Club and *sekolahalam* Minangkabau under supervision of **langit**selatan.

TSE 2016 also afforded large media coverage of **langit**selatan activities. This has helped **langit**selatan to establish itself as a prominent astronomy communicator in Indonesia. In continuation, **langit**selatan will provide more education material through the website to help teachers and students to learn astronomy.

Acknowledgement

langitselatan would like to thank Ministry of Tourism, Sriwijaya Air, SMKN 1, ITB'85, Andi Yudha Asfandiyar, Cheung Sze-leung (IAU OAO), Hidehiko Agata (NAOJ), Pamela L. Gay (Cosmoquest), Sydney Observatory, Surveyors New South Wales Inc. and all the local people who support us during site survey and TSE trip.

References

[1] NASA 2016 *Five Millennium Catalog of Solar Eclipses* http://eclipse.gsfc.nasa.gov/SEcat5/SE2001-2100.html (accessed on September 9th, 2016)

A perspective about the total solar eclipse observation from future space settlements and a review of Indonesian space researches

D Sastradipradja[1,2], F M Dwivany[3] and L Swandjaja[3]

1 Member of AIPI, Indonesian Academy of Sciences – Commission for Basic Sciences, Medan Merdeka Selatan 1, Jakarta Pusat 10110, Indonesia
2 Retired Professor of Bogor Agricultural University, Indonesia
3 School of Life Science and Technology, Institut Teknologi Bandung, Indonesia

E-mail: dngawur@yahoo.com, fenny@sith.itb.ac.id, l_swandjaja@yahoo.com

Abstract. Viewing astronomy objects from space is superior to that from Earth due to the absence of terrestrial atmospheric disturbances. Since decades ago, there has been an idea of building gigantic spaceships to live in, i.e., low earth orbit (LEO) settlement. In the context of solar eclipse, the presuming space settlements will accommodate future solar eclipse chasers (amateur or professional astronomers) to observe solar eclipse from space. Not only for scientific purpose, human personal observation from space is also needed for getting aesthetical mental impression. Furthermore, since space science indirectly aids solar eclipse observation, we will discuss the related history and development of Indonesian space experiments. Space science is an essential knowledge to be mastered by all nations.

1. Introduction

The Sun provides energy for life on Earth through light (radiation) and heat. A total solar eclipse (TSE) gives us an opportunity to see close to the Sun. Solar studies are carried out using filtered telescopes and cameras and multiple tools to look at the solar surface and corona, the elements produced in different parts, etc. Observations are made from Earth, but watching from space is superior. It is anticipated that in LEO extra-terrestrial settlements, solar eclipse chasers who live in it will improve their observational methods, not only direct methods but also robotics due to the space hazards to human health. Solar eclipse chasers are amateurs or professionals collecting new knowledge of the Sun as well as enjoying viewing the spectacle. Above all, personal viewing to see the TSE is needed in particular for artistically impression.

2. Solar eclipse observations from Earth

Since centuries ago, people have successfully predicted and made a list of upcoming solar eclipses complete with date, time, and type of eclipse and where it will appear. The effort is very useful enabling people to make an observation program in advance. Since solar eclipses can be accurately predicted, many countries where the Moon umbra falls make efforts to invite solar observers to visit their country during the eclipse for touristic and scientific reasons. A solar eclipse is shaped when the Sun, the Moon as the obscuring object, and the Earth are aligned in a straight line and the shadow of the Moon will fall and cross the Earth [1]. The shadow covers a certain area with the center part having a TSE. In other parts of the shadow there will be a partial eclipse (PSE). Two more types of solar eclipse are known: the annular (ASE) and the hybrid (HSE) types. An ASE occurs when the Moon is too far away from Earth to completely cover the Sun's disk. A HSE occurs when the eclipse is between TSE and ASE and are very rare occasions [2]. Those events cause a group of people, the 'solar eclipse chasers', emerge who will travel to remote locations on Earth to observe predicted solar eclipses and they are not frightened experiencing difficulties. TSE can be seen from Earth or from space, in example from futuristic LEO space settlements because the angular size of the Sun seen from these points in space is about the same visual size of the Moon so that the Moon can completely obscure the Sun. TSE may exist until circa 1.4 billion years from now. By that time the Moon will be positioned too far away from Earth with an increase distance of 30,400 km due to yearly increase of 2.2 cm. This is caused by tidal acceleration of Moon's orbit around Earth. The angular size of the Moon will be too small to cover the Sun entirely and no natural TSE exist ever since [2].

3. Solar eclipses as seen from space

LEO satellites have a distance to the Sun in reasonable reach from Earth, around 250 – 1200 km from the surface of the Earth [3]. Human has always been living on Earth, but it is predicted that one day human will need more space. Living in other space beyond the Earth has been a dream of mankind since the first successful space mission. This dream led human to plan for settlements in other satellites, planets or other celestial objects. These days, LEO satellites have been constantly manned for certain mission. Hence, solar eclipse may be experienced by the crew in those satellites. The advantage of observing solar eclipses from space is that the observation setup is freed from the disturbing weather and cloud conditions. We have until to date just a few astro-cosmonauts spending a short time, a few years at the most, in space. Only these astro-cosmonauts have had the opportunity watching solar eclipses from space. A few notable events [4] are attested briefly below. The crew of Gemini XII observed a beautiful total solar eclipse from space in 1966 [5]. A theoretical possibility is making an artificial object to obscure the sunlight, but there is no way at present to make such large artificial satellite to give the "Moon caused eclipse effect". At the altitude of the International Space Station (ISS), an object would need to be about 3.35 km across to blot the Sun out entirely [1]. During the Apollo–Soyuz Test Project of July 1975, Apollo and Soyuz space crafts docked together for 44 hours to conduct collaborative space manouvers followed by separation and use of Apollo to create an artificial solar eclipse allowing Soyuz to take photographs of the solar corona [4,5]. Robotics approaches may be used due to the health hazards for life of the space environment. Much have been achieved on solar knowledge, but human presence in space is also needed for manual monitoring and repair purposes of equipment and for getting mental artistic impressions.

4. Present and future developments

If space tourism [6] will becoming less expensive and on its turn will create active space related activities, more people can enjoy traveling into space to catch the opportunity of watching solar eclipses and other celestial phenomena. Future activities will expand readily if extraterrestrial settlements are emerging. We will soon come into a situation in which the keyword of futuristic development of human civilization is making the space fit for a living habitat. It is not too early now to seriously start thinking about joining other advanced nations in preparing for the future off-Earth living habitats for mankind. The growing activities will begin with the efforts to get material needed for life such as water, oxygen, minerals, sand, building materials, primary products, and carbon, from other sources like other planets, the Moon and asteroids than from Earth. Gradually, mankind will require other services for other needs like food, housing, education and recreation as sports-entertainment. Here is where space tourism should develop. One of the most basic arguments that calls for human settlement off-Earth as a first-order priority is the survival of human civilization by creating alternative living space [7].The space dwellers are the ones who will do the space activities, among them will be the future solar eclipse observers.

NASA plans to send manned mission to Mars by 2030s [8] and subsequent building a colony SpaceX on Mars. It would take 400 years to reach a settlement of 80,000 inhabitants to ensure the required safe human genetic diversity to be established. The Mars environment will need paraterraforming which is equally [9] needed for other planets and space habitats. It costs a lot, but building a LEO space city will be more expensive; although ultimately in the future LEO cities will also be built. Experiencing TSE from Mars is not possible though, as Luna can only be seen as a tiny spot in the sky of Mars. This planet knows solar eclipses by one of its own moons Phobos which is of an ASE type having a broad ring of unobscured Sun [10]. Following the example of Appolo 12, a Mars orbiting satellite may make Mars as the object to obscure the Sun and a TSE may be seen by the viewers in the orbiting vehicle.

5. Present state of solar studies and space research in Indonesia

Research in astronomy including activities of solar studies in Indonesia have been routinely carried out facilitated by Bosscha observatory since 1923 and coordination by LAPAN (Indonesian Nationmal Space Agency). LAPAN also make a statement of its commitment to enhance space research [11]. Our discussions will continue on efforts doing space research. Interest in space research in Indonesia started around 1985, when NASA offered Indonesia to send its astronaut into space within NASA program of sending manned spaceships into space, carrying with him/her some space experiments to be conducted in space [12]. Regretfully, this plan was discontinued due to technical, financial and other constraints. Only just in the first decade of this century, the real Indonesian space experiment started to be executed.

5.1. LAPAN endorsed researchers and academics to send research materials to outer space

Due to its duty to develop space science, LAPAN is involved in a cooperation between Indonesia with JAXA in a KIBO program related to Biological Sciences. Indonesia has hosted the workshop on "The 1st Space Exploration and KIBO Utilisation for Asia", the first workshop related to utilisation space exploration for research in many fields of human activities. The Space Environment Utilization Working Group (SEUWG) plans to optimize the use of KIBO, the Japanese Experiment Module on ISS, as the way to facilitate effective joint utilization of the space environment among Asia-Pacific countries [13]. The Space Seeds for Asian Future (SSAF) program is one of the activities which utilise KIBO-AB and initiated by the Asia-Pacific Regional Space Agency Forum (APRSAF).

5.2 NASA supports space experiments by Indonesian high school students

Yeast and rice from Indonesia out into space. Two high school teams were selected to create micro labs to be sent to the ISS for research purposes. The micro labs were transported to the ISS on a Cygnus cargo ship, and subsequently the micro labs were put to use and able to be monitored from Earth in real time. The purpose of the micro labs are to study the alcohol production in yeast and paddy cultivation in zero-gravity environment [14]. The results will then be able to be used for future research. Overview of the research is that 'NanoRacks-Surya Institute Indonesia-*Oryza sativa* Growth in Microgravity Conditions' (NanoRacks-SII-Plant Growth) determines how the growth of *Oryza sativa* and its roots differs with that on Earth and affected by microgravity (MG) and lightless conditions. It also explores the methods on how to grow plants in space, specifically a staple food such as rice, which requires lots of water. This investigation explores the different needs and feasibility of growing rice in weightless, confined conditions as opposed to that on Earth. The benefit of this research is to acquire knowledge in how to cultivate foodstuffs and plants in a confined, weightless environment that is useful in supporting astronauts for planned, long-term missions. In addition, another benefit is the knowledge gained being useful in selecting suitable foods for cultivation in space. The impacts of gravity and light towards plant growth are also better explored. This knowledge leads people to develop more efficient methods of cultivating staple food like rice.

5.3 Research using simulated microgravity (MG)

ITB has built a clinostat which is subsequently used for studies with plant and seeds [15] in a weightless environment. The research was jointly carried out by ITB and LAPAN. Sastradipradja [16] built a simple rotating wall vessel cell incubator system using a kymograph which is setup horizontally carrying cell incubation tubes attached to the rotating drum. A test trial using growing goat's lymphocytes demonstrated that the system supports 3-D growth of cells.

5.4 Research soon to be initiated on metabolomics in simulated MG environment

It is the intense wish of the first author (*pers. reflections* 2016) of the present article to see an umbrella research project to be conducted by the younger generation to do research on 'Metabolomics of animals in simulated MG environment (dry immersion in water)". The young researchers are well informed of animal calorimetry techniques [17]. Metabolism is an essential function of life. Data on energy metabolic rates under MG environment can be related to functional metabolic body size and an allometric relationship between metabolic rate and body size can be formulated to give deeper understanding of the phenomenon of scaling of diverse organisms under MG environments. Metabolic studies may deepen our understanding of the catabolic reduction effect of MG on life by quantitating its value and revealing the cause of and how to manipulate them further for better bodily benefit.

6. Conclusions

Natural total solar eclipses by the Moon will continue to happen, observed by viewers on Earth and its immediate surrounding space. Viewing solar eclipses from space is superior to observation from Earth. Robotics approaches may and still continue to be employed. However, human personal observation from space is needed as a mind-stretcher. This fact coincidentally happen parallel with extra-terrestrial off-Earth accommodation. NASA's prospective plan to send manned space mission to Mars by the 2030s will find different way of eclipse (by its own moon) or only occultation. It is imperative that all nations be involved developing space science to make a space fit for life. Currently developing countries have less chance to do space-related mission or research. Nevertheless, we can use another approach to do space-related research. For example, the research on microgravity can be done using simulation on a clinostat, rotating wall vessel cell incubators or water immersion approaches.

7. Acknowledgement

The authors would like to thank Dra. Clara Yono for providing information on vision & Mission of LAPAN. Thanks are also due to Prof. J.W. Saputra, Indonesian Research Funds, Ir. Srisetiowati Seiful, MBA and Ir. Syailendra Harahap, SSI, both from the Surya Institute for briefing on the NASA sponsored Indonesian Highschool student's research. We would also like to thank the ISSEL editor and reviewer for the help and opportunity to publish this paper in proper form.

References

[1] Zeilik M 2002 *Astronomy: The Evolving Universe* (USA: Cambridge University Press)
[2] Littmann M, Espenak F and Willcox K 2008 *Totality: Eclipses of the Sun* (New York, USA: Oxford University Press)
[3] Gorman A 2005 The Archaeology of Orbital Space *Aust. Sci. Conf.* 338-357
[4] Brooks C G, Grimwood J M, and Swenson L S Jr. 1980 *Chariots for Apollo: A History of Manned Lunar Spacecraft* Isis **71** 4 (December 1980)
[5] Letherbridge C 2001 *Gemini 12 Fact Sheet* http://www.spaceline.org/flightchron/gemini12.html
[6] Heppenheimer T A 1977 *Colonies in Space* www.nss.org/settlement/coloniesinspace
[7] O'Neil D, Bekey I, Makins J, Rogers T F and Stallmer E W 1997 *General public space travel and tourism – volume 1 executive summary* (NASA)

Journal of Physics: Conference Series **771** (2016) 012043 doi:10.1088/1742-6596/771/1/012043

[8] Chang K and Victor D 2016 *Can the U.S. really get astronauts to Mars by 2030?* New York Times (print on October 12, 2016) A18
[9] Pershakov V M and Petrova T O 2012 *Terraforming of Mars planet* UDC 624:001.76:523(045) 9.53-6
[10] Fogg M J 1988 *Elsevier* **22** **3** 415-20
[11] Djamaluddin T 2016 Indonesia serius membangun luar angkasa *Majalah Sains Indonesia* **54**
[12] Satradipradja D 2015 *30 Tahun Perjalanan Gagasan INSPEX (Indonesian Space Experiment)* AIPI-KIPD
[13] Mathers N *et al* 2014 *Trans. JSASS Aero. Tech. J.* **12** **29** 1-5
[14] *NanoRacks-Surya Institute Indonesia-Oryza Sativa Plant Growth in Microgravity Conditions (NanoRacks-SII-Plant Growth) - 07 20 2016* http://www.nasa.gov/mission_pages/station/research/experiments/2389.html#publications
[15] Swandjaja L, Esyanti R R. Khairurrijal, Latief C, Dwivany F M 2014 *JSASS Aero. Tech. J.* **12** **29** 5-10
[16] Sastradipradja D 2007 *J. Med. Vet. Ind.* **11** 22-35
[17] Sastradipradja D 2015 Metabolisme: Arah dan Perkembangannya. (Denpasar, Bali: Penerbit Buku Arti)

Total solar eclipse education for young generation at Palangkaraya, Central Kalimantan

S Fatima[1], Widyanita[1], H Fahriyah[1], A K Rhodiyah[1], C D Satrya[2], M Hilmi[1], G E Ramadhania[1], L Naufal[1], F A M Mulki[1] and D Herdiwijaya[1]

[1]Astronomy Department, Institut Teknologi Bandung, Jalan Ganesha no. 10, Bandung, Indonesia.
[2]Physics Department, Institut Teknologi Bandung, Jalan Ganesha no. 10, Bandung, Indonesia.

Email: fatimas61@yahoo.com

Abstract. The path of Total Solar Eclipse (TSE) on March 9th 2016 passed through several cities in Indonesia and one of them is Palangkaraya, Central Kalimantan. The TSE natural phenomenon provided a special moment and gave unforgettable and lifelong experiences for children who live in Palangkaraya. Some miss-information and a bad impression can be felt by children who do not understand about TSE that causes momentary darkness during totality phase. Therefore we designed a children education programs about the TSE that as follow: (1) socialization about TSE, (2) Popular astronomy seminar, (3) How to observe the Sun? (4) writing competition about TSE and (5) TSE observation. The events were held on March 8th – 9th 2016. More than 200 representatives of elementary school students and teachers throughout Palangkaraya have actively participated. The keynote speaker was an Indonesian expert astronomer with help from alumni of astronomy olympiad in order to provide inspiration for the participants, especially to the students. We conclude that students as young generation of the nation may have more motivation to work in science by direct learning from natural phenomena.

1. Introduction

Interest in learning astronomy in Indonesia is not very high. Many people in Indonesia still have wrong perception about astronomy. This could be seen when astronomy events happen in Indonesia. Some people do not care what was happening. Many news that contain wrong information about the events are spreading very fast in Indonesian society. Therefore, Indonesian astronomers need to introduce astronomy further to the society. Astronomy events in Indonesia open good chance to introduce astronomy and correct many wrong news about the event. The special one is Total Solar Eclipse (TSE) happened on March 9th 2016. Many people do not know what is SE and how to observed it safely.

The path of Total Solar Eclipse (TSE) on March 9th 2016 passed through 12 provinces in Indonesia. One of them is Central Kalimantan, with its capital city Palangkaraya. The number of children in Palangkaraya (with age <15) is greater than a half of its population [1]. Elementary education in Palangkaraya quite good, as indicated by Angka Partisipasi Kasar (Gross Participation Ratio) at 110.95% and Angka Partisipasi Murni (Net Participation Ratio) at 96.47% [1], as well as the number of Public Elementary School many (127 schools) and Private Elementary School (32 schools) in 2015 [2]. Thus, because of the rarity of the TSE event and the last TSE happened in Indonesia over 10 years ago; we believe that most of elementary school students have not ever seen the TSE. Beside that, astronomy can be used to attract young people to sciences and technology [3]. And we know that amateur astronomers (and professional astronomers, of course) can also make important contributions to education [4]. Therefore, we (the public outreach committee that include: professional astronomers, amateur astronomers, and local education stakeholder of Palangkaraya) intend to hold educational event and public outreach of TSE March 2016 in Palangkaraya with the main target for elementary school children.

2. Method and outreach activities

We held some activities to appoarch the goal to encourage Palangkaraya people, specially young generation to love science by experiencing the TSE phenomena. Contructive learning that is learning new ideas by direct experiences [5] was applied for the participants in this TSE event. We used educating, writing and observing as the methods. We educated participants by socialization about TSE and popular astronomy seminar. We held writing competition for elementary school students to evaluate the education achievement of TSE and astronomy seminar. And also gave chances the participants to experience the TSE and solar observation using telescope. Participants of the activities are elementary school students, their teachers, representative of education government and parents. The main location of activities were in SDN 4 Menteng and SDN Percobaan Palangkaraya.

2.1 Socialization about TSE

Socialization was held at SDN Menteng 4 started at 8.00 am. The activity held on March 8th, that was preparation to TSE observation in the next day. We presented information about TSE before the event using some movies about the TSE. We gave many information about TSE that had occured on March 9th for about 2 minutes 30 seconds (Figure 1). The information are about phases of solar eclipse: contact time in Palangkaraya, location to observe the eclipse, and how to observe the eclipse.

2.2. *Popular astronomy seminar*

This activity was held at SDN Menteng 4, in parallel classes. Guest speaker at this seminar was Prof. Bambang Hidayat (Indonesian expert astronomer). At the seminar, Prof. Bambang reviewed a movie about basic astronomy. Some children asked Prof. Bambang about the movie and they discussed many aspects of astronomy and also about the TSE (Figure 2).

2.3. *Solar observation*

This activity was held at SDN Percobaan which is near to the SDN 4 Menteng. By using the portable telescope and special eclipse-glasses, students of SDN Percobaan observed the Sun. The students were very enthusiastic when observing the Sun, they got into long queue repeatedly to see the Sun with telescope or eclipse glasses.

Figure 1. Participants payed attention to the speaker of TSE socialization.

Figure 2. One of the student asked Prof. Bambang Hidayat about astronomy.

Figure 3. The student when participated writing competition. They looked happy but serious when made poem or short story about TSE.

Figure 4. Prof. Bambang and committee gave prize to three students who are the winner of short story competition about TSE.

2.4. TSE writting competition

We invited elementary school students grades 1-4 to make a short story about TSE and grades 5-6 to create a poem about TSE. Participants who followed the event are 135 representative students from elementary schools around Palangkaraya (figure 3). We announced three winners for each writing poetry and short story competition (figure 4). We choosed the winners based on three main criteria, i.e. conformity of science facts, plotting and aesthetic of the words they used.

2.5. TSE observation

The activity was held at SDN Percobaan on 9th of March 2016. Around 6 am, observation was started. Even though, because of cloudy day and a light rain, we just observed with a pair of eclipse-glasses. When TSE in phase 1, we could not see the sun clearly because of black thin cloud. Phase 1 occured from 6:23 am, after that an eclipse of the Sun by the Moon happened (figure 5). We could observed the Sun with eclipse-glasses from 6:13 am until 6:30 am. Before the totality, we had an opportunity to see a little light of Sun until the totality happened. The part of the Sun covered by black cloud during the totality was hard to see, therefore we could not see TSE process completely. When the landscape fell into darkness, the temperature dropped suddenly and nature became silent, insects and birds falling silent and all people shouting a big WOW! The cricket sang and children shouted repeatedly being astonished to feel the darkness in the morning like in the night. The totality happened at 7:28 am around 2 minutes 30 seconds. After the totality, the Sun just covered by much thin cloud, so that we could observed without eclipse-glasses. We could see the eclipse with cloud as the filter. Moreover, the children were enthusiastic to see eclipse in the last phase.

Figure 5. There was so much cloud around Sun direction during TSE process.

Figure 6. Participants observed TSE using special sun glasses. The glasses made from ND5 filter.

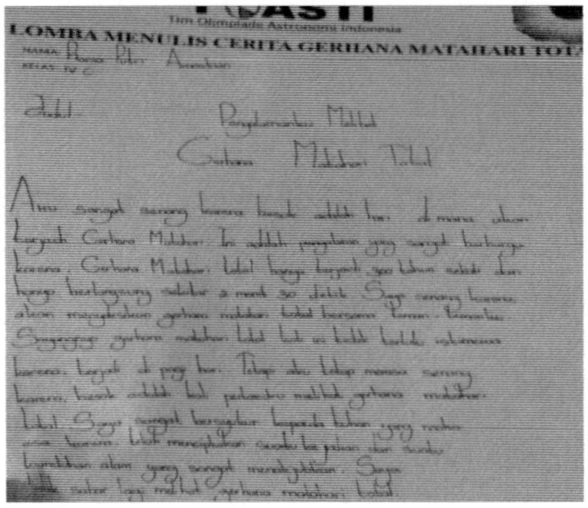

Figure 7. One sample of TSE short story made by Rania Putri. In the story she said: "My experience observe TSE, I am very happy that tomorrow is the day to observe TSE. This is valuable experience because the TSE only happen about 300 years once in same place and take time only about 2^m 30^s. I am happy that I will observe TSE with my friends. Although this TSE will happen in the morning, I am still happy because this is my first experience. I thank God because He creates miracle and wonderful phenomena. I cannot wait any longer to see TSE".

3. Program reception

The TSE socialization was attended by more than 200 participants i.e. elementary school students representative from around Palangkaraya, their teacher and their parents. The participant enthusiasm can be seen from their passionate activity during this section. Moreover, the participant have high curiosity about the TSE which can be seen from so many questions they asked the speakers.

One of education achievement of TSE can be seen from the students work. About 79.1% students who participated in writing competition using TSE facts for their poem or short story (figure 7) i.e. the period of TSE, the danger of observing TSE in wrong way, and the duration of TSE. This high precentage show us there is small enough of misconceptions or inconsistent comprehension [6] about TSE that they got from our education programs. Furthermore, on March 9th, the participant followed the rules for the TSE observation. Moreover, participants were so interested that no one was frightened during the solar eclipse. They were very excited, counting down together and gave big applause followed by 'wow' expression at the time of totality TSE. Other important result that is we got 6 students as the champions of the writing competition of TSE. They could inspire and encourage others to be champions also.

4. Discussion

Education and motivation event by direct experiences will give unforgettable memories especially for very young generation. Children can have wonderful imagination after they experienced the unique phenomena, i.e. TSE. And also, it was a good chance to push them to create a big dream, especially in science, by getting inspiration from great role model from astronomy.

Acknowledgement

The activities were helped by many people and elements. We thank to Intan Ophelia Binti as coordinator of the activities and TOASTI (community of astronomy olympian alumni), Prof. Bambang Hidayat, Education Service of Central Kalimantan, Liberti Hia from Kelas Inspirasi of Central Kalimantan, Tabengan Daily Newspaper, SDN 4 Menteng and SDN Percobaan.

References

[1] Pemerintah Kota Palangka Raya 2016 *Jumlah Penduduk* https://www.palangkaraya.go.id/statis-15-jumlahpenduduk.html (accessed on May 10th 2016)
[2] Pemerintah Kota Palangka Raya 2016 *Pendidikan* http://palangkaraya.siap.web.id/data-sekolah/data-daftar/ (accessed on May 10th 2016)
[3] Percy J R 2006 *JAAVSO* 248 253.
[4] Percy J R 1997 *New Trends in Teaching Astronomy*, ed. D. McNally *et al.* (Cambridge: Cambridge Univ. Press).
[5] Kavanagh C L Agan and C Sneider 2005 *Astronomy Education's Review 4* **19** 52
[6] Comins N 2001 *Heavenly Errors: Misconceptions about the Real Nature of the Universe* (New York: Columbia University Press).

Effect of microgravity simulation using 3D clinostat on cavendish banana (*Musa acuminata* AAA Group) ripening process

Fenny Martha Dwivany [1,2], Rizkita R. Esyanti [1,2], Adeline Prapaisie [1], Listya Puspa Kirana [1], Chunaeni Latief [3] and Ari Ginaldi [3]

[1]School of Life Science and Technology, Institut Teknologi Bandung, Indonesia;
[2]Biosciences and Biotechnology Research Center, Institut Teknologi Bandung, Indonesia
[3]Indonesian National Institute of Aeronautics and Space (LAPAN), Bandung, Indonesia

E-mail: fenny@sith.itb.ac.id

Abstract. The objective of the research was to determine the effect of microgravity simulation by 3D clinostat on Cavendish banana (*Musa acuminata* AAA group) ripening process. In this study, physical, physiological changes as well as genes expression were analysed. The result showed that in microgravity simulation condition ripening process in banana was delayed and the *MaACO1*, *MaACS1* and *MaACS5* gene expression were affected.

1. Introduction

According to several researches, microgravity conditions may lead to the limitation of air convection, resulted in a change in the composition of O_2 and CO_2 in the surface around the plant [1,2]. The condition can lead to a stagnant air layer that will thicken around plant surface, meanwhile the air from outside was difficult to diffuse, causing unbalanced in composition of CO_2 and O_2. CO_2 will increase as a result of respiration, while O_2 will decrease due to the limitations of air convection [2]. Changes in the composition of the air can change some metabolic processes in plants, for example ethylene biosynthesis.

Ethylene is a gaseous plant hormone that is naturally produced. One function of ethylene is a signal in the process of fruit ripening. In climacteric fruit such as banana, ethylene production will increase along with the increase of the rate of respiration. Increased ethylene does not occur in non-climacteric fruit. Along with increase of ethylene, the fruit will undergo maturity, changes in physiology, biochemistry, molecular and organoleptic characters [3].

Ethylene is synthesized from methionine and the reaction is catalyzed by two enzymes, ACC synthase and ACC oxidase [4]. Methionin is changed into S-adenosyl-methionin (SAM), and ACC synthase catalyze reaction of SAM into ACC (1-Aminocyclopropane-1-carboxylic acid). Finally, ACC oxydase catalyze ACC conversion into ethylene and this step really depends on O_2 and accelerated by fruit ripening [5]. It has been known that ACC synthase is encoded by multigene families ACS, whereas ACC oxidase is encoded by multigene families ACO. According to previous research, *MaACO1, MaACS1, MaACS5* were expressed higher in ripening bananas [6,7,8].

In this study, the effect of microgravity simulation performed by 3D clinostat and closed chamber condition on banana fruit ripening process were analysed based on physical, physiological and gene expression study. As control, banana was placed at open air condition as well closed chamber on ground and clinostat open condition. Until recently, report related to effect of microgravity on fruit ripening process was limited. Understanding the effect of space condition on plant response to various signals from the environment can also materialize to the idea of establishing a greenhouse/laboratory to minimize effect of microgravity for further space farming. In addition, understanding of fruit maturation and decay from this experiment can also be applied to design a better postharvest technology.

2. Methods

Mature green Cavendish Bananas, (*Musa acuminata* AAA Group), were used as samples after 24 hours treatment by ethylene. Banana samples were wrapped in paper tissues containing active charcoal to avoid physical injury and to absorb water from respiration process. Three replicates of samples were then placed into four groups. Experiment was conducted using 3D clinostat [9] over seven days then samples were harvested at day 0, 1, 4, 5 and 7 and stored at -80°C for further molecular analysis. Physical and physiological analyses were performed to observe changes in peel colour and to measure peel to pulp ratio. In addition, gene expression analysis was also performed, a total of RNA was isolated from 1 gram sample [10], then qPCR was performed [8,11]. The qPCR result was analysed using relative analysis method with equations, $\Delta\Delta Ct = (Ct_{target\ gene} - Ct_{GAPDH})T_x - (Ct_{gene\ target} - Ct_{GAPDH})\ T_0$, where T was the treatment time [12].

3. Results

Effect of 3D clinorotation on fruit ripening was analysed using physical and physiological approaches. Peel colour changed gradually from green to yellow during seven days of fruit ripening process. The peel colour at closed chamber on clinostat changed slightly and took longer time compared to control and other treatments (figure 1(a)). Mass ratio of peel to pulp fruit ripening bananas was also measured, ripening process at closed chamber on clinostat treatment was slowest and resulted highest ratio of peel to pulp (figure 1(b)).

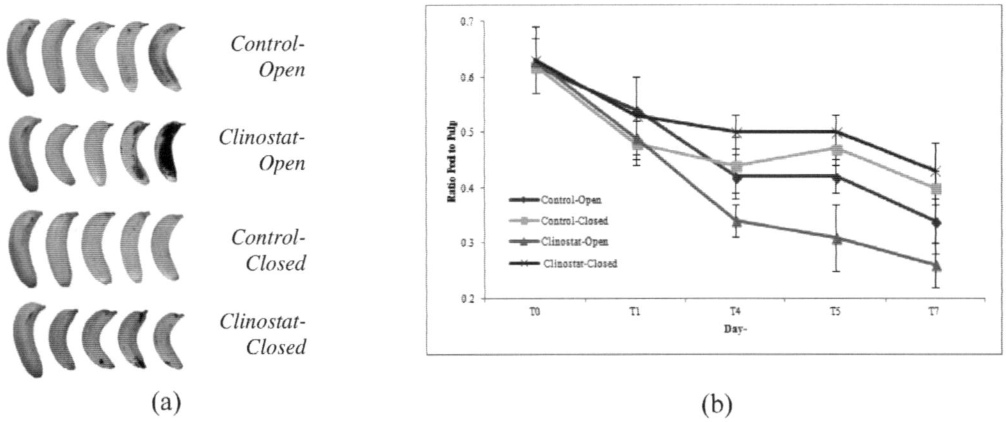

Figure 1. (a) Peel colour changes, (b). Mass ratio of peel to pulp

In addition to physical and physiological analyses, analysis at molecular level was also performed using gene expression analysis. Total RNA was isolated from all banana samples. The expression of fruit ripening-related (*MaACS1*, *MaACS5* and *MaACO1*) genes expressions were characterized by qPCR method using gene specific primers. The result of *MaACS1* expression analysis during fruit ripening showed that gene expression increased in all conditions from day 0 until day 7 (2A). Lower gene expression of *MaACS1* was shown in all treated samples compared to control. On the other hand, the expression of *MaACS5* in control condition increased from day 0 and reached the highest at day 5, whereas at other treatment still increasing after day 5 (figure 2(b)).

The expression of *MaACO1* gene during fruit ripening was shown in figure 2(c). The highest gene expression at closed chamber on clinostat was at day 1, then it decreased whereas in other treatments reached the highest expression later at day 4 then decreased. However, gene expression at open chamber on clinostat increased again at day 7.

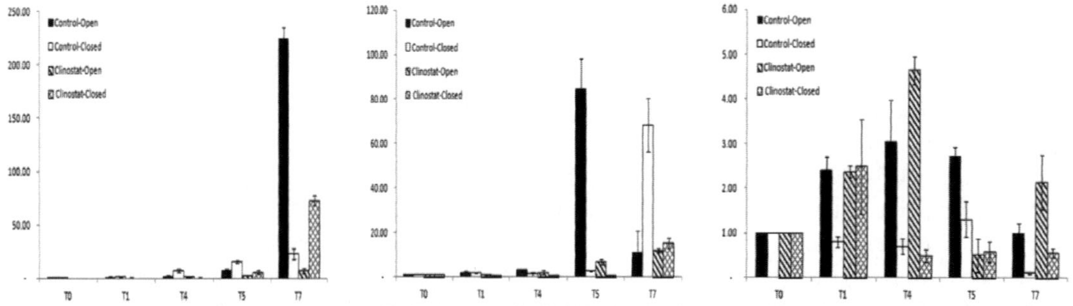

(a) (b) (c)

Figure 2. Relative quantification analysis of gene expressions from banana pulp during ripening process in four storage conditions: control open, clinostat open, control closed, and clinostat close chamber. (a). *MaACS1*, (b). *MaACS5* and, (c). *MaACO1*.

4. Discussion

According to the results, longest self-life was obtained when banana was stored at microgravity simulation condition (closed chamber on clinostat), followed by closed chamber and open chamber on clinostat, and last was control.

In this study, changes in peel colour were slowest in banana placed at closed chamber on clinostat compared to other treatment. This phenomenon was confirmed by the pattern of chlorophyll content at the same treatment which was also the lowest (data not shown). The result might be related to the limited oxygen content around the fruit. Peel colour of banana commonly changed gradually from green into yellow during ripening process, which was caused by chlorophyll degradation [13].

Mass ratio of fruit peel to pulp is an indicator of bananas fruit ripening [14]. The mass ratios of peel to pulp for all conditions in this study were decrease during ripening process. However, closed chamber on clinostat treatment showed the highest ratio and slowest ripening process. The decrease in the mass ratio of peel to pulp might be associated with changes in water levels on both tissues.

Molecular analysis showed similar patterns of *MaACS1* gene expression found in all conditions. *MaACS1* showed an increase in gene expression but the averages of gene expression were lower in other conditions compared to control. On the other hand, *MaACS5* gene expression patterns in treated conditions were different to control. In control condition, the gene expression increased and reached the peak at day 5 then decreased, whereas gene expression at treated conditions has not reached the peak and increased continually after day 5. Further study to obtain the peak of *MaACS5* expression in treated conditions was needed. The gene expression pattern of *MaACO1* showed that the peak of gene expression at control was attained earlier compared to other condition.

This study showed that *MaACS1*, *MaACS5* and *MaACO1* gene expression pattern were affected by different chamber condition during fruit ripening. Analysis on *MaACS1* showed that lower gene expression resulted in slower ripening process compared to control. Delayed activity of *MaACS5* gene expression might correlate with slower ripening process at treated condition compared to control. Interestingly, longest shelf-life of bananas stored at closed chamber on clinostat condition might correlate with *MaACO1* gene expression. In this condition, the gene expression of *MaACO* decreased drastically after day one, while in other conditions increased continually until day 4. Since these genes belong to gene families which encode ACC synthase and ACC oxidase, their activity all together might affect fruit ripening. It has been known that the decrease of O_2 concentration in environment result in inhibition of ACC oxidase activity [5].

Our previous study showed that banana fruit ripening in bamboo closed chamber was also delayed compared to control [15]. It was reported that banana stored at closed chamber filled with 5% carbon dioxide showed lower *MaACS5* and *MaACS6* gene expressions [8]. Several other studies also reported a correlation between delay in banana ripening and high carbon dioxide composition around the fruit [16,17,18]. It has been known that ethylene synthesis from ACC is induced by the presence of oxygen and is inhibited at high concentrations of carbon dioxide [5].

5. Conclusion

Simulated microgravity conditions using closed chamber on clinostat condition resulted in delaying of banana fruit ripening. In this study, we also found that *MaACS1*, *MaACS5* and *MaACO1* gene expression were affected when banana was placed in such condition.

Acknowledgement

Author would like to thank LAPAN, JAXA, Prof. Khairurijal, Dr. Sony Suhandono and Banana Group from ITB for their supports to this project. This research was funded by ITB Research Innovation Grant to Rizkita R. Esyanti and ITB International Research Grant to Fenny M. Dwivany.

References

[1] Musgrave ME, Kuang A and Matthews SW 1997 *Planta* **203** 1-19
[2] Monje O, Stutte GW Goins GD and Porterfield GE 2003 *Adv. Space Res.* **31(1)** 151-67
[3] Prassana V, Prabha TN, and Tharanathan 2007 *Critic. Rev. in Food Sci. and Nut.* **47** 1-19
[4] Yang SF and Hoffman NE 1984 *Ann. Rev. of Plant Physiol.* **35** 155-89
[5] Taiz L and Zeiger E 2002 *Plant Physiology* (Los Angeles: Sinauer Associates, Inc.)
[6] Liu X, Shiomi A, Nakatsuka Y, Kubo R, Nakamura and Inaba A 1999 *Plant Physiol.* **121** 1257-65
[7] Choudhury SR, Roy S, Saha PP, Singh SK, and Sengupta DN 2008 *Plant Cell Rep.* **27** 1235-49
[8] Dwivany F, Esyanti RR, Robertlee J, Paramaputra IC, Permatadewi RK, Tambun DH, Handayani RU, Pratiwi AS and Zaskia H 2014 *AIP Conference Proceedings* ***1589***, 285 doi:10.1063/1.4868801
[9] Swandjaja L, Esyanti RR, Khairurrijal, Dwivany FM, and Latief C 2014 *Trans. JSASS Aerospace Tech. Japan.* **12** ists 29 Th 5-10
[10] Zhuang J, Su J, and Chen W 2006 *Mol. Plant Breeding* **4(1)** 143-6
[11] Handayani RU and Dwivany F 2014 *J. Math Fund. Sci.* **46(2)** 194-3
[12] Livak KJ, and Schmittgen TD 2001 *J. Methods* **25** 402-8
[13] Yang X, Pang X, Xu L, Fang R, Huang X, Guan P, Lu W and Zhang Z 2009 *J. Experimental Bot.***60(14)** 4051-62
[14] Dadzie BK and Orchard JE 1997 *Routine postharvest screening of banana/plantain hybrids criteria and methods* International network for banana and plantain (Inibap) technical guidelines
[15] Pratiwi A, Dwivany F, Larasati D, Islamia HC and Martien R 2015 *AIP Conference Proceedings* ***1677***, 100005 doi:10.1063/1.4930763
[16] Mapson LW and Robinson JE 1966 *Food Tech.* **1** 215-25
[17] Kanellis AK, Solomos T, and Mattoo AK 1989 *Plant Phy.* **90** 251-25
[18] Kittur FS, Saroja N, Habibunnisa, and Tharanathan RN 2001 *Eur. Food Res. Tech.* **213** 306-11

Physical, chemical and biological characteristics of space flown tomato (*Lycopersicum esculentum*) seeds

Rizkita R. Esyanti[1,2], Fenny M. Dwivany[1,2], Maria Almeida[1], Leonita Swandjaja[1]

[1]School of Life Science and Technology, Insitut Teknologi Bandung, Indonesia
[2]Biosciences and Biotechnology Research Center, Institut Teknologi Bandung, Indonesia

E-mail: rizkita@sith.itb.ac.id

Abstract. Several research showed that space flown treated seeds had a different characteristic with that of ground treated seed, which eventually produced a different characteristic of growth and productivity. Research was conducted to study the physical, chemical and biological properties, such as the rate of germination and the growth of tomato (*Lycopersicum esculentum*) space flown seeds compared with that of control one. Observations of physical properties using a SEM showed that there were pores on the surface of some tomato space flown seeds. Observations using a stereo and inverted microscope showed that the coat layer of space flown seeds was thinner than control seeds. The total mineral content in the control seeds (22.88%) was averagely higher than space flown seeds (18.66%), but the average carbohydrate content in control seed was lower (15.2 ± 2.79 %) than the space flown seeds (9.02 ± 1.87 %). The level of auxin (IAA) of control seeds (142 ± 6.88 ppm) was averagely lower than the space flown seeds (414 ± 78.84 ppm), whereas the level of cytokinins (zeatin) for the control seeds (381 ± 68.86 ppm) was higher than the space flown seeds (68 ± 9.53 ppm), and the level of gibberellin (GA3) for the control seeds (335 ± 10.7 ppm) was higher than the space flown seeds (184 ± 7.4 ppm). The results of this study showed that the physical and chemical properties of tomato space flown seeds were generally different compare with that to control seeds, so that it might also be resulted in different germination and growth characteristic. The germination test showed that space flown seeds had lower germination rate compare to control. The growth pattern indicated that planted space flown seeds generally grew better than control. However, those data were more homogenous in control seeds compare to that in space flown tomato seeds.

1. Introduction

Tomato (*Lycopersicum esculentum*) is a member of Solanaceae which has high nutrient content in addition to its high economic value. The average of tomato production in Indonesia in 2000 was 80 kg/ha. This number is considered low compare to the potential of the production, 150 kg/ha [1]. Productivity can be improved by selecting improved crops. One of the methods used in producing improved crops is radiation exposure [2]. Radiation can cause genetic variation in the seed. This variation can be used as an approach to produce improved crops [3]. Experiments in China showed that space flown seeds have produced new kinds of rice, wheat and vegetables which are more productive, larger, and more nutritious than their earthly treatment [4]. Recent publication also noted that space flown seed program in China also produced cucumbers and tomatoes which are oversize with remarkably high sugar content as a result of cosmic radiation and microgravity. However, similar research in the US showed non comparable results. Other research in Alfalfa showed that the rate of germination of the space-flight seed and the growth of the seedlings of Alfalfa was higher compare to control on Earth [5].

The purpose of this study was to compare the rate of germination, growth, and productivity of space-flight tomato seeds to the control seeds on Earth, in addition to the evaluation on physical and chemical properties of the seeds.

2. Methods

The tomato seeds (*Lycopersicum esculentum* cv. Arthaloka) used in this experiment had been sent to the International Space Station (ISS) by Japanese Exploration Agency (JAXA) in 2010 and stored in Japanese Experiment Module (JEM) which also known as *Kibo* (hope) for three months before it was sent back to Indonesia.

2.1. Physical observation.

The Scanning Electron Microscope (SEM) imaging was done in *Pusat Penelitian dan Pengembangan Geologi Kelautan* (PPGL) Bandung. The sample seeds were coated by palladium-gold (PdAu) in Fine Coat Sputter JFC-1100. The image was observed with SEM JEOL JSM 6063LS and the image was captured by Back Electronic Scatter (BES) 10 kV. The seeds were also observed by stereo microscope after they were soaked in water.

2.2. Chemical content of the seeds

2.2.1. Mineral content measurement. The mineral content was measured by Energy Dispersive x-Ray Spectroscopy (EDS) in PPGL Bandung.

2.2.2. Carbohydrate content. Carbohydrate content of the seeds was measured using Luff Schroll method [6]. The known volume was converted to mass based on Luff Scroll table, the carbohydrate content was then calculated using the equation (1) below.

$$Carbohydrate\ content = \frac{mass\ of\ glucose\ x\ dillution\ factor\ x\ 0.95}{mass\ of\ sample} \tag{1}$$

2.2.3. Auxin, Cytokinin and Gibberellin contents. 1 gram of seed sample was extracted in 25 ml of 80% methanol. The methanol was evaporated in room temperature. The pH of the extract was then lowered to 2.5 before the addition of 5 ml ethyl acetate. The water and ethyl acetate phase were separated. The water phase was evaporated. The dried sample was added by 1 ml methanol before it was analysed [7,8]. The sample was analysed in high performance liquid chromatography (HPLC). The eluent was methanol and water (7:3). The flow rate was 0.8 ml/minute. The standard used for auxin was IAA, zeatin for cytokinin and GA3 for gibberellin.

2.3. Germination rate in soil

The germination test was conducted by germinating the seeds in the pots (5 cm x 10 cm). The soil surrounding the seeds was kept moist until the seeds were fully germinated.

2.4. Plant growth and productivity

The seedlings with 3-4 leaves were transferred to polybags. The seedlings were watered regularly up to 14 weeks. For maintenance, the plants were fertilized and treated with fungicide. Every week the height of the plants was measured.

3. Results and discussion

The image from SEM observation showed some pores (see arrow in figure 1) were formed on the surface of space-flight seed, while that of control seed had no pore. This observation was supported by the research on Alfalfa seeds which showed pores formed after cosmic radiation during the space flight [5]. Tomato seeds exposed to gamma ray radiation would have morphological damage, such as pore formation on the seed coat surface [9]. The gamma ray would interact with the water molecule in the cells and formed free radical. The energy released during the free radical formation would go around the seed surface and induce cell damage. Another research on space-flight tomato seeds showed similar result [10]. The pore produced due to radiation would cause water and air enters the seeds faster during germination and could speed up the germination process.

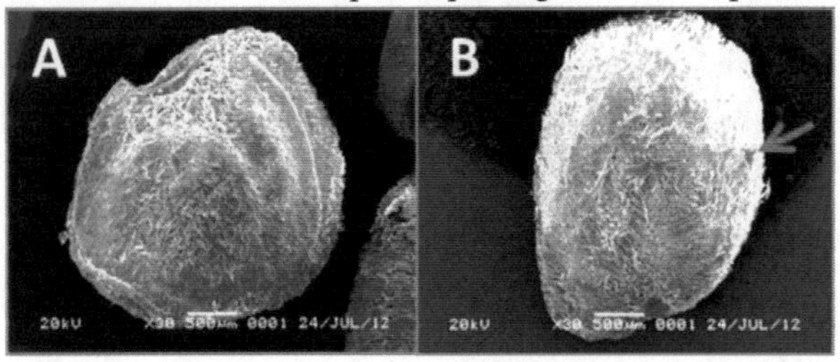

Figure 1. (A). Control seed had no pore. (B). Space-flight seed had a pore (arrow) in the seed coat.

Visual observation on the seed surface of space-flight seeds showed thinner seed coat compare to control (figure 2). Other research on rice seeds showed that cosmic radiation could change the morphology of the seeds [11]. The changes varied based on the dose and duration of gamma ray exposure.

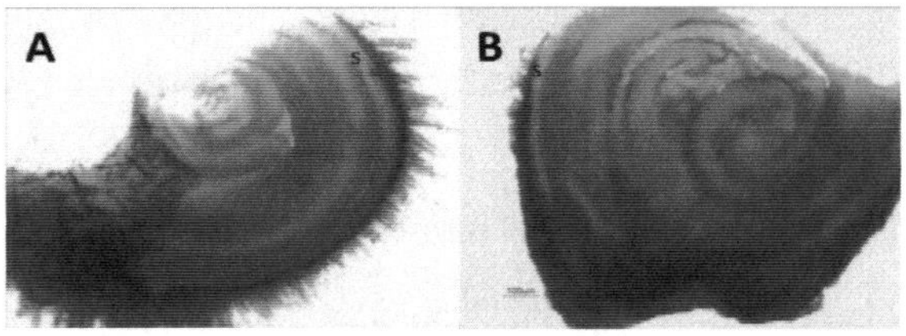

Figure 2. Control seed (A) has thinner seed coat (s) compare to space flight seed (B).

The total mineral content of the space flight seed was also generally lower (18.66%) than control seed (22.88%). This result was similar to Alfalfa seeds flown to the space, which were then resulted in softer surface than control seed [5] even though the cause has yet to be found. This result might also be supported by the calcium content of the space flight seed, which had lower calcium content (6.27%) compare to control seed (7.44%). The carbohydrate content of the space flight seed (9.02±1.87%) was lower than the control seed (15.2±2.79%). The result was similar to the result of the research on the hypocotyl of the *Arabidopsis,* which showed that space treated hypocotyl contained less cellulose and xyloglucan [12]. The cytokinin and gibberellin content of the space flight seeds were lower, but the auxin content was higher than control (table 1). Low cytokinin concentration could prevent mitosis [13]. Lower gibberellin content would slow down the germination process in the space flight seeds, while high concentration of auxin in the seeds would promote plant growth [4].

Table 1. The average concentration of auxin, cytokinin and gibberellin per gram tomato seeds.

Hormone	Control (ppm)	Space flight seed (ppm)
Auxin	142.055± 6.88	414.445 ± 78.84
Cytokinin	381.605± 68.86513	68.885 ± 9.53887
Gibberellin	335.01 ± 10.7	184.465 ± 7.4

The germination test showed that space seeds had lower rate of germination (73%) compare to control (83%). The pore on the seed coat could probably increase the water imbibition into the space flight seeds, but the low level of gibberellin might slow the process of germination. This result was supported by another research on *Arabidopsis* seeds [15], which stated that radiation disturbed the radicle and root formation in the seedlings.

The observation of the growth rate was presented in figure 3. After the fifth week, there was no significant difference between the space flight plants compare to the control. The higher growth rate during the sixth to the tenth week in space flight plant was probably caused by the higher auxin concentration in the seeds, which then recovered to the same growth rate after plant could produce its own auxin.

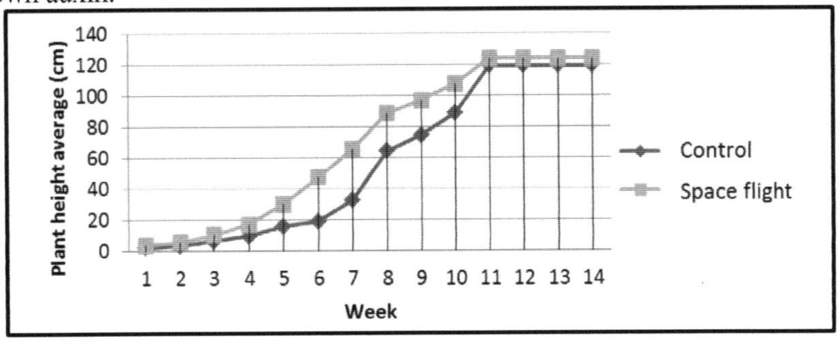

Figure 3. Tomato plant height during 14 weeks of field trials.

4. Conclusion

The physical characteristic of the space flight seeds support water imbibition but chemical content in the seed slowed the germination rate. The plant growth from the space flight seed was initially better than control, but at the end there was not significantly different.

Acknowledgements

Author would like to thank LAPAN, JAXA and Banana Group from ITB for their supports to this project. This project was funded by Research Grant ITB granted to Dr. Rizkita R. Esyanti.

References

[1] Setijo P 2009 *Benih Tomat* 5th Ed. (Yogyakarta: Kanisius)
[2] Guo LJ, Min LS, and Wang P 2000 *J.of Space Med. and Med. En.* **13**(2) 114-18
[3] Nechitailo GS, Jinying L, Huai X, Yi P, Chongqin T, and Min L 2005 *Adv.in Space Res.* **36** 1329-33.
[4] Hancock, T 2011 *Space Seeds' bring bigger harvests, Chinese scientists claim* http://www.zdnet.com/article/space-seeds-bring-bigger-harvests-chinese-scientists-claim/
[5] Ren W, Zhang Y, Deng B, Guo H, Cheng L, and Liu Y 2010 *African J. Biotech.* **9**(43) 7273-79.
[6] Sutadi D and Hanafi 2004 *Penuntun Praktikum Kimia Pangan* (Bogor: Departemen Perindustrian dan Perdagangan, Akademi Kimia Analis)
[7] Genkov T, Ivanov I and Ivanova I 2002 Laboratory of Bioorganic Synthesis *J. Plant Physiol.* **22** 3-4 and 94-104
[8] Karadeniz A, Topcoug F and Inan S 2005 *World J. of Microbiol. Biotech.* **22** 1061-64
[9] Norfadzin F, Ahmed OH, Shaharudin S, and Abdulrahman 2007 *J. of Agric. Res. Int.* **2**(7) 620-25
[10] Hammond EC, Bridgers K, and Berry FD 1996 *J. of Radiation Meas.* **26**(6) 851-6
[11] Cheng Z, Liu M, Zhang M, Hang X, Lei C, and Sun Y 2007 *Adv. Space Res.* **40**, 540-49.
[12] Hoson T, Komisaka S, Wakabayashi K, Sogamura Y, Mori R, Tanimoto E, Takeba G, Nishitani K, Izumi R, Ishioka N *et al* 2000 *J. of Biol. Sci. in Spaces* **14** (2) 75-96
[13] Staden JV, Zazimalova E and George EF 2008 *Plant Growth Regulators II in Plant Propagation by Tissue Culture 3 rd Ed.* (Netherlands: Springer) p. 213
[14] Campbell NA, Reece JB and Mitchell LG 2003 *Biologi Jilid 2 Fifth Edition* (Jakarta: Erlangga)
[15] Kranz A, Bork U, Bucker H, and Reitz G 1990 *J. of Radiation Appl. and Inst.* Elsevier **17** (2) 155-65

Relation between cloud thickness-cloud number concentration differences and rain occurrence based on Koren-Feingold model

R Sulistyowati[1,2], S Viridi[3], R Kurniadi[3] and W Srigutomo[1]

[1] Physics of Earth and Complex Systems Research Division, Physics Department, Faculty of Mathematics and Natural Sciences, Institut Teknologi Bandung, Jal an Ganesha 10, Bandung 40132, Indonesia
[2] Physics Education Department, Faculty of Teacher Training and Science Education, Universitas PGRI Palembang, Jalan Ahmad Yani, Lrg. Gotong Royong 9/10 Ulu, Palembang, Indonesia
[3] Nuclear Physics and Biophysics Research Division, Physics Department, Faculty of Mathematics and Natural Sciences, Institut Teknologi Bandung, Jalan Ganesha 10, Bandung 40132, Indonesia

E-mail: ritarahman08@gmail.com

Abstract. Koren-Feingold (KF) model, a model that relates growth of cloud thickness (H) and cloud number concentration (N) is discussed and analyzed in this work. Two boundary conditions are required by this model, where the first is cloud thickness potential H_0 and the second is aerosol concentration N_0. The initial conditions are simply $H(0) = 0$ and $N(0) = 0$. Several pairs of (H_0, N_0) values are chosen in calculating the precipitation. Three categories of rainfall are used in this work, which are no drizzle D0, light drizzle D1, and drizzle D2. As H evolves in time t, it produces maximum cloud thickness H_{max} and saturation cloud thickness H_{sat}, as also N does N_{max} and N_{sat}. Two kinds of cloud thickness difference, $\Delta H_1 = H_0 - H_{max}$ and $\Delta H_2 = H_0 - H_{sat}$ are proposed and also for cloud concentration difference ΔN_1 and ΔN_2 with similar definitions. Pairs of (ΔH_1, ΔN_1) and (ΔH_2, ΔN_2) are used in analyzing simulation results. The first pair can be used as prediction of rainfall occurrence, while the second pair is more for confirmation and understanding the relation between cloud thickness and cloud concentration in producing rainfall. It is observed that $H < H_0$ and $N < N_0$ are always fulfilled. Rainfall in category D2 will have significant differences in H and N, while in category D0 will not. Typical differences for H are about 80 % and 30-50 % for both cases. Deeper discussion about (ΔH_1, ΔN_1) and (ΔH_2, ΔN_2) is presented in this work.

1. Introduction

Aerosol-cloud interaction is a very complex process. It is an important key in understanding weather change, including the occurrence of rain [5, 1, 6]. However, the effect of aerosol particles on clouds and precipitation is still under debate [4, 3] and poorly described. Numerous researches, based on observation and numerical, have been intensively carried to improve better understanding about this interaction. One of the numerical model that can explain the phenomenon is a model initiated by Koren-Feingold [2]. This model consists of equations that describe the coupled interactions between cloud thickness (H) and cloud number concentration (N) at rain occurances. In this study, Koren-Feingold

(KF) model is used to calculate changes in cloud thickness (ΔH) and aerosol concentrations (ΔN) at different rainfall events. An analysis of the values of ΔH and ΔN are expected to explain the behavior of changes in cloud thickness and aerosol concentrations due to rain.

2. Methodology

KF model will be solved as equation (1), (2) and (3) below [2]:

$$\frac{dH}{dt} = \frac{H_0 - H}{\tau_1} + \dot{H}_r(t-T) \quad \text{with} \quad \dot{H}_r = -\frac{1}{c_1 H} R = -\frac{\alpha H^2}{c_1 N} \tag{1}$$

$$\frac{dN}{dt} = \frac{N_0 - N}{\tau_2} + \dot{N}_r(t-T) \quad \text{with} \quad \dot{N}_d = -c_2 NR \tag{2}$$

$$R(t) = \frac{\alpha H^3 (t-T')}{N(t-T')} \tag{3}$$

First equation described the growth rate of cloud, first term of right side indicates cloud growth before rain falls, while the second term serves the same but after rain falls. c_1 is a constant value as a function of cloud–base temperature and pressure; for warm adiabatic cloud $c_1 \cong 2 \times 10^{-3}$ m^{-1} and α is a constant related to warm stratocumulus clouds. H_0 shows carrying capacity of cloud, namely potential cloud thickness. So that if there is no rain, after constant time (τ_1), maximum amount of cloud growth H could reach asymtotic limit of H_0. In the equation (2), N_0 is assumed to be aerosol background concentration belong to the system (carrying capacity). τ_2 is constant time of aerosol growth characteristic and $c_2 \cong 3 \times 10^{-1}$ m^{-1} [2].

The used input is H_0 and N_0. Some input pair H_0 and N_0 are employed to calculate some rain (R) occurences and generate values of rainfall with several categories, namely: No Drizzle (D0), Light Drizzle (D1), and Drizzle (D2). For the initial conditions, $H(0) = 0$ and $N(0) = 0$ are set. As H evolves in time t, it produces maximum cloud thickness H_{max} and saturation cloud thickness H_{sat}, as also N does N_{max} and N_{sat}. Some parameters were used to analyze cloud thickness change and cloud number concentration (N) change, before and after rain. Those are defined as shown at the table 1 below:

Table 1. Parameter and definition of cloud thickness (ΔH) and cloud number concentration (ΔN) change

Parameter	Definition
ΔH_1	$\Delta H_1 = H_0 - H_{max}$
ΔH_2	$\Delta H_2 = H_0 - H_{sat}$
ΔN_1	$\Delta N_1 = N_0 - N_{max}$
ΔN_2	$\Delta N_2 = N_0 - N_{sat}$

3. Results and discussions

From the calculation of the KF model equation, we obtained the relationship between ΔH and ΔN as seen in figures 1 and 2 below

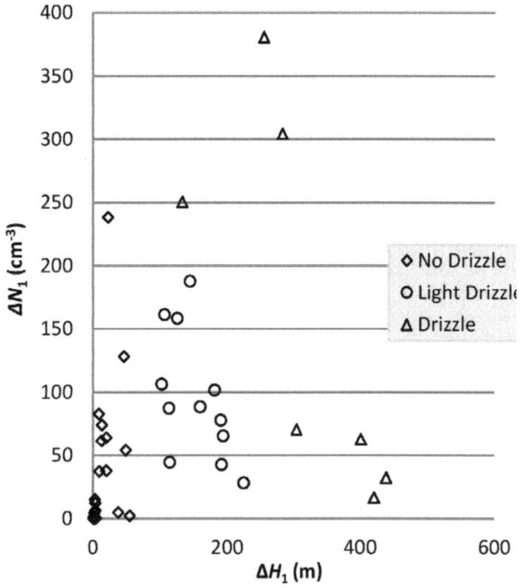

Figure 1. Relation between ΔH_1 and ΔN_1

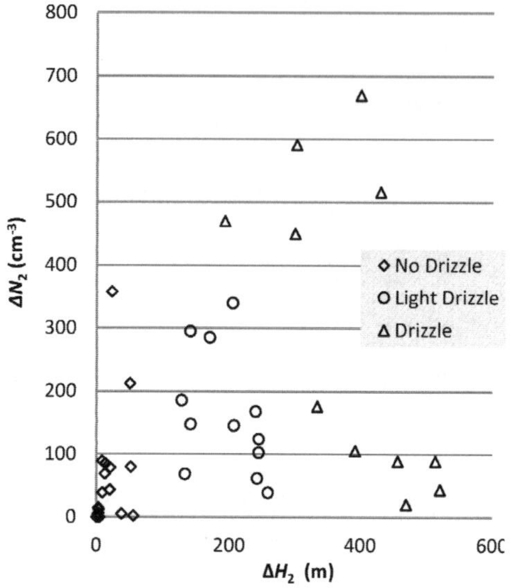

Figure 2. Relation between ΔH_2 and ΔN_2

Figure 1 shows the coupled relationship between ΔH and ΔN in different rain events. The condition of No Drizzle (D0) generally occurred in ΔH, both ΔH_1 and ΔH_2, less than 100 m. Light Drizzle (D1) condition occurs in ΔH between 100-200 m; while Drizzle (D2) condition occurs in ΔH>200 m. While, for all condition of the same ΔH, either D0, D1 or D2 generally occure at all range of ΔN. Based on this analysis, it can be stated that in prediction of the presence/absence of rain, ΔH_1 and ΔH_2 value have contribution more than the value ΔN_1 and ΔN_2.

Further, the change of cloud thickness and cloud number concentration at numerous rain occurences could be analyzed from percentage of N_{sat} decrease to N_{max}.

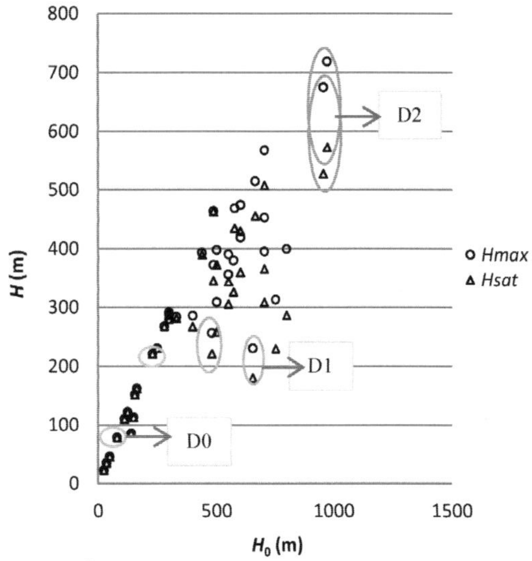

Figure 3. Plot of H_{max} and H_{sat} for some rain occurences

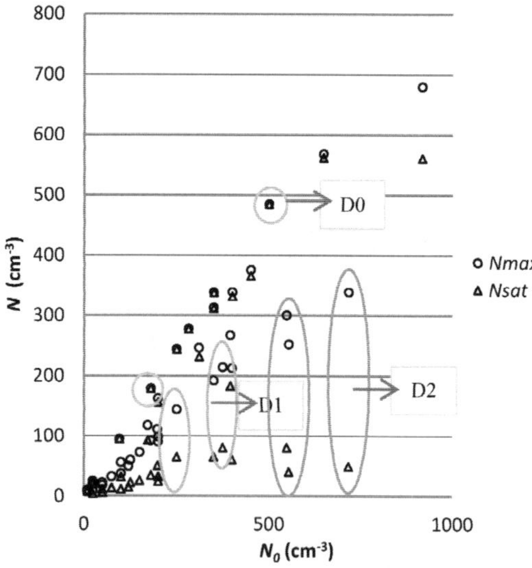

Figure 4. Plot of N_{max} and N_{sat} for some rain occurences

In the condition of No Drizzle (D0), there are no significant differences between the value ΔH_1 and ΔH_2, as well as between ΔN_1 and ΔN_2. Conversely, for the condition of Light Drizzle (D1) and Drizzle (D2), the difference in value ΔH_1 and ΔH_2, ΔN_1 and ΔN_2, increased. At conditions of Light Drizzle, the concentration of cloud number saturation (N_{sat}) due to rain occurences may decrease as much as 30-50% of the value N_{max}. While the condition of Drizzle, N_{sat} value may decrease until it reaches 80% of the value N_{max} that can be achieved.

4. Conclusion

KF models can be used to identify some rain occurences through the analysis of the value of ΔH and ΔN. For various categories of rain occurences, both value (ΔH_1, ΔN_1) or (ΔH_2, ΔN_2) showed the same pattern that increases linearly. Rain occurences are more likely to be influenced by ΔH than ΔN. Parameter ΔH_1 more relevant for the prediction of the presence/absence of rain (before the rain occurs) while ΔH_2 more relevant to investigate the character/kind of rain because the calculation of ΔH_2 requires H_{sat} value that can only be known after the rain.

Acknowledgments

This research is funded by Institut Teknologi Bandung Research Program 2016 (Division Research B).

References

[1] Houghton J T et al 2001 *IPCC Third Assessment Report Climate change 2001: The scientific basis* (Cambridge: Cambridge Univ. Press) p 994
[2] Koren I and Feingold G 2011 Aerosol–cloud–precipitation system as a predator-prey problem *Proceedings of the National Academy of Sciences* **108** 30 12227-32
[3] *NASA planning document. Studies of Emissions and Atmospheric Composition, Clouds and Climate Coupling by Regional Surveys* (SEAC4RS) 2012 https://www.espo.nasa.gov/pub/SEAC4RS-overview- 21OCT2010.pdf.
[4] National Research Council (NRC) 2005 *Radiative forcing of climate change: Expanding the concept and addressing uncertainties* (Washington DC: The National Academies Press)
[5] Ramanathan V *et al* 2001 *Science* **294** 5549 2119-24
[6] Tao W K et al 2012 Impact of aerosols on convective clouds and precipitation *Rev. Geophys.* **50** 2

The Effect of atmospheric humidity level to the determination of Islamic Fajr/morning prayer time and twilight appearance

Nihayatur Rohmah

Institut Agama Islam (IAI), Ir. Soekarno 99 Beran Ngawi East Java Indonesia

nihayaturrohmah1@gmail.com

Abstract. Islamic prayer times are based on the astronomical position of the Sun in the sky. One of them is the Fajr prayer. It is marked by some indicators in the morning twilight which is white light spread in the Eastern horizon. However, determining the true time of twilight can be difficult. One of the reasons is the effect of atmospheric humidity to the appearance of morning twilight. The higher the humidity, the redder twilight sky appearance. This paper discusses this effect. It is shown that despite of the same Sun's position, sky color can vary considerably. Observations of various solar dip angle have been made to study this effect. Visibility for different angle can change accordingly. We obtained that the average solar dip for Fajr prayer by morning twilight images was -18°39'29.4".

1. Introduction

The twilight is divided into civil, nautical and astronomical twilight corresponding to solar elevation angle of -6, -12 and -18 degrees, respectively [1]. Islamic Fajr time is related to the dawn or morning twilight. However, no clear definition of the dawn. If it is interpreted as the time of the first light, then the dawn will correspond to -12 to -18° solar elevation angle, but it is not possible to be more precise. The rise of morning twilight for the beginning of fajr (namely the true fajr or al-fajr as-sadiq) could be discerned on the horizon by the appearance of a light, which ascends towards the sky and resembles a tail of the fox--this is known as al-fajr al-kazib (the false dawn). Then the light spreads on the horizon (and becomes like a white cotton) in such a way that whenever you look towards it. In other words, it is the spreading of the light on the horizon after it had been ascending towards the sky [2].

In meteorology, visibility is a measure of the distance at which an object or light can be clearly discerned. Visibility is also affected by local air pollution, which strongly corresponds with aerosol. Visibility degradation is caused by the absorption and scattering of light by particles, aerosol and gases in the atmosphere. A further reduction of visibility occurs as these aerosol particles grow. Moreover, much of the aerosol that contributes to is hygroscopic. This means that that in conditions of high humidity the aerosols grow by absorbing water vapor, although not to the extent that occurs in fog formation. Dry aerosol particles scatter light, and this scattering increases above the deliquescence point, where solid particles absorb water and turn into fine droplets, at around 70-80% relative humidity. Hence, visibility may be low on days with high aerosol loading and when humidity is high. Since rain scavenges aerosol particles from the atmosphere, some of the best visibility days occur after strong rain events. Therefore, color appearance of the sky and transparency or visibility in the atmosphere could be connected to relative humidity, temperature, and atmospheric pressure [3][4].

The purpose of this paper is to determine the corresponding position of the sun in relation to the aforementioned definitions of dawn; and to construct color model examples at the time twilight appears, at various levels of atmospheric humidity.

2. Method

Morning twilight observations were carried out with the naked eyes and by digital camera or images recording. For naked eyes observations, the observer needs to have the eyes in full pupil dilation and as accommodated to low-light (dark) as possible. For image recordings, we need to specifically regulate the camera's shutter speed (1/4 - 5 sec.), ISO levels (400 - 800), aperture (f/2 - f/5), resolution, focal length, and so on, such that the camera can record subtle light changes in a very dark sky. Several locations have been selected for the purpose of comparison of different levels of altitude, local relative humidity and visibility that were Juwiring, Pati, Yogyakarta (Central Java), Kaibon (East Java), and Bandung (West Java). Images of the sky were taken in the east and near horizon direction. Timing has been made just before sunrise to sunrise ahead with no moonlight interference.

3. Results and discussions

Figure 1 shows the splendid colors of twilight with different measured relative humidity. Based on the image data, we found a relation between humidity and dawn color. From left to right, the humidities are 50%, 80%, and 95% respectively. The higher humidity, the more pronounced are the orange-reddish colors. Colors are caused by scattering of sunlight by the particle and gas constituents in the atmosphere. Colors come about because the amount of scattering varies with wavelength. Since scattering characteristics depend on particle density, size and material, different atmospheric conditions are responsible for numerous varieties of twilight color displays.

Figure 1. The dark horizon and orange sky tint seen above from left to right: a.) 50% level of humidity, 23 August 2010 b.) 80% level of humidity, 19 August 2013, c.) 95% level of humidity, 29 July 2013.

Table 1 gives summaries of observational results from morning twilight images. Determination of the time of true fajr (column 7) is obtained by images analysis with an indicator of light that spreads along the horizon. We have applied simple procedures to detect intensity variation near horizon from array of pixels as pixels profile. We took a note for the time of increasing intensity. Elevation angle of the Sun (column 8) at the time of increasing intensity is calculated by using Accurate Time version 5.1, a freeware developed by Mohammad Odeh of the Jordanian Astronomical Society (JAS). The minimum and maximum of solar dip were resulted by -15°30'46" and -24°53'20", respectively. The average value of the angle of elevation of the Sun can be obtained as -18°39'29.4". It is slightly deeper dip than defined astronomical twilight.

Table 1. The effect of humidity on the Sun's elevation angle of the true dawn

No 1	Date 2	Location 3	Elevation 4	Temperature (°C) 5	Humidity (%) 6	Time of fajr (UT+7) hms 7	Altitude (dms) 8
1.	11/10/2010	Juwiring	111 m	17.7	86	04:02:09	-19:9:05
2.	21/07/2010	Juwiring		18.6	89	04:31:10	-19:33:50
3.	23/07/2010	Juwiring		18.1	85	04:37:44	-18:02:08
4.	27/07/ 2013	Juwiring		18.9	89	04:25:36	-20:52:29
5.	29/07/ 2013	Juwiring		19.3	94	04:22:03	-21:41:48
6.	01/08/ 2013	Juwiring		17.9	89	04:33:01	-19:02:45
7.	19/08/ 2013	Juwiring		18.5	83	04:05:16	-24:53:20
8.	20/08/ 2013	Juwiring		18.2	81	04:34:21	-17:43:47
9.	22/07/ 2012	Pati	1 m	20.3	90	04:27:03	-19:50:38
10.	23/07/ 2012	Pati		21.6	79	04:37:23	-17:25:34
11.	22/09/ 2012	Pati		22.6	79	04:19:56	-17:03:02
12.	23/09/ 2012	Pati		22.6	79	04:16:49	-17:41:27
13.	19/08/2010	Kaibon	10 m	28.4	71	04:19:27	-20:41:44
14.	20/08/2010	Kaibon		27.9	66	04:29:07	-18:15:55
15.	23/08/2010	Kaibon		27.0	77	04:20:17	-20:10:07
16.	07/02/2011	Kaibon		26.8	88	Thick cloud	-
17.	09/02/2011	Sedoro		25.3	88	Thick cloud	-
18.	06/07/2011	Merbabu	3100 m	18.8	91	04:39:04	-17:33:27
19.	18/09/2012	Yogjakarta	1 m	27.0	74.5	04:20:58	-18:03:55
20.	11/10/2010	Bandung	1200 m	20.7	86	04:36:10	-14:02:34
21.	12/10/2010	Bandung		20.7	86	04:28:15	-15:51:27
22.	15/10/2010	Bandung		20.7	86	04:28:10	-15:30:46

4. Conclusions

Based on the image analysis of morning twilight from several locations with different relative humidity, we can conclude that the higher humidity, the twilight sky colors became orange-reddish. Also, visibility was influenced by the level of humidity. The beginning of Fajr prayer can be derived by averaging solar elevation angle of defined the true dawn was -18°39'29.4".

Acknowledgements

The author would like to thank Prof. Dr. Thomas Djamaluddin (LAPAN, Indonesia), Dr. Dhani Herdiwijaya (Astronomy, ITB Indonesia), and the anonymous reviewers for their valuable comments and suggestions.

References

[1] United States Naval Observatory Astronomical Applications Department 2005 *Multiyear interactive computer almanac, 1800–2050* (USA: Willmann-Bell)
[2] Rozenberg G V 1966 *Twilight: A Study in Atmospheric Optics* (New York: Plenum Press) p 160
[3] Brimblecombe P 1981 Long-term trends in London Fog Science of the Total Environment **22** 19
[4] Seinfeld J H, Pandis S N 2006 *Atmospheric Chemistry and Physics - From Air Pollution to Climate Change* 2nd ed. (John Wiley and Sons) ISBN 0-471-82857-2

Analysis of high altitude clouds in the martian atmosphere based on Mars Climate Sounder observations

L Puspitarini[1], A Määttänen[2], T Fouchet[3], A Kleinboehl[4], D M Kass[4], J T Schofield[4]

[1]Bosscha Observatory and FMIPA Institut Teknologi Bandung Indonesia
[2]LATMOS/IPSL, UVSQ Université Paris-Saclay, UPMC Univ. Paris 06, CNRS, Guyancourt, France
[3]LESIA Observatoire de Paris, Meudon, France
[4]Jet Propulsion Laboratory, Caltech, Mail Stop 169-237, 4800 Oak Grove Dr, Pasadena, CA 91109

E-mail: lucky.puspitarini@gmail.com

Abstract.
High altitude clouds have been observed in the Martian atmosphere. However, their properties still remain to be characterized. Mars Climate Sounder (MCS) aboard Mars Reconnaissance Orbiter (MRO) is an instrument that measures radiances in the thermal infrared, both in limb and nadir views. It allows us to retrieve vertical profiles of radiance, temperature and aerosols. Using the MCS data and radiative transfer model coupled with an automated inversion routine, we can investigate the chemical composition of the high altitude clouds. We will present the first results on the properties of the clouds. CO_2 ice is the best candidate to be the main component of some high altitude clouds due to the most similar spectral variation compared to water ice or dust, in agreement with previous studies. Using cloud composition of contaminated CO_2 ice (dust core surrounded by CO_2 ice) might improve the fitting result, but further study is needed.

1. Introduction

Martian climate has been an issue of scientific curiosity for centuries and that clouds exist in Martian atmosphere has been known for some considerable time. In the past 10 years, surprising high altitude clouds (50-100 km) above the Martian surface have been discovered (see [1, 2, 11, 10, 7]). These clouds are most likely composed of carbon dioxide (CO_2) (see [10]), but their chemical properties still remain to be characterized.

Mars Climate Sounder (MCS) aboard the Mars Reconnaissance Orbiter (MRO) spacecraft (see [8, 13]), is an instrument that measures radiances in nine different channels in the thermal infrared, both in limb and nadir views (see Fig. 1). It has been able to retrieve vertical profiles of temperature and aerosols (see [6]). The high altitude clouds have also been observed using MCS (see [5, 12]). We will describe the first results on the properties of the high altitude clouds based on MCS data and using radiative transfer model coupled with an automated inversion routine developed by [4].

Table 1. MCS Channels, band passes, band center, and main absorbers (Table is adapted from [8, 6]).

Channel	Band Pass (cm^{-1})	Band Center (μm)	Main Absorbers
A1	595-615	16.5	CO_2
A2	615-645	15.9	CO_2
A3	635-665	15.4	CO_2
A4	820-870	11.8	H_2O ice
A5	400-500	22.2	dust
A6	3300-33000	1.65	
B1	290-340	31.7	dust
B2	220-260	41.7	H_2O vapor, H_2O ice
B3	230-245	42.1	H_2O vapor, H_2O ice

2. Mars Climate Sounder (MCS) observations

The MCS observes in 9 channels across the visible and infrared ranges of the electromagnetic spectrum (0.3 - 45 μm). Those channels, from two telescopes of the MCS, are A1-6, and B1-3 (see Fig. 1 and Table 1). A1-3 channels are sensitve to CO_2 gas, A4 is to water ice, A5 is to dust, whereas B1-3 are to dust, water vapor, and water ice. As previously mentioned, the MCS acquires nadir and limb observations. We used results only from the limb observation (the vertical profiles). Each channel of MCS is composed of 21 detectors. In limb observation, each detector has a vertical resolution of $\sim$ 5 km. Thus, it can observe from the surface to the high altitude. Position of a detector corresponds to a certain altitude in Martian atmosphere.

We used data from the first MCS observations that are during Mars Year (MY) 28, Season 4 or Solar Longitude 115-133. From $\sim$ 200 selected vertical profiles, we detected $\sim$ 50 observations with evidence of clouds. The Martian clouds are characterized by a local maximum at high altitude in the radiance profiles due to the emission and scattering process by the clouds (see Fig. 2).

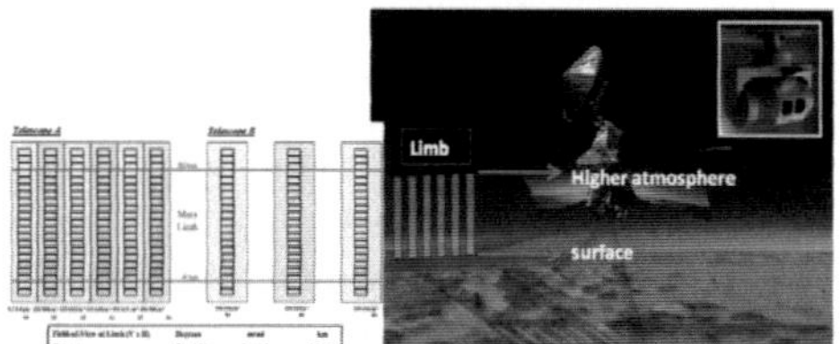

Figure 1. The nine channels of MCS Instrument A1-6 and B1-3. Each channel has 21 detectors. In limb observation, position of a detector corresponds to a certain altitude (image is taken from [8]).

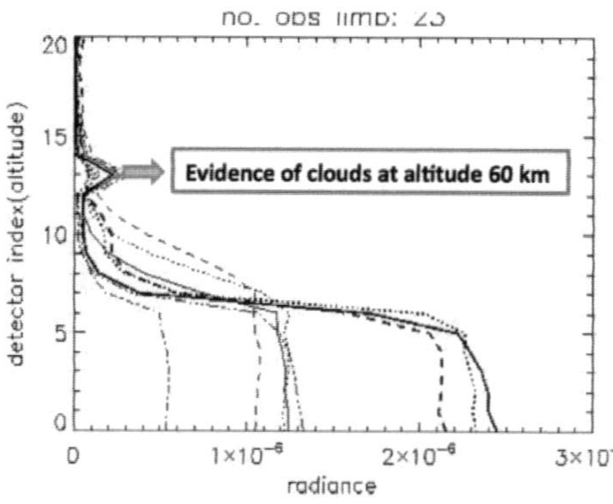

Figure 2. An evidence of the high altitude clouds is a local maximum in the vertical profile of radiance from limb observation of MCS. Different line types correspond to different channels.

Journal of Physics: Conference Series **771** (2016) 012049 doi:10.1088/1742-6596/771/1/012049

3. Method

To analyze the properties of the clouds, we use the MCS data together with radiative transfer model, coupled with automated inversion routine (see [4]). There are two codes in the analysis: (1) radiative transfer code (direct model) to calculate the radiance profile, and (2) inversion code that adjust the radiance profile to retrieve temperature and aerosols profiles. We started with temperature and aerosols (dust and water ice only) profiles from Mars Climate Database (MCD) (see [3]) and added a pre-defined CO_2 ice profile to get a realistic input for the inversion routine (*a priori* profile). Optical parameters (extinction factor, scattering albedo, etc) were obtained using T-Matrix code (see [9]). We selected a radiance profile from MCS data that showed the signature of the high altitude clouds. We ran the inversion code that calculated synthetic radiance profile and compared it with the observed radiance profile. The code can vary the temperature and dust profiles to adjust the radiance profile. Aerosols information in the inversion routine is noncompulsory. However, since aerosols make a significant contribution to the radiance profile, they should be included. From the inversion, we can obtain the new vertical temperature and aerosols profiles. We used these new profiles from the inversion and added with cloud layer (CO_2 ice, water ice, or dust profil at high altitude) to adjust the radiance at the cloud altitude.

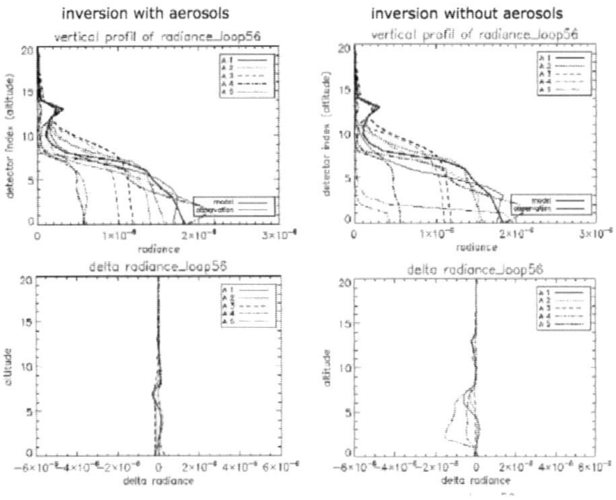

Figure 3. A test of the inversion routine. Upper figures are the vertical radiance profiles (detector index vs. observed and synthetic radiances of different channels shown in different line types; the synthetic radiances are shown in bold lines). Lower figures show the difference between the observed and the synthetic radiance. The need to include aerosols in the inversion can be clearly seen from this figure (see lower right figure).

4. Results and discussions

- An example of the fitting from the inversion routine can be seen in Fig. 3 (upper left).
- We tested the inversion routine with and without taking aerosols into account. We remark that aerosols play important role in modifying thermal structure of martian atmosphere, especially at the lower altitude (see Fig. 3). We noticed strong differences between the observed and the obtained radiance profile of channel A4 and A5, showing the non-negligible quantity of water ice and dust at the lower altitude (see Fig. 3, lower right).
- CO_2 ice is the best candidate to be main component of the high altitude clouds due to the most similar spectral variation compared to water ice or dust (see Fig. 4).

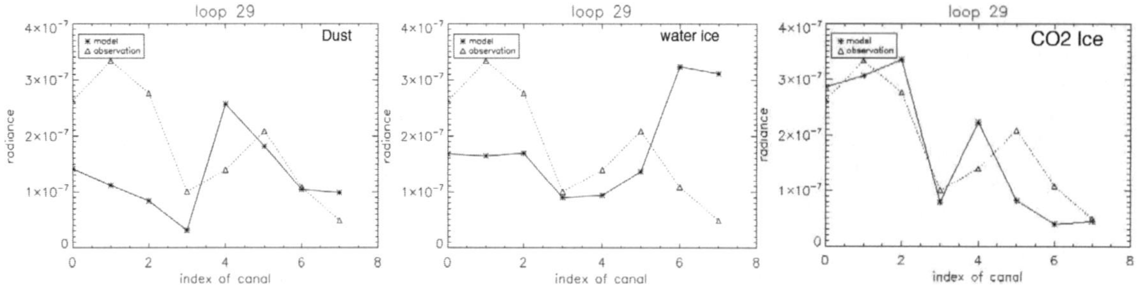

Figure 4. Spectral variations of radiance at the altitude of the clouds as a function of the MCS channels (A1-5, B1-3). The figure shows a test of the routine for searching the most probable composition of the high altitude clouds. The CO_2 ice has the most similar spectral variation to the observed variation and thus is our primary candidate for the cloud composition in this observation.

- From the spectral variations of radiance at the altitude of the cloud layer (see Fig. 4, right), we remark that our fitting results tend to be under- or overestimate in some channels. This suggest that properties of the clouds might be slightly different from our prediction. For example, they might not be pure CO_2 ice. Dust core surrounded by CO_2 ice might improve the fit, but further study is needed.
- Because of the poor fits to the radiances, we could not derive effective radius (r_{eff}) of the cloud, but approximations with $r_{eff} = 1.5$, 2.3 and 3 μm show that the radiance model is quite similar to the observed radiance.

To improve our results, we need a more complex calculation to estimate optical properties of the component of the cloud.

References

[1] Clancy R T and Sandor B J 1998 *Geophys. Res. Lett.* **25 (4)** 489-492
[2] Clancy R T, Wolff M J, Whittney B A, Cantor B A, Smith, M D 2007 *Journal of Geophysical Research: Planets* **112 E04004**
[3] Forget F, Hourdin F, Fournier R, Hourdin C, Talagrand O, Collins M, Lewis S R, Read P L, Huot J-P 1999 *Journal of Geophysical Research: Planets* **104 (E10)** 24155-24176.
[4] Fouchet T *et al* 2007 *Icarus* **190** 32-49
[5] Kass D M, Abdou W A, McCleese D J, Schoeld J T, Määttänen A 2011 *MCS climatology of detached, localized haze layers In: Forget, F., Millour, E. (Eds.) Fourth International Workshop on the Mars Atmosphere: Modelling andObservations* 400-403
[6] Kleinboehl A *et al* 2009 *Journal of Geophysical Research* **114** 1-30
[7] Määttänen A *et al* 2010 *Icarus*, **209**, 452-469
[8] McCleese D J, Schofield J T, Taylor F W, Calcutt S B, Foote M C, Kass D M, Leovy C B, Paige D A, Read P L, Zurek R W 2007 *Journal of Geophysical Research* **112** 1-16
[9] Mischenko M I, Travis L D, Lacis A A 2002 *Scattering, Absorption, and Emission of Light by Small Particles* (Newyork: Cambridge University Press)
[10] Montmessin F, Gondet B, Bibring J-P, Langevin Y, Drossart P, Forget F, Fouchet T 2007 *Journal of Geophysical Research* **112 (E11)**
[11] Montmessin F *et al* 2006 *Icarus* **183** 403-1410
[12] Sefton-Nash E, Teanby N A, Montabone L, Irwin P G J, Hurley J, Calcutt S B et al. 2013, *Icarus* **222** 342
[13] Zurek R W and Smrekar S E 2007 *Journal of Geophysical Research* **112 (E5)**

Journal of Physics: Conference Series **771** (2016) 012050 doi:10.1088/1742-6596/771/1/012050

Perturbation of circumsolar dust ring on stability of Sun-Earth triangular libration points

B Dermawan

Astronomy Research Division, Faculty of Mathematics and Natural Sciences, Bandung Institute of Technology, Jalan Ganesha 10, Bandung 40132, Indonesia

E-mail: budider@as.itb.ac.id

Abstract. Asteroid collisions and cometary outgassing produce grains and dust that cause to fill the interplanetary space. The particles are temporarily trapped into orbit of the Earth to develop a circumsolar dust ring. The ring encloses the Sun-Earth equilateral triangular libration points in terms of Restricted Three-Body Problem (RTBP). On the other hand, regions close to the triangular points are preferable as locations for placing astronomical satellite. Here we study the planar-circular RTBP of Sun-Earth system with considers the oblateness of Earth and the presence of a circumsolar dust ring. Perturbation of the dust ring on linear stability of the points is discussed and we find that the points are still stable. However, in general the presence of the circumsolar (resonant) dust ring decreases value of the critical mass parameter.

1. Introduction

It has been known that Earth has a circumsolar dust ring of asteroidal and cometary origin [1]. The dust and grains are temporarily trapped into orbit of the Earth (figure 1). The ring embeds Earth along orbiting the Sun. From observations taken by on board infrared instruments of some satellites there is an asymmetry zodiacal cloud that envelops the inner solar system. The asymmetry of the zodiacal cloud can be accounted for this dust ring. The cloud has micrometer-sized particles and can be removed by drag forces (Poynting-Robertson and solar wind) that eventually spiral in the particles towards the Sun.

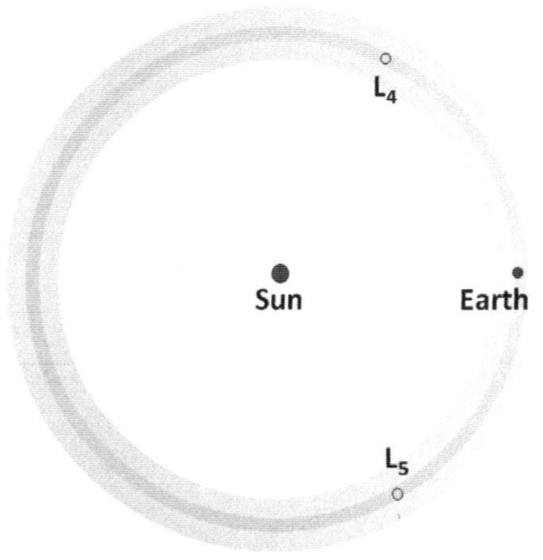

Figure 1. Illustration of circumsolar dust ring. Darker color represents higher number particle density.

The dust ring is located in Earth's orbit and consequently, it covers the Sun-Earth equilateral triangular libration (Lagrangian) points L_4 and L_5 (see figure 1) with respect to the Restricted Three-Body Problem (RTBP). The triangular points are considered as future potential and promising locations for hovering satellites such as that for Earth-Moon system [2]. There are at least two scientific interests for placing satellites around the Sun-Earth triangular points. First, the points are ideal locations for monitoring the space-weather away from the Sun-Earth line; and second, the L_4 and L_5 regions may harbor asteroids [3] that of significant interest to small body experts. Recent study [4] describes an attractive behavior and transfer mechanism between the triangular points.

Because the triangular points may efficiently harbor satellite for various astronomical purposes, understanding accurate locations of the triangular points becomes important. Singh and Taura [5,6] considered effects of radiation, oblateness, and the presence of circular cluster of material points on the equilibrium points in RTBP. They found that locations of the triangular points are influenced by the additional effects. Here we study the Sun-Earth triangular libration points in Planar-Circular RTBP under the influences of oblateness of Earth and the presence of circumsolar dust ring.

2. Equations of motion

Classical RTBP studies motion of a third mass-less body under the gravitational forces of the point-mass primaries (massive bodies). Orbits of the primaries around their mutual center of mass are circular and the third body has a planar orbit. In this work the classical RTBP is modified by including additional effects. Potential due to the dust ring [7] is:

$$\frac{m_r}{\sqrt{r^2+d^2}},$$

where m_r is the total mass of the dust, r the radial distance of the third infinitesimal body, and d the parameter that represents the flatness and the core size of the dust ring.

In rotational coordinate system, locations of the Sun, Earth, and the third infinitesimal body are, respectively, (ξ_1,η_1), (ξ_2,η_2), and (ξ,η). Under the effects of oblateness and circular dust ring, equations of motion of the third body in barycentric and dimensionless coordinate system $\left(\bar{\xi}=\xi/1\,\text{au}, \bar{\eta}=\eta/1\,\text{au}\right)$ are:

$$\frac{d^2\bar{\xi}}{df^2}-2\frac{d\bar{\eta}}{df}=\frac{d\bar{V}}{d\bar{\xi}},\quad \frac{d^2\bar{\eta}}{df^2}+2\frac{d^2\bar{\xi}}{df}=\frac{d\bar{V}}{d\bar{\eta}},$$

where the potential

$$\bar{V}=\frac{n^2}{2}\left(\bar{\xi}^2+\bar{\eta}^2\right)+\frac{(1-\mu)}{r_1}+\frac{\mu}{r_2}\left(1+\frac{A_E}{2r_2^2}\right)+\frac{m_r}{\sqrt{r^2+d^2}}, \tag{1}$$

with

$$n^2=1+\frac{3}{2}A_E+\frac{2m_r r_c}{\left(r_c^2+d^2\right)^{3/2}},\quad \mu=\frac{m_E}{m_S+m_E},$$

$$r_i^2=\left(\bar{\xi}-\bar{\xi}_i\right)^2+\left(\bar{\eta}-\bar{\eta}_i\right)^2,\ i=1,2;$$

A_E is Earth's oblateness coefficient derived from its equatorial and polar radii [5], n is known to be the mean motion, r_c is radial distance of the third body in the classical RTBP, μ is the mass parameter with m_E is Earth mass and m_S is mass of the Sun.

3. Location of triangular equilibrium points

Conditions of equilibrium are satisfied when the partial derivatives of components of the potential (1) are zero: $\overline{V}_{\overline{\xi}} = \overline{V}_{\overline{\eta}} = 0$. After some algebra, the conditions produce coordinate of the triangular equilibrium points $(\overline{\xi}_0, \overline{\eta}_0)$, i.e. $\overline{\xi}_0 \neq 0$ and $\overline{\eta}_0 \neq 0$:

$$\overline{\xi}_0 = \frac{1}{2} - \mu - \frac{A_E}{2},$$
$$\overline{\eta}_0 = \pm\sqrt{3}\left[\frac{1}{2} - \frac{A_E}{6} - \frac{2m_r(2r_c - 1)}{9(r_c^2 + d^2)^{3/2}}\right]. \tag{2}$$

(+ for L_4 and – for L_5)

In classical RTBP coordinates of the triangular points are $\left(\frac{1}{2}, \pm\frac{\sqrt{3}}{2}\right)$. From (2) it is clear that effects of Earth oblateness and the dust rings shift locations of the triangular points of classical RTBP. Earth oblateness affects both abscissa and ordinates of the triangular points. While the mass parameter changes only the abscissa of the points, the dust ring (mass and physical properties) influences only on the ordinates.

4. Linear stability and critical mass parameter

To examine the linear stability of the triangular equilibrium points, small displacements to the points are added: $u = \overline{\xi} - \overline{\xi}_0$, $v = \overline{\eta} - \overline{\eta}_0$. By taking linear part of Taylor expansion, the equations of motion become

$$\ddot{u} - 2\dot{v} = \overline{V}_{\overline{\xi}\overline{\xi}} u + \overline{V}_{\overline{\xi}\overline{\eta}} v, \quad \ddot{v} + 2\dot{u} = \overline{V}_{\overline{\xi}\overline{\eta}} u + \overline{V}_{\overline{\eta}\overline{\eta}} v, \tag{3}$$

where dots represent derivatives with respect to time (t). Arranging $u = C_1 e^{\lambda t}$ and $v = C_2 e^{\lambda t}$, where C_1 and C_2 are constant coefficients, and second partial derivatives of the potential (1) with respect to its components, yields the corresponding characteristic equation

$$\lambda^4 + \left(4 - \overline{V}^0_{\overline{\eta}\overline{\eta}} - \overline{V}^0_{\overline{\xi}\overline{\xi}}\right)\lambda^2 + \overline{V}^0_{\overline{\xi}\overline{\xi}}\overline{V}^0_{\overline{\eta}\overline{\eta}} - \left(\overline{V}^0_{\overline{\xi}\overline{\eta}}\right)^2 = 0,$$

that the second partial derivatives are evaluated at the triangular points $(\overline{\xi}_0, \overline{\eta}_0)$ to obtain the characteristic's roots (λ). Distinct pure imaginary roots reveal stable motion of the third body around the triangular equilibrium points.

By taking the discriminant of the characteristic equation equals zero, the critical mass parameter (μ_c) can be derived,

$$\mu_c = \frac{1}{2}\left(1 - \sqrt{\frac{23}{27}}\right) - \frac{1}{9}\left(1 - \frac{13}{\sqrt{69}}\right)A_E + \left[\frac{3}{2} - \frac{83 + 12r_c^2}{6\sqrt{69}} + \frac{(76 - 8r_c)(r_c^2 + d^2)}{27\sqrt{69}}\right]\frac{m_r}{(r_c^2 + d^2)^{5/2}}. \tag{4}$$

Motion of the third infinitesimal body is stable when $0 \leq \mu \leq \mu_c$.

From (4), by inserting values of A_E, r_c, m_r, and d, we obtain value of μ_c. Comparing value of μ for Sun-Earth system to the μ_c, we find that μ is much less than μ_c. Therefore, motion of the third body, such as a satellite, close to the Sun-Earth triangular equilibrium points remains stable under the influence of the Earth oblateness and a circumsolar dust ring. In general, the denser the dust ring, the lower value the critical mass parameter.

5. Conclusions

Linear stability of the Sun-Earth triangular equilibrium points has been studied by including effects of Earth oblateness and circumsolar dust ring. Both effects shift coordinate of the triangular points compared to that of the classical one. Particularly, effect of the circumsolar dust ring changes slightly the ordinates of the points. We also find that the denser the dust ring, the lower value the critical mass parameter. Although under the influence of a circumsolar dust ring, motion of the third infinitesimal body, such as a satellite, remains stable around the triangular equilibrium points.

References

[1] Dermott S F, Jayaraman S, Xu Y L, Gustafson B A S, Liou J C 1994 *Nature* **369** 719
[2] Catlin K A and Craig A. McLaughlin C A 2007 *Journal of Spacecraft and Rockets* **44** 660
[3] Connors M, Wiegert P and Veillet C 2011 *Nature* **475** 481
[4] Llanos P J, Hintz G R, Lo M W and Miller J K 2013 *Journal of Earth Sci.and Eng.* **3** 515
[5] Singh J and Taura J J 2014a *Astrophys. Space Sci.* **349** 681
[6] Singh J and Taura J J 2014b *Astrophys. Space Sci.* **350** 127
[7] Miyamoto M and Nagai R 1975 *Publ. Astron. Soc. Japan* **27** 533

Study of stability of mean-motion resonances in multiexoplanetary systems

M Handayani[1] and B Dermawan[2]

[1] Undergraduate Program in Astronomy, Faculty of Mathematics and Natural Sciences, Institut Teknologi Bandung, Jalan Ganesha 10, Bandung 40132, Indonesia
[2] Astronomy Research Division, Faculty of Mathematics and Natural Sciences, Institut Teknoologi Bandung, Jalan Ganesha 10, Bandung 40132, Indonesia

E-mail: budider@as.itb.ac.id

Abstract. Many exoplanetary systems have been found to harbour more than one planet. Some of them have commensurability in orbital periods of the planets (resonant-planet pair). The aim of this work is to analyse the stability of resonant-planet pair configuration in two multiexoplanetary systems which have two planets in near mean-motion resonances, i.e. Kepler-9 and HD 10180 systems. This work considers numerical and comparative empiric-analytical studies. Numerical studies are performed using the integrator package SWIFT with an integration time of 10 Myr. Results from numerical integrations indicate that all orbital solution sets of the systems are stable. Further numerical explorations also demonstrate that the systems are stable for small perturbations in the orbital elements and mass variations. Analyses of stability based on comparative empiric-analytical are done by applying a known stability criterion to all systems. We find that all systems tend to be stable.

1. Introduction

Exoplanets are known common throughout the universe. Some of them belong to multiplanetary systems. Study about their dynamical behaviours becomes one of the topic of interests in celestial mechanics, which are associated with orbit and stability of the systems. Orbital studies of planetary systems show commensurabilities among their orbits, known as resonances, which are also common in the Solar System. Near-resonances phenomena are also found in multiexoplanetary systems [1]. One of these resonances is two-planet mean-motion resonance (resonant-planet pair). Analyses of resonant-planet pair provide an idea about prediction of planets' orbital configuration with respect to its stability.

There are two approaches to study the stability of resonant-planet pair configuration. First is numerical study using N-body integration. Numerical study calculates orbital evolution of a system during a given integration time or a time-scale, so that its stability can be decided. The stability of resonant-planet pair is indicated by evolution of orbital elements together with libration/circulation of the resonant arguments [2]. Second, the stabillity conditions of resonant-planet pair can be determined by applying a known empiric-analytical solution of the stability of three-body problem [3].

There are many known exoplanetary systems in near-resonances. This study focuses on Kepler-9 and HD 10180 systems. Kepler-9 is a single-star (1.07 Solar mass), known to have three planets in the system. Two outer planets (b and c) are claimed to be in 2:1 near-resonance [1]. Larger system, like HD 10180 system (having six confirmed planets) has two planets (d and e) which are in 3:1 near-resonance

[1]. The HD 10180 star is also known a single-star (1.06 Solar mass). With current technology and methods, undiscovered planets in Kepler-9 and HD 10180 systems are believed to have smaller masses. Therefore, their perturbations to orbits of the known planets should be small.

2. Numerical studies

The SWIFT package [4] is commonly used in celestial mechanics studies. The package is designed to integrate a set of mutually gravitationally interacting bodies together with a group of test particles which feel the gravitational influence of the massive bodies but do not affect each other or the massive bodies. We used exoplanetary data consisting of the usual a set of six Keplerian orbital elements (semi-major axis a, eccentricity e, inclination i, longitude of ascending node Ω, argument of pericenter ω, mean anomaly M) that was then transformed into a set of Cartesian orbital elements to suit the SWIFT codes. We set values of Keplerian orbital elements when they were not available.

2.1. Data

Data for orbital elements and physical parameters of Kepler-9 and HD 20180 systems were taken from some references [5, 6]. Using TRADES software [5] two sets of orbital solutions for Kepler-9b and c are provided. The initial integration conditions for both of them were different, leading to the differences of their derived planet's mass. As for set solution 1, they used TRADES with initial parameters and Transit Times from [7], while set solution 2 using TRADES+PSO+LM [5] with Transit Times from [8]. The given solutions include planetary mass m in Jupiter mass (M_{Jup}), orbital period T, e, Ω, ω and M. We eventually derived value of a using the Kepler Third's Law.

A set of orbital solutions for six confirmed planets in HD 10180 system are given [6]. Because of using radial velocity method, the planetary masses are considered as minimum ($m \sin i$). True masses are typically higher by about 15% due to the geometric effects of inclination. We derived value of M using the given epoch and time of pericenter passage.

The parameters are presented in Table 1. Inclinations for both systems are unknown, so we set the values to zero. Unknown values of Ω are also set to zero.

Table 1. Orbital solutions and masses of Kepler-9 and HD 10180 systems.

Planets	a (au)	e	Ω(°)	ω (°)	M (°)	m ($M_{\rm Jup}$)
Kepler-9b#	0.144	0.131 (0.058)	-	18.9 (356.1)	333.8 (3.8)	0.246 (0.137)
Kepler-9c#	0.230	0.119 (0.068)	1.6 (~0)	102.9 (167.6)	7.5 (307.4)	0.169 (0.094)
HD 10180d*	0.129	0.131	-	325.0	37.5	0.038**
HD 10180e*	0.270	0.051	-	147.0	168.8	0.081**

Values in parenthesis are for the set solution 2 [5]. # [5], * [6], ** $m \sin i$

2.2. Integration scheme

We did not include the rotational motions of the bodies and only took planets that have been confirmed in http://exoplanetarchive.ipac.caltech.edu/ on January 28, 2016. The inner planet's mass in each systems was added to mass of the host star for the efficiency of computing times. We have done numerical integrations for both systems and set an integration time of 10 Myr. For each system one set of the planets' orbital solutions was integrated. For the purpose of numerical explorations, we use the second solution [5] for Kepler-9 system, while for HD 10180 system we added 15% mass to the known planetary mass.

3. Empiric-analytical study

Among empiric-analytical studies for determining stability of multiplanetary system, including Hill's stability, there is a new solution to clarify the stability of resonant-planet pair system, considering that the system follows three-body problem [3]. The criterion is given in (1)

$$\Gamma = \frac{a_2(1-e_2)}{a_1(1+e_1)} - 2.4[\max(\mu_1, \mu_2)]^{1/3}\left(\frac{a_2}{a_1}\right)^{1/2} + 1.15 > 0, \quad (1)$$

where Γ is a stability parameter, μ is ratio of planetary to the star mass, and indices 1 and 2 refer to inner and outer planets. This criterion discriminates stable and unstable regions, although numerical studies [3] still found a "grey" region between them, meaning that it can be stable or unstable. The above criterion has been applied successfully [9] to other systems, i.e. one two-planet sytem and one multiplanetary system. We applied Petrovich criterion (1) to examine whether or not the resonant-planet pairs in Kepler-9 and HD 10180 systems are stable.

4. Results and discussion

4.1. Evolution of orbital elements and resonant arguments

We describe evolution of orbital elements and the resonant arguments (θ and Φ) of the resonant-planet pairs. Following [2] the resonant arguments for 2:1 resonance: $\theta_{1,2} = 2\lambda_2 - \lambda_1 - \varpi_{1,2}$, $\Phi = 4\lambda_2 - 2\lambda_1 - \Omega_1 - \Omega_2$, and for 3:1 resonance: $\theta_{1,2} = 3\lambda_2 - \lambda_1 - 2\varpi_{1,2}$, $\Phi = 3\lambda_2 - \lambda_1 - \Omega_1 - \Omega_2$, with $\lambda = \varpi + M$ and $\varpi = \omega + \Omega$. Subscripts 1 and 2 refer to inner and outer planets of the pair.

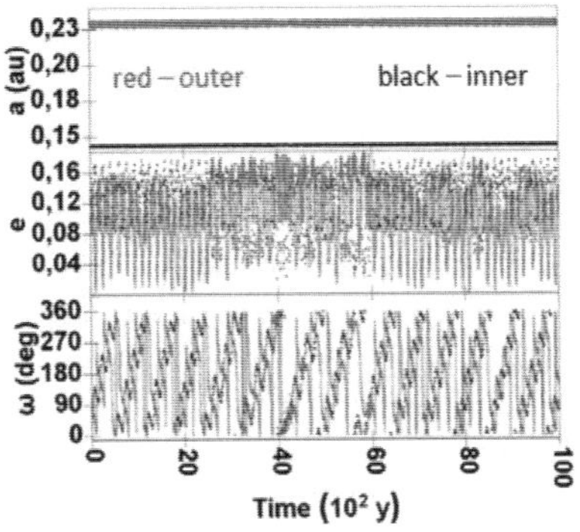

Figure 1. Evolution of orbital elements of Kepler-9 system set solution 1.

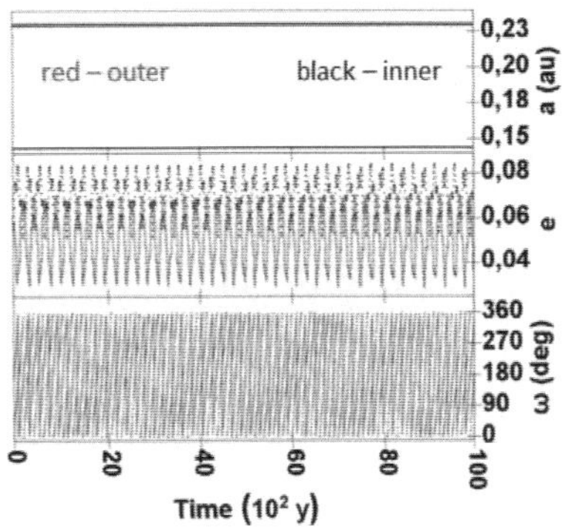

Figure 2. Evolution of orbital elements of Kepler-9 system set solution 2.

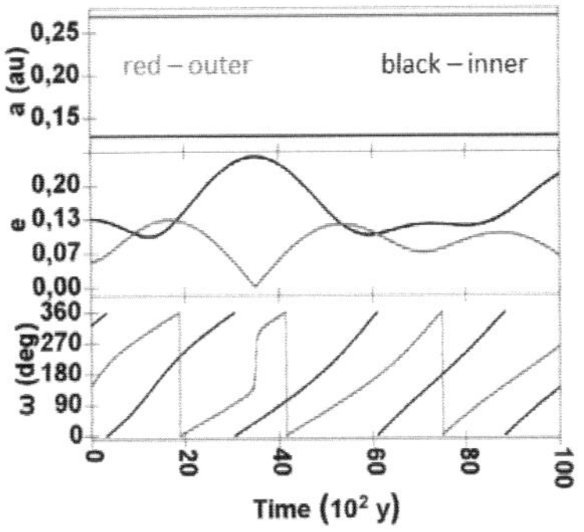

Figure 3. Evolution of orbital elements of HD 10180 system.

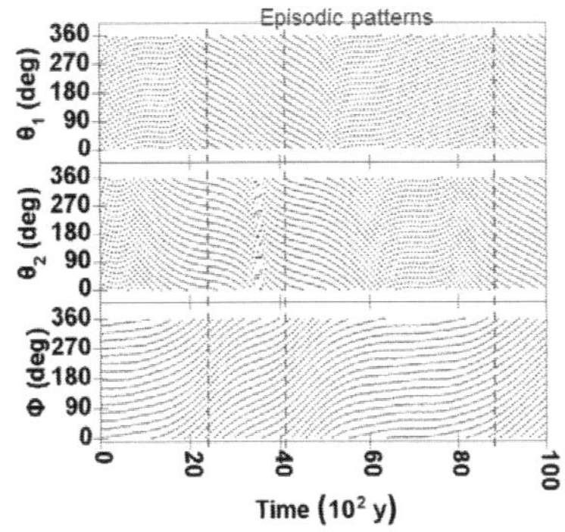

Figure 4. Evolution of resonant arguments of HD 10180 system.

Figures 1 and 2 show orbital evolutions of the resonant-planet pairs of Kepler-9 set solutions 1 and 2, while figure. 3 is that for HD 10180 system. We see that the resonant-planet pairs remain "bound" to their host stars. The evolution of their orbital elements are limited in narrow range values, indicating that the systems are indeed in orbital commensurability. Argument of pericenters circulate as the nodes of pericenter of the planet pairs evolve regularly. We found that their configurations remain unchange for 10 Myr. Set solution 2 of Kepler-9 is more stable than that for set solution 1. Additional 15% planetary masses to the HD10180 system do not much affect the system. Figure 4 depicts evolution of resonant arguments of HD 10180 system which shows clear episodic patterns. Less noticeable patterns are taken place for Kepler-9 system. This means that HD 10180 system is more dynamic.

4.2. Petrovich stability

We applied Petrovich criterion (1) to Kepler-9 system set solutions 1 and 2, also to HD 10180 system with minimum mass solution. Kepler-9 system set 2 and HD 10180 system have Γ-values greater than zero, but Kepler-9 system set 1 has value below zero (see figure 5). According to Petrovich's work, the inner resonant planet of Kepler-9 system set 1 might be plunged into the host star as $\mu_1 > \mu_2$.

It is obvious in figure. 5 that HD 10180 system is located in stable region with 3:1 resonance line, while Kepler-9 system is located in "grey" region with 2:1 resonance line. However, Kepler-9 system set 2 tends to be stable ($\Gamma > 0$).

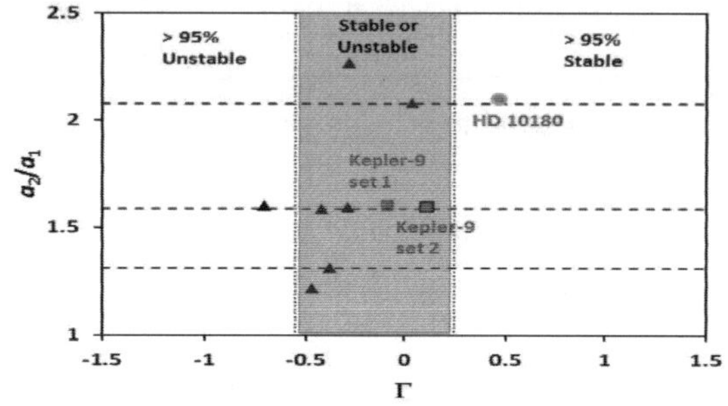

Figure 5. Petrovich stability plots of multiplanetary systems. Circular shape denotes HD 10180 system, the rectangles represent Kepler-9 set solution1 1 and 2. The triangle shapes stand for other known resonant-planet pair systems.

5. Conclusions

Our work finds that resonant-planet pairs in Kepler-9 set 2 and HD 10180 systems are dynamically stable for at least 10 Myr. Numerical simulations showed that Kepler-9 set 2 appears more "bound" than that for set 1. Verification using Petrovich criterion supports this numerical result. In addition, HD 10180 systems is more dynamic, considering the clearly seen of episodic patterns in the evolution of resonant arguments. Additional planetary masses do not much affect the dynamics of HD 10180 system.

Acknowledgments

The authors acknowledge Faculty of Mathematics and Natural Sciences, Institut Teknologi Bandung for financial support.

References

[1] Ghilea M C 2015 Statistical distributions of mean motion resonances and near-resonances in multiplanetary sytems *Preprint* arXiv: 1410.2478v3
[2] Barnes R, Deitrick R, Greenberg R, Quinn T R and Raymond S N 2015 *Astrophysical. Journal* **801** 101
[3] Petrovich C 2015 The stability and fates of hierarchical two-planets systems *Preprint* arXiv: 1506.05464v2
[4] Levison H F and Duncan M J 1994 *Icarus* **108** 18
[5] Borsato L, Marzari F, Nascimbeni V, Potto G, Granata V, Bedin L R and Malavolta L 2014 *Astronomy & Astrophysics* **571** A38
[6] Kane S R and Gelino D M 2014 *Astrophysical Journal* **792** 111
[7] Holman M J 2010 *Sci.* **330** 6000
[8] Mazeh T 2013 *Astrophysical Journal Supplement Series* **208** 16
[9] Mia R.and Kushvah B S 2016 *MNRAS* **457** 1089

Journal of Physics: Conference Series **771** (2016) 012052 doi:10.1088/1742-6596/771/1/012052

The locations of triangular equilibrium points in elliptic restricted three-body problem under the oblateness and radiation Effects

Ibnu Nurul Huda[1,2] **and Budi Dermawan**[3]

[1] Master Program in Astronomy, Faculty of Mathematics and Natural Sciences, Bandung Institute of Technology, Jalan Ganesha 10, Bandung 40132, Indonesia
[2] Master Program (M2) in Dynamic of Gravitational System, Paris Observatory, 61 Avenue de l'Observatoire 75014, Paris
[3] Astronomy Research Division, Faculty of Mathematics and Natural Sciences, Bandung Institute of Technology, Jalan Ganesha 10, Bandung 40132, Indonesia

E-mail: ibnunurulhuda@yahoo.com

Abstract. The Restricted Three-Body Problem (R3BP) considers motion of a third infinitesimal object under the gravitational influences of the primaries (bigger and smaller massive bodies) whose orbits are around the center of mass. If the orbits are elliptical, this belongs to Elliptic R3BP (ER3BP). In planar case it possesses five equilibrium points consisting of three collinear (L_1, L_2, and L_3) and two triangular (L_4 and L_5). To mimic a better astrophysical R3BP, such as motion of a satellite in star-planet system, the classical problem can be generalized by considering the effects of radiation pressure and oblate spheroid shape on the primaries. We study analytically the locations and the stability of L_4 and L_5 equilibrium points in the frame of ER3BP with incorporating the effects of radiation for bigger primary and oblateness for smaller primary. Our study suggests that the oblateness factor (A_2) and the radiation factor (q_1) shift the positions of L_4 and L_5 points compared with the classical ones. We also find that there is a stability limit for motion of the third body around these points.

1. Introduction

The Circular Restricted Three-Body Problem (CR3BP) considers motion of a third infinitesimal object under the gravitational influences of the primaries (bigger and smaller massive bodies) whose orbits are circular around the centre of mass. In planar case it possesses five equilibrium points consisting of three collinear (L_1, L_2, and L_3) and two triangular points (L_4 and L_5). The classical problem of CR3BP was modified by including additional forces. The effect of radiation pressure in CR3BP was first studied by Radzievskii in 1950 for the solar problem [3]. In both papers, however, Radzievskii did not consider the linear stability. The linear stability of equilibrium point was discussed by Chernikov in 1970 [3] who considered the relativistic Poynting Robertson effect which may be treated as perturbation, and Simmons in 1985 [4] with the effect of radiation pressure for all range of value.

This treatment may also be applicable to Elliptical Restricted Three-Body Problem (ER3BP). Singh and Umar in 2012 [1] considered the motion under the influences of an oblate spheroid in bigger primary and a radiation pressure in smaller primary. Kumar and Narayan in 2012 [2] studied the existence and stability of collinear equilibrium points in ER3BP under the effects of oblate spheroid primaries with the bigger primary has radiation pressure. Here we study the locations of L_4 and L_5 in ER3BP with bigger primary contain the radiation and smaller primary is an oblate spheroid.

2. Equations of motion

We assume that m_1 and m_2 are masses of the primaries, f is the true anomaly of system, r is the distance between the primaries, e is the eccentricity, and a is the semi-mayor axis of the primaries. In the case of three-body problem, it is more convenience to introduce the third body in the rotational (ξ, η, ζ) rather than in the inertial (x, y, z) coordinate system. The unit of time is chosen to make the gravitational constant $G = 1$. Introducing a transformation to a dimensionless coordinate system $\bar{\xi} = a\xi / r$, $\bar{\eta} = a\eta / r$, equations of motion of the third object in planar case are

$$\frac{d^2\bar{\xi}}{df^2} - 2\frac{d\bar{\eta}}{df} = \frac{dV}{d\bar{\xi}}, \quad \frac{d^2\bar{\eta}}{df^2} + 2\frac{d\bar{\xi}}{df} = \frac{dV}{d\bar{\eta}}.$$

The potential function V is defined by

$$V = (1 + e\cos f)^{-1}U \qquad U = \frac{1}{2}(\bar{\xi}^2 + \bar{\eta}^2) + \frac{1}{n^2}\left(\frac{(1-\mu)}{r_1} + \frac{\mu}{r_2}\right),$$

where n is mean motion of elliptical orbit, $\mu = m_2 / (m_1 + m_2)$ with $\mu \le 1/2$ and

$$r_i^2 = (\bar{\xi} - \bar{\xi}_i)^2 + (\bar{\eta} - \bar{\eta}_i)^2 \quad (i = 1, 2).$$

with (ξ_1, η_1) and (ξ_2, η_2) are the primaries coordinate.

3. Additional forces

3.1. Radiation pressure

The radiation pressure force (F_p) changes with distance by the same law as the gravitational attraction force of m_i (F_{gi}) and acts opposite to it. It is possible that this force will lead to a reduction of the effective mass of the massive particle. We use q_i to represent radiation pressure coefficients of the primaries. We can write the total force as

$$F_{gi} - F_{pi} = q_i F_{gi} \qquad q_i = 1 - \frac{F_{pi}}{F_{gi}}$$

3.2. Oblate spheroid

We use A_i to represent oblateness coefficients of the primaries respectively, such that $0 < A_i << 1$ and

$$F_{gi} = \frac{m_i}{r_i} + \frac{3m_i}{2r_i^4}\left(\frac{AE^2 - AP^2}{5R^2}\right) \qquad A_i = \frac{AE_i^2 - AP_i^2}{5R_i^2},$$

where AE_i and AP_i are the dimensional equatorial and polar radii of m_i. R_i is radius of m_i which assume the primary as spherical object.

To mimic the extrasolar planetary system with single star and planet, we assume that only the bigger primary is the source of radiation ($q_1 \neq 1$, $q_2 = 1$) and only the smaller primary is oblate spheroid ($A_1 = 0$, $A_2 \neq 0$).

4. Location of triangular points

Conditions of triangular equilibrium points are given as follows

$$\bar{V}_{\bar{\xi}} = \bar{V}_{\bar{\eta}} = 0. \qquad (1)$$

Following [5] from (1) we obtain the location of equilibrium points $(\bar{\xi}_0, \bar{\eta}_0)$

$$\bar{\xi}_o = \frac{1}{2} - \mu + \varepsilon_1 - \varepsilon_2, \quad \bar{\eta}_o = \pm\frac{\sqrt{3}}{2}\left[1 + \frac{2}{3}(\varepsilon_1 + \varepsilon_2)\right].$$

where

$$\varepsilon_1 = -\frac{1}{2}A_2 - \frac{1}{3}(1 - q_1), \quad \varepsilon_2 = 0.$$

When the primary is oblate $r_i = 1 + \varepsilon_i$ with $\varepsilon_i << 1$. The positif value of $\bar{\eta}_0$ is the position of L_4 and the negative value of $\bar{\eta}_0$ is the position of L_5.

Figure 1 shows the displacement of triangular points under the effect of radiation and oblateness. The gray dot represents the positition of m_2. The yellow dot represents the positition of m_1. The positions of infinitesimal object are represented by the blue dots (for classical problem) and the red dots (for the presence of radiation and oblateness effects) with L_4 is located above axis and L_5 is located below axis. The triangular points shift drawn toward the bigger primary with increment of A_2 and reduction of q_1.

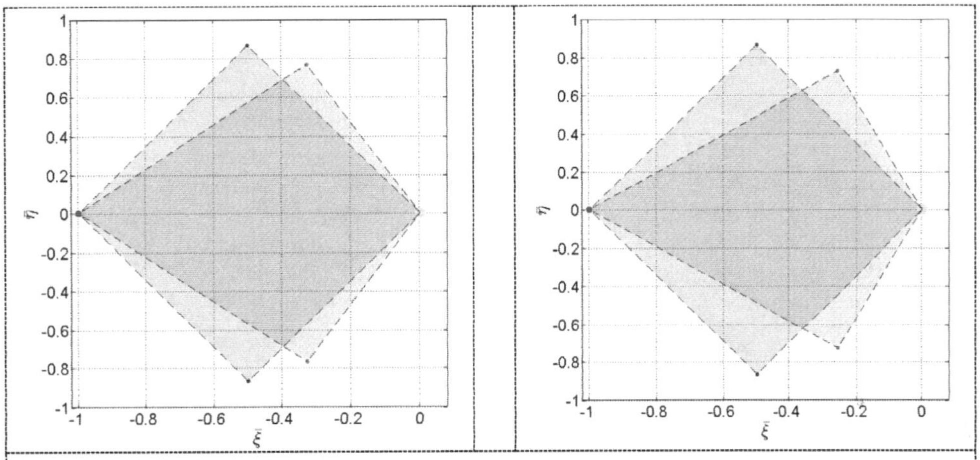

Figure 1. Variation of L_4 and L_5 locations for $\mu = 0.002$: (left) $q_1 = 0.5$; $A_2 = 0.01$, (right) $q_1 = 0.5$; $A_2 = 0.15$.

5. Linear stability

Define the displacement of triangular points by:

$$u = \bar{\xi} - \bar{\xi}_o, \quad v = \bar{\eta} - \bar{\eta}_o,$$

the equation of motion become

$$\ddot{u} - 2\dot{v} = \bar{V}_{\bar{\xi}\bar{\xi}}u + \bar{V}_{\bar{\xi}\bar{\eta}}v, \quad \ddot{v} + 2\dot{u} = \bar{V}_{\bar{\xi}\bar{\eta}}u + \bar{V}_{\bar{\eta}\bar{\eta}}v. \tag{2}$$

Substitution $u = \alpha e^{\lambda f}$ and $v = \beta e^{\lambda f}$ to (2), with α and β are constant and λ is root of characteristic equations, characteristic equation can be derived

$$\lambda^4 + (4 - \bar{V}_{\bar{\eta}\bar{\eta}} - \bar{V}_{\bar{\xi}\bar{\xi}})\lambda^2 + (\bar{V}_{\bar{\xi}\bar{\xi}}\bar{V}_{\bar{\eta}\bar{\eta}}) - \bar{V}_{\bar{\xi}\bar{\eta}}^2 = 0$$

The stability of the system depends on the form of λ. If all roots are pure imaginary the system is stable. Fig. 2 shows the ER3BP's stability regions. It shows that the region is smaller when we consider q_1 and A_2.

Journal of Physics: Conference Series **771** (2016) 012052 doi:10.1088/1742-6596/771/1/012052

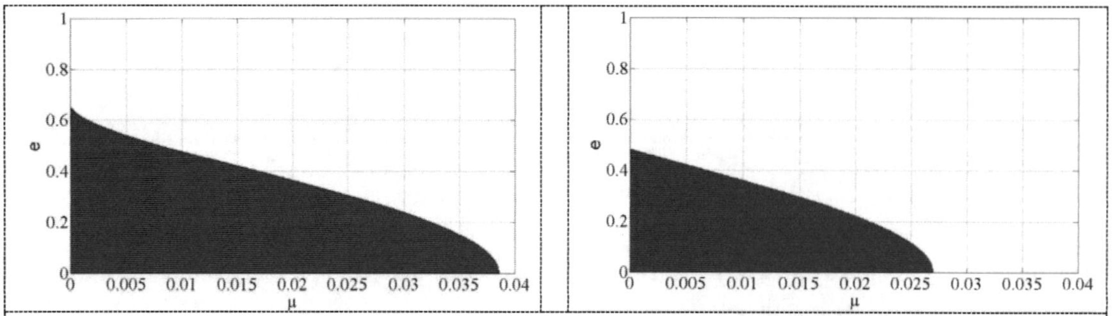

Figure 2. Stability curves for (left) $q_1 = 1.0$; $A_2 = 0.0$, (right) $q_1 = 0.9$; $A_2 = 0.01$.

6. Conclusion

The position of triangular equilibrium points in ER3BP under the effect of radiation and oblateness primaries have been studied. We conclude that the positions are shift when it recognizes the radiation factor (q_1) and oblateness coefficient (A_2). The stability regions are changed due to these effects.

References

[1] Singh J and Umar A 2012 *Astronomical Journal* **143** 5
[2] Kumar R C and Narayan A *Int'l. J. Pure and Appl. Math.* **80** 477-494.
[3] Chernikov Yu A 1970 *Soviet Astronomy* **14** 176-181, Jul.-Aug. 1970
[4] Simmons J F L, McDonald A J C, and Brown J C 1985 *Celestial Mechanics* **35** 145-187
[5] Usha T, Narayan A, and Ishwar B *Astrophys Space Sci.* **349** 151-164

Update on the impact of the proton radius on the neutron star radius

T Mart[1,2] and A Sulaksono[1]

[1] Departemen Fisika, FMIPA, Universitas Indonesia, Depok 16424, Indonesia
[2] Indonesian Academy of Sciences (AIPI), Jl. Merdeka Selatan No. 11, Jakarta 10110, Indonesia

E-mail: terry.mart@sci.ui.ac.id

Abstract. We present the new result of our investigation on the extraction of proton radius and the impact of different proton radii on the radius of neutron star, after correcting the mistake in the previous calculation of the proton radius. The new value of the extracted proton radius is 0.864 fm. The effect of this correction on the calculated neuron star radius is trivial.

1. Introduction

It has been long believed that the standard and precise measurement of the proton radius would be obtained by using the elastic electron-proton scattering $e^- + p \to e^- + p$, for which the cross section reads (see Ref. [5] for further explanation)

$$\frac{d\sigma}{d\Omega} = \left(\frac{d\sigma}{d\Omega}\right)_{\rm point} \frac{1}{(1+\tau)} \left[1 + \frac{\tau}{\epsilon}\mu_p^2\right] G_{E,p}^2(Q^2, \Lambda) \ . \tag{1}$$

Note that the radius is obtained from the relation $r_E \equiv \langle r_E^2 \rangle^{1/2} = [-6dG_{E,p}(Q^2)/dQ^2|_{Q^2=0}]^{1/2}$. The latest measurement at MAMI yields $r_E = 0.879(5)_{\rm stat.}(4)_{\rm syst.}(2)_{\rm model}(4)_{\rm group}$ fm [1], in accordance with the standard CODATA value [2], $r_E = 0.8768(69)$ fm, obtained from measurements of the Lamb shift in hydrogen atom. However, a recent measurement of the Lamb shift in muonic hydrogen atom [3] leads to a problem, since this measurement yields a smaller proton charge radius, i.e. $r_E = 0.84184(67)$ fm. In the previous work we try to solve this problem by assuming that the radii of protons are not identical. As a consequence, the cross section given by Eq. (1) must be averaged over all proton sizes, i.e., averaged over the form factor cut-off Λ. Therefore, the proton form factor must be also averaged in the squared form,

$$\langle G_{E,p}^2(Q^2, \Lambda_1) \rangle = \frac{1}{2\Delta\Lambda} \int_{\Lambda_1 - \Delta\Lambda}^{\Lambda_1 + \Delta\Lambda} G_{E,p}^2(Q^2, \Lambda) d\Lambda \ , \tag{2}$$

where $2\Delta\Lambda$ represents the range of the cut-off variation near the middle value Λ_1.

2. Extraction of the proton radius

By using Eq.(2) and assuming a dipole for $G_{E,p}(Q^2, \Lambda)$ we fitted the lowest Q^2 form factor data given in Ref. [4]. The lowest value of χ^2/N was obtained for $\Delta\Lambda/\Lambda_1 = 21.5\%$ and $\Lambda_1 = 0.8203$

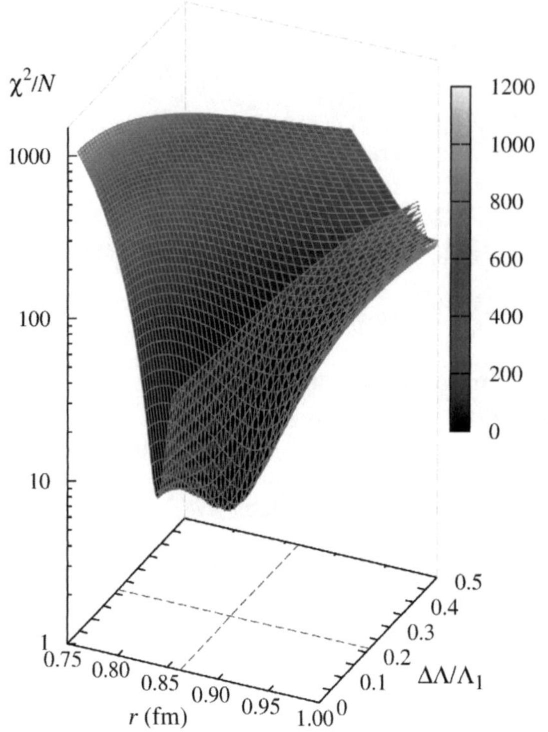

Figure 1. The value of χ^2/N as functions of the proton charge radius r and $\Delta\Lambda/\Lambda_1$. The position of smallest χ^2/N is indicated by the intersection of the two dashed (blue) lines.

GeV, which corresponds to the relative variation of proton radius $\Delta r_E/r_E \approx 21.5\%$ and the proton radius of $r_E = 0.833$ fm [5]. The result was surprising and immediately received a comment [6], which concerns on the accuracy of the extraction. A reply to this comment has been addressed [7]. Very recently, a new comment has appeared [8], which concern on the extracted proton radius value. We have also addressed a reply [9] and admitted that we found a mistake in the extraction process, i.e., in obtaining the radius we used the original dipole form factor $(1 + Q^2/\Lambda_1^2)^{-2}$, instead of the averaged one, i.e., Eq. (5) of Ref. [5]. We have revised our calculation and found the correct value of proton radius is 0.864 fm. This value is larger than our previous result, as well as the recent measurement of the Lamb shift in muonic hydrogen atom [3]. The χ^2/N value as functions of the proton radius and the ratio of cut-off range is depicted in Fig. 1, where we can see that the lowest value of χ^2 is obtained for $r_E = 0.864$ fm and $\Delta\Lambda/\Lambda_1 = 21.5\%$. Applying the relativistic correction explained in Ref. [11], the radius decreases to 0.844 fm [12], which is very close to the recent muonic hydrogen atom measurement [3].

3. Radius of the neutron star

As described in our previous work [5] the energy density of the neutron star matter depends on the densities of nucleon and scalar particles. Therefore, the effect of different nucleon radii enters the neutron star calculation via the nucleon and scalar densities

$$\rho_i = A\bar{\rho}_i \ , \qquad \rho_{s,i} = A\bar{\rho}_{s,i} \ , \tag{3}$$

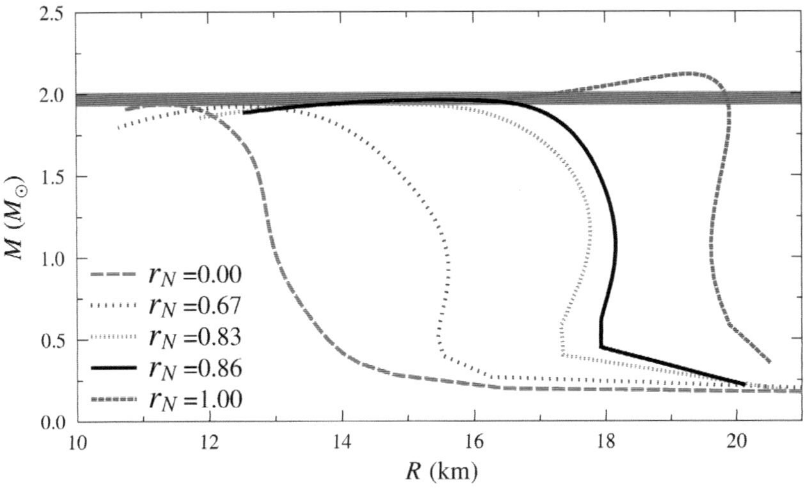

Figure 2. The neutron star mass as a function of its radius obtained by assuming five different nucleon radii r_N. The result has been obtained by using $\beta = 0.01$ in Eq. (5). The gray horizontal band shows the mass of the SRJ164-2230 pulsar, which is believed to be the heaviest observed neutron star [10].

where $\bar{\rho}_i$ and $\bar{\rho}_{s,i}$ indicate the densities of the ith nucleon and scalar particles if they were point particles. In Eq. (3) the proportionality constant A reads

$$A = \frac{1}{1 + V_p\bar{\rho}_p + V_n\bar{\rho}_n}, \tag{4}$$

where V_p and V_n are the proton and neutron volumes, respectively, and we have assumed that $V_p = V_n \equiv V_N = \frac{4}{3}\pi r_N^3$.

The dependency of the nucleon radius r_N on the density of matter is parameterized by using

$$r_N(\rho) = r_N(0)\left\{1 + \beta\left(\frac{\rho}{\rho_0}\right)^2\right\}^{-2}, \tag{5}$$

where $\rho = \rho_p + \rho_n$ and ρ_0 is the value of ρ at saturation point. The ratio between r_N and $r_N(0)$ as a function of the matter density ρ for different values of β is shown in Fig. 6 of Ref. [5].

To obtain the neutron star mass as a function of its radius we solve the Tolman-Oppenheimer-Volkoff (TOV) equation with different particle densities in the neutron star core. The result is shown in Fig. 2, where we compare the effect of different nucleon radii on the mass-radius profile of the neutron star. Note that the new calculation is represented by $r_N = 0.86$ fm in Fig. 2, i.e., the solid black line.

Obviously, the calculated radius of the neutron star depends on the value of nucleon radius r_N, while the maximum mass of neutron star is determined by the dependence of the nucleon radius on the matter density (β). As shown in Ref. [5], the choice of β is very decisive here, the use of $\beta = 0.0005$ could overshoot the experimental data. Figure 2 also shows that the corrected proton radius $r_N = 0.86$ fm only slightly shifts the mass-radius relation from our previous result, which used $r_N = 0.83$ fm. Nevertheless, the discrepancy between the present result and the point-particle approximation increases and becomes more apparent.

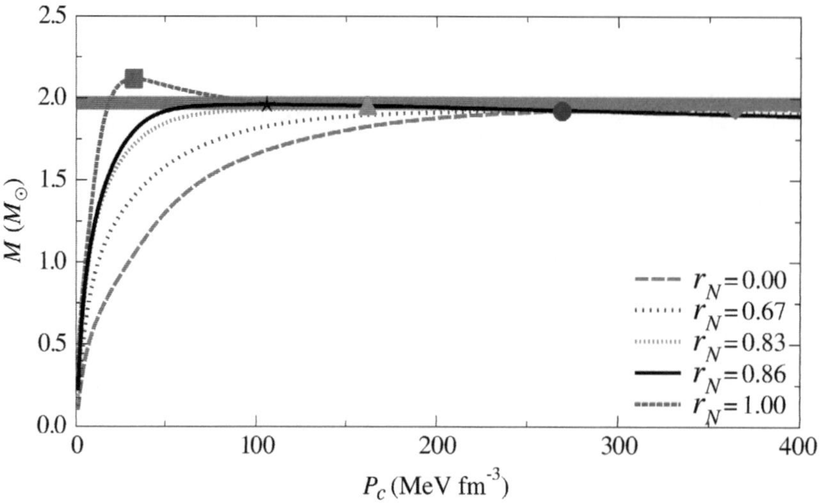

Figure 3. As in Fig. 2, but for the neutron star mass as a function of its center pressure. Solid square, star, triangle, circle, and diamond indicate the maximum masses with the corresponding center pressures.

Figure 3 displays the relation between the neutron star mass and its center pressure. Clearly, the effect of the correction on proton radius is quite subtle in this case. As a conclusion, we may say that the effect of the correction on proton radius is very mild in the calculation of the neutron star. Given that the real result would be obtained by averaging the uncertainty in the extracted proton radius, i.e., about 20%, it is save to say that the effect of the correction is trivial in this case.

4. Conclusion

We have corrected our calculation of the proton radius extraction and found a new radius of 0.864 fm, instead of 0.833 fm. Without the relativistic correction this value is larger than the new measurement obtained from the muonic Hydrogen atom, but still smaller than the standard value. This correction does not, however, disturb our result of the neutron star calculation.

Acknowledgments

This work has been partly supported by the Research-Cluster-Grant-Program of the University of Indonesia, under contract No. 1862/UN.R12/HKP.05.00/2015.

References

[1] Bernauer J C *et al.* 2010 *Phys. Rev. Lett.* **105** 242001
[2] Mohr P J, Taylor B N and Newell D B 2008 *Rev. Mod. Phys.* **80** 633
[3] Pohl R *et al.* 2010 *Nature* **466** 213
[4] Bernauer J C 2010 *Ph.D. Thesis,* Universität Mainz
[5] Mart T and Sulaksono A 2013 *Phys. Rev.* C **87** 025807
[6] Downie E J, Briscoe W J, Gilman R, and Ron G 2013 *Phys. Rev.* C **88** 059801
[7] Mart T and Sulaksono A 2013 *Phys. Rev.* C **88** 059802
[8] Arrington J 2016 *Phys. Rev.* C **93** 039801
[9] Mart T and Sulaksono A 2016 *Phys. Rev.* C **93** 039802
[10] Demorest P B, Pennuci T, Ransom S M, Roberts M S E and Hessels J W T 2010 *Nature* **467** 1081
[11] Bedoya Fierro D, Kelkar N G and Nowakowski M 2015 *JHEP* **1509** 215
[12] Kelkar N G, Mart T and Nowakowski M 2016 submitted to *Makara J Sci*

The spectral evolution of nebular phase from Nova V5668 Sgr

Robiatul Muztaba[1], Hakim L. Malasan[1], and Akira Arai[2]

[1]Astronomy Division and Bosscha Observatory, Institut Teknologi Bandung 40391, Bandung, Indonesia
[2]Koyama Astronomical Observatory, Kyoto Sangyo University, Motoyama, Kamigamo, Kita-ku, Kyoto, Kyoto 603-8555, Japan

E-mail: muztaba.r@gmail.com

Abstract. The spectral evolution of the Nova V5668 Sgr has been observed on 12 June, 23 June, and 15 August 2015 using NEO-R1000 spectrograph attached to the C-11 telescope at Bosscha Observatory, Lembang, Indonesia. The observed spectra indicate that the nova has been entering the nebular phase. The spectra during this phase showed strong forbidden lines of [O I] λ6300, 6364, [O II] λ7320, and [O III] λ5007, [N II] λ5755. The spectra also showed emission line of Fe II λ4924, 5169, 5317, 6149. We classify Nova V5668 Sgr into the Fe II type. We present and discuss the variation of nebular abundance $N(X)/N(H)$ of the Nova V5668 Sgr in the light of its evolutionary stages.

1. Introduction

Nova is an attractive astrophysical object to be comprehensively studied. The instability of the state of the accretion disc will create variability of long or short periods of photometry to reach the minimum apparent brightness [1]. The change in the light curve of nova indicates differences in the physical conditions. This, consequently also has an impact on the variation of its spectral line. When the nova passes peak of maximum luminosity, absorption lines will be weaken, while the emission lines begin to strengthen. This indicates the existence of large and hot region, commonly called the nebulae [3]. The appearance of striking features of forbidden lines [OIII] λ5007 and [NII] λ5755 indicates that the nova is under nebular phase.

Nova Sgr 2015 No. 2 (V5668 Sgr) was first discovered by John Seach (Chatsworth Island, NSW, Australia) on March 2015. Its position is RA (J2000.0) = 18h36m56.84s and Dec (J2000.0) = -28°55′39.8″. Here, we present our spectroscopic study of V5668 Sgr which shows the evolution of spectra from June to August 2015.

2. Optical spectroscopic observations

The observations were allocated during nearly three months since 9 June until 29 August 2015 at Bosscha Observatory, Lembang. Unstable sky conditions prevailing at Bosscha Observatory has hampered monitoring observational mode. Observations were successfully carried out and we obtained 3 spectra of the nova taken on 12 June, 23 June, and 15 August 2015. We used the Celestron-11 Telescope (f/10) combined with a low-resolution spectrograph NEO R-1000 (spectral coverage 3548-7977 Å, and resolution, $R = \lambda/\Delta\lambda = 1000$, slit width 65 μm, and grating 600 groove/mm) and CCD ST-8 XME (1530x1020 pixels, 9 micron/pixel, On-chip Binning 1x1 and 1x2). We used Image Reduction and Analysis Facility (IRAF) to produce wavelength and flux calibrated spectra. For wavelength

Journal of Physics: Conference Series **771** (2016) 012054 doi:10.1088/1742-6596/771/1/012054

calibration we used spectral atlas for FeNeAr, whereas for the flux calibration, spectrophotometric standard stars HR 7001 and HR 7596 were employed [2].

3. Results

The results of spectra which have been calibrated for each observation date can be seen in figure 1. In spite of low S/N ratio, emission lines appear very strongly and the continuum is also detected. The spectrum shown in green taken on 15 August suggests that the nova envelope became optically thin, since ratios of [N II] $\lambda5755$, [O II] $\lambda7320$ from continuum light are stronger than the spectrum shown in blue taken on 12 June. The spectrum shown in red taken on 23 June and the spectrum shown in green taken on 15 August, show that these fluxes become weak. The broad emission feature around 4600 Å is Bowen blended band (C III/N II). Bowen band along with CIII, N II lines indicate the existence of the hard UV radiation from the photosphere of V5668 Sgr. On the other hand, [O III] $\lambda5007$ and He II $\lambda4686$ are not so strong yet but clearly detected.

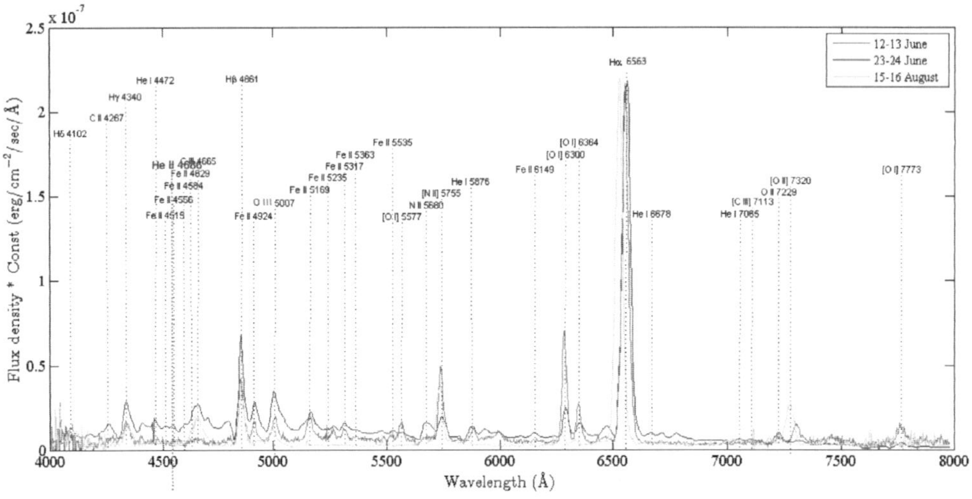

Figure 1. Low-resolution spectra evolution of V5668 Sgr obtained at Bosscha Observatory

Nova classification based on spectrum depends on its own dominant spectrum of chemical elements. According to [5], spectrum of V5668 Sgr is categorized as Fe II type. Expansion velocity of this nova has been measured from its P-Cygni profile in Hß of the blue spectra on 12 June. The expansion velocity was determined to be 2087 ± 179 km/s.

4. Discussion and summary

The changes of light curve and spectral evolution could explain the physical condition of the nova. The changes in physical conditions that occur in V5668 Sgr lead to differences electron temperature and density value of total hydrogen as shown by table 1.

Table 1. The changes of hydrogen density and electron temperature.

Date	6-12-2015	6-23-2015	8-15-2015
N(H) Total Hydrogen density	1.08×10^{11} cm^{-3}	3.15×10^{8} cm^{-3}	4.05×10^{8} cm^{-3}
Electron Temperature (K)	26388 K	15528 K	15528 K

Figure 2 shows the change of light curve for V5668 Sgr from pre-validated AAVSO database. The observations carried out on June 12, 2015 showed a magnitude range of 8-9, while on June 23, 2015 showed a magnitude 11-12. On 12 and 23 June, the V5668 Sgr was in the formation of dust and nebular phase. This is shown by the declining light curve that indicates the formation of a new shell. At this phase there is a significant expansion of shell [4]. From the analysis of P-Cygni profile in Hβ λ4861, it is obtained that the expansion velocity is 2087 ± 179 km/s. The high expansion velocity is capable to carry materials far out along the radial direction. The process of shell expansion which is fast moving make the electron temperature decreases.

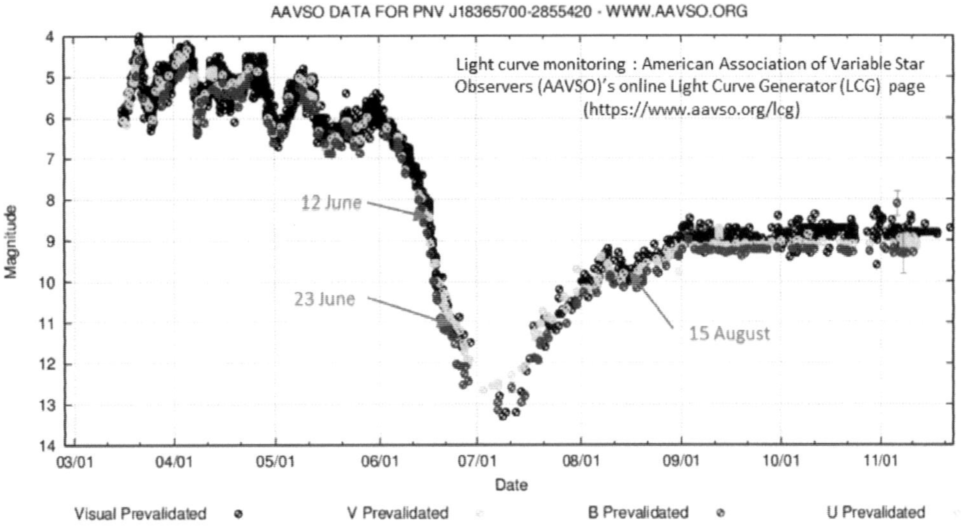

Figure 2. Visual-magnitude light curve of V5668 Sgr from pre-validated AAVSO database

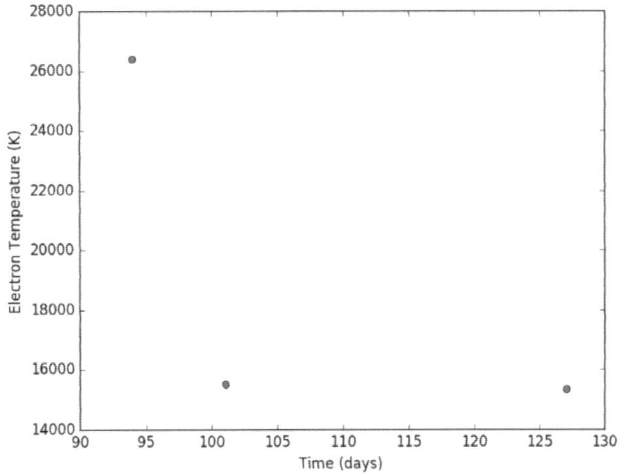

Figure 3. Changes of electron temperature after outburst

Figure 3 shows on 94-days after the outburst, physical condition nova still has a thick shell, so that the intensity is absorbed by the existing material at the shell, heating throughout the region. Thus this increases the electron temperature which cause recombination processes become dominant. There is an increase of neutral hydrogen in the shell. So that the contribution of neutral hydrogen atoms will produce the forbidden lines by collisional excitation and fluorescence resonance processes. Therefore, this condition allows the V5668 Sgr is in nebular phases and produce a nebular spectra. The increase process of a hydrogen neutral is evidenced by the decline in the ratio of the density profile like [O I] λ6300, 6364 to the total amount of Hydrogen which is getting lower on 101-days and 127-days after the outburst as shown in figure 4.

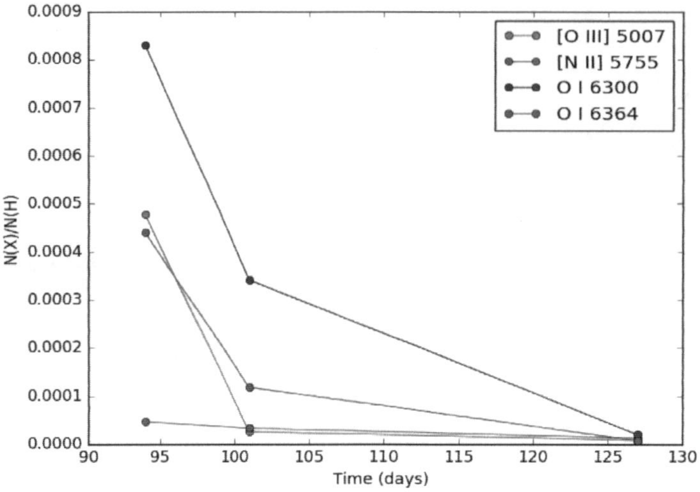

Figure 4. The changes ratio of element nebulae to the total amount of Hydrogen

From this nova spectroscopy observation we can conclude that the spectra during nebular phase showed strong forbidden lines of [O I] λ6300, 6364, [O II] λ7320, and [O III] λ5007, [N II] λ5755. The spectra also showed emission line of Fe II λ4924, 5169, 5317, 6149. Nova clasification based on spectrum depends on its own dominant spectrum of chemical elements. According to [5], the spectrum is included to Fe II type.

Acknowledgments
This work was supported by Bosscha Observatory of Institut Teknologi Bandung and financial supported by Institut Teknologi Bandung. We acknowledge with thanks the light curve observations that used in this research from AAVSO International Database contributed by observers worldwide. This paper is based on one of the author's thesis (RM) submitted to the graduate school of Institut Teknologi Bandung in partial fulfilment of the requirement for the degree Master of Science.

References

[1] Bode M F and Evans A 2008 *Classical Nova* (UK: Cambridge University Press)
[2] Hamuy M, Suntzeff N B, Heathcote S R, Walker A R, Gigoux P and Philips M M 1994 *Publications of the Astronomical Society of the Pacific* **106** 566-589
[3] Osterbrock D E 1989 *Astrophysics of Gaseous Nebulae and Active Galactic Nuclei* (USA: University Science Books)
[4] Williams R E, Woolf N J, Hege E K, Moore R L and Kopriva D A 1978 *Astrophysical Journal* **244** 171-181
[5] Williams R E 1992 *Astronomical Journal* **104** 725-733

Evolution effect of BD+60°2522 to Bubble Nebula NGC 7635

Aprilia and I A Arfianty

Department of Astronomy, Faculty of Mathematics and Natural Sciences, Institute Technology of Bandung, Jalan Ganesha 10, Bandung Indonesia

E-mail: aprilia@as.itb.ac.id, iraalfi@gmail.com

Abstract. Bubble Nebula is a bubble formed by the interaction between the stellar wind of BD+60°2522 with ambient interstellar gas. We use a web-based stellar evolution code, the EZ-web, to construct the evolution of BD+60°2522. From the evolution, we obtain the age of the system needed for the interstellar bubbles model. Then from the model, we determine parameters such as radius, expansion velocity, luminosity, temperature, and density of the Bubble.

1. Introduction

Bubble Nebula is an interstellar bubble located in the Perseus Arm of the Galaxy, and cataloged as NGC 7635. First discovered in 1787 by a British astronomer William Herschel, it is an enormous bubble being formed by the interaction between the stellar wind of a hot massive star, BD+60°2522, with its surrounding. This star is an O6.5IIIef star [4] with mass $45M_\odot$ [5], radius $15R_\odot$, luminosity $398{,}000L_\odot$, and effective temperature 37,500 K (data from Simbad). The image of Bubble Nebula taken by NASA's Hubble Space Telescope was chosen as the mark of the 26th anniversary of the launch of Hubble telescope into the Earth orbit on April 24, 1990 (figure 1 in this paper).

Figure 1. Image of Bubble Nebula NGC 7635 taken by NASA's Hubble Space Telescope as the mark of the 26th anniversary of the launch of Hubble telescope into the Earth orbit (https://www.nasa.gov/).

As a massive star, BD+60°2522 has already experienced mass loss since its Zero Age Main Sequence (ZAMS) stage. The mass loss rate is $10^{-5.76}$ $M_\odot$/year [3], and believed to form the bubble. According to

[3], the expanding velocity and radius of NGC 7635 are 36 km/s and 1.5 pc. This makes age of the bubble as 4 x 10^4 years. This number is different to the result of [5] for the age of BD+60°2522, i.e. 2 x 10^6 years. This shows that NGC 7635 is not formed at the beginning of the evolution of BD+60°2522. The first bubble has already expanded and merged with the ambient interstellar medium [5].

In this work, we calculate physical parameters of Bubble Nebula, such as radius, expanding velocity, luminosity, temperature, and gas density, based on the evolution of BD+60°. We use web-based stellar evolution code, EZ-web (http://www.astro.wisc.edu/~townsend/), for the evolution, and models of the formation of interstellar bubble from [2, 11, 10, 1].

2. Data and method

The formation of NGC 7635 is affected by the evolution of its central star, BD+60°2522. We adopt metallicity $Z = 0.02$ and 61 $M_{\odot}$ as the initial mass. The effect of convective overshooting is not included. We obtain present parameters for BD+60°2522: the age of 4.53 x 10^4 years, luminosity 5.42 x 10^5 $L_{\odot}$, and effective temperature 4.64 x 10^4 K.

For the age of BD+60°2522 in main sequence stage, we obtain 3.37 x 10^6 years. Comparing this to the result from [5] for the age of the star in main sequence stage, i.e. 4 x 10^6 years, it is not different significantly.

3. Results and discussion

According to a model from [11] a bubble was formed by the interaction between stellar wind and the interstellar gas which has constant atom density ($n_0 = 1$ cm^{-3}). As is shown in figure 2, the interaction yields four regions: stellar wind region (a), inner shocked stellar wind region (b) with radius R_1, outer shocked interstellar gas region (c) with radius R_2, and ambient interstellar gas (d). [10] and [1] applied this model to the formation of Bubble Nebula NGC 7635, as is shown in figure 3.

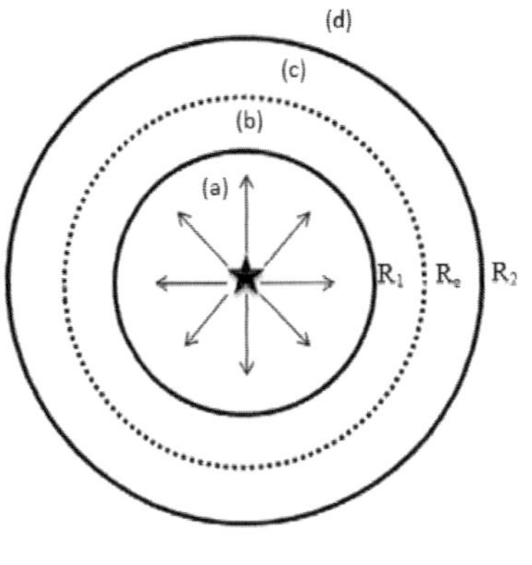

Figure 2. Schematic view for the formation of bubble around a massive star [11]. Expanding stellar wind is indicated in (a), inner shocked stellar wind in (b), outer shocked interstellar gas in (c), and ambient interstellar gas in (d). Region (a) and (b) is separated by R_1, (b) and (c) by R_c, while (c) and (d) by R_2.

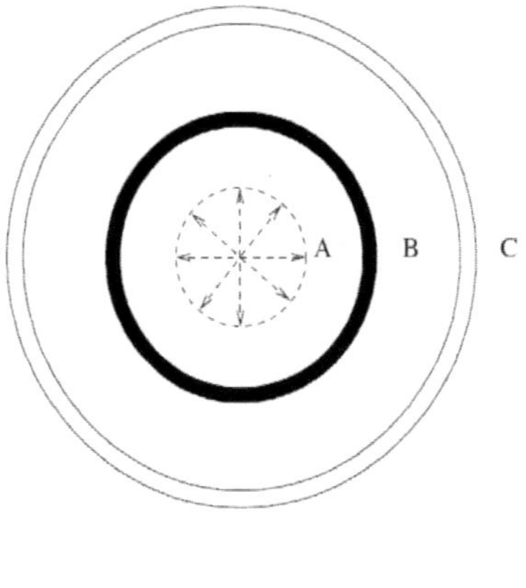

Figure 3. Schematic view for the formation of Bubble Nebula NGC 7635 [10]. Expanding wind hits the inner shock (dashed line), then kinetic energy is converted into thermal energy, which pushes a shell (thick line) into H II region (B), then pushes a shell (double line) into the interstellar medium (C).

To obtain the Bubble's physical parameters, such as the outer shell radius (R_2), expanding velocity (V_2), and stellar wind luminosity (L_w), we use equations from [2,11] written below,

$$R_2(t) = 27L_{36}^{\frac{1}{5}}n_0^{-\frac{1}{5}}t_6^{\frac{3}{5}}\ \text{pc} \tag{1}$$

$$V_2(t) = 16L_{36}^{\frac{1}{5}}n_0^{-\frac{1}{5}}t_6^{-\frac{2}{5}}\ \text{km/s} \tag{2}$$

$$L_w = \frac{1}{2}\frac{dM_w}{dt}V_w^2 \tag{3}$$

Here, n_0 is the density for interstellar gas ($n_0 = 1\ cm^{-3}$), t_6 is the age of BD+60°2522 in main sequence stage, in 10^6 years (t_6 = 3.37 x10^6 years), $\frac{dM_w}{dt}$ is the mass loss rate for BD+60°2522 ($\frac{dM_w}{dt} = 10^{-5.76}\ M_{\odot}/\text{years}$ [3]), V_w is stellar wind velocity and L_{36} is stellar wind luminosity in 10^{36} erg/s. Table 1 shows parameters determined from equations (1) – (3) for different stellar wind velocities from some references.

Table 1. Parameters obtained from equations (1) – (3) for different stellar wind velocities of BD+60°2522

Stellar wind velocity, V_w (km/s)	References	Stellar wind luminosity, L_w (10^{36} erg/s)	Bubble's shell radius (pc)	Bubble's expanding velocity (km/s)
1800	Johnson (1980) [6]	1.7745	60.5228	10.6425
2200	Johnson (1982) [7]	2.6507	65.5812	11.5320
2500	Johnson (1982) [7]	3.4230	69.0218	12.1370
2700	Leitherer (1988) [9]	3.9925	71.1796	12.5165

Using parameters in table 1, we calculate temperature, density, and luminosity of Bubble Nebula using equations from [2] which are

$$T = 1.6 \times 10^6 \left(L_{36}^{8/35} n_0^{2/35} t^{-6/35}\right)\ \text{K} \tag{4}$$

$$n = 0.01 \left(L_{36}^{6/35} n_0^{19/35} t_6^{-22/35}\right)\ \text{cm}^{-3} \tag{5}$$

$$L = 3.8 \times 10^{32} \left(L_{36}^{37/35} n_0^{18/35} t_6^{16/35}\right)\ \text{erg/s} \tag{6}$$

The results are shown in table 2 for different stellar wind velocities as in table 1.

Table 2. Parameters obtained from equations (4) – (6) for different stellar wind velocities of BD+60°2522

Stellar wind velocity, V_w (km/s)	References	Bubble's temperature (10^6 K)	Bubble's gas density (cm^{-3})	Bubble's luminosity (10^{33} erg/s)
1800	Johnson (1980) [6]	1.4967	0.0057	1.3335
2200	Johnson (1982) [7]	1.6405	0.0061	2.0383
2500	Johnson (1982) [7]	1.7392	0.0064	2.6708
2700	Leitherer (1988) [9]	1.8015	0.0065	3.1428

According to [8], interaction between stellar wind and ambient interstellar medium is in snowplow phase (like a snowplow piles up snow in front of it as it moves forward), which consists of two condition, energy conserving and momentum conserving modes. Energy conserving mode is reached when L_b <

Journal of Physics: Conference Series **771** (2016) 012055 doi:10.1088/1742-6596/771/1/012055

L_w, while $L_b > L_w$ for momentum conserving mode. We obtain $L_b < L_w$ for Bubble Nebula, which means that Bubble Nebula is now in energy conserving mode of snowplow phase.

4. Conclusion

From the calculation based on model from [11], we obtain radius of Bubble Nebula NGC 7635 is 60.52 – 71.18 pc, which is expanding with velocity 10.64 – 12.52 km/s. This condition is reached at the age of the central star BD+60°2522 of 3.37 x10^6 years, i.e. age of its main sequence stage. The temperature of NGC 7635 is (1.5 – 1.8) x 10^6 K with gas density 0.006 cm^{-3}.

References

[1] Anand M Y, Kagali B A and Murthy J 2009 *Bull. Astr. Soc. India* **37** 1
[2] Castor J I, McCray R and Weaver R 1975 *Astrophysical Journal* **200** L107
[3] Christopoulou P E, Goudis C D, Meaburn J, Dyson J E and Clayton C A 1995 *Astrophysical Journal* **295** 509
[4] Conti P S and Leep E M 1974 *Astrophysical Journal* **193** 113
[5] Dawanas D N, Wardana A and Malasan H L 2007 *Astrophys. Space Science* **312** 23
[6] Johnson H M 1980 *Astrophysical Journal* **235** 66
[7] Johnson H M 1982 *Astrophysical Journal Suppl. Ser.* **50** 551
[8] Lamers H J G L M and Cassinelli J P 1977 *Introduction to Stellar Winds* (Cambridge: Cambridge University Press)
[9] Leitherer C 1988 *Astrophysical Journal* **334** 626
[10] van Marle A J, Langer N and Garcia-Segura G 2004 *RevMexAA (Serie de Conferencias)* **22** 136
[11] Weaver R, Castor J I and McCray R 1977 *Astrophysical Journal* **218** 377

Constraining cosmological parameter with SN Ia

A N Indra Putri and H R Tri Wulandari

Astronomy Department, Faculty of Mathematics and Natural Sciences, Bandung Institute of Technology, Indonesia

E-mail: annisa.novia@as.itb.ac.id

Abstract. A type I supernovae (SN Ia) is an exploding white dwarf, whose mass exceeds Chandrasekar limit (1.44 solar mass). If a white dwarf is in a binary system, it may accrete matter from the companion, resulting in an excess mass that cannot be balanced by the pressure of degenerated electrons in the core. SNe Ia are highly luminous objects, that they are visible from very high distances. After some corrections (stretch (s), colour (c), K-corrections, etc.), the variations in the light curves of SNe Ia can be suppressed to be no more than 10%. Their high luminosity and almost uniform intrinsic brightness at the peak light, i.e. $M_B \sim -19$, make SNe Ia ideal standard candle. Because of their visibility from large distances, SNe Ia can be employed as a cosmological measuring tool. It was analysis of SNe Ia data that indicated for the first time, that the universe is not only expanding, but also accelerating. This work analyzed a compilation of SNe Ia data to determine several cosmological parameters (H_0, Ω_m, Ω_Λ, and w). It can be concluded from the analysis, that our universe is a flat, dark energy dominated universe, and that the cosmological constant Λ is a suitable candidate for dark energy.

1. Introduction

The universe form from an event called Big Bang almost 14 billion years ago. In its early, universe is very hot and dense. When the universe expands, temperatures decrease enough to atom and then galaxies, stars, and planet because gravitation from giant cloud. Stars end their life in many ways. One of them is supernovae. Supernovae are stars that have reached the end of their life in very dramatic fashion with high explosion. Supernova Ia discovered in binary system consist of massive star and white dwarf. Massive star will transfer its energy to white dwarf until limit Chandrasekar ($\sim 1.44\ M_\odot$) and exploded as Supernovae Ia (SN Ia).

Observations of SN Ia at low redshift ($z < 1$) provide indication that the expansion of the universe at the present time is accelerating [4,5,6]. This expansion has been attributed to a dark energy component with negative pressure and thus cause accelerated expansion. The results from SN Ia observation can constraint cosmological parameters like Hubble constant (H_0), matter density (Ω_m), cosmological constant (Ω_Λ), and equation of state (w). These results indicate that dark energy is the dominant component in the matter-energy budget of the universe [4,5,6].

Many surveys take observation for Supernova Ia to obtain more Supernova Ia data. Adding Supernova Ia data can constraint cosmological parameter with better precision. The goal of this work is to constraint cosmological parameters using SN Ia data, and combination of SN Ia + Cosmic Microwave Background (CMB) data from WMAP9 and Planck mission. The steps are fitting light curve, constructing Hubble diagram from SN Ia data, and constraining cosmological parameter with SN Ia, SN Ia + CMB from WMAP9, and Planck mission.

2. Dataset and fitter

2.1. Dataset

This work uses SN Ia sample data from [3] which refer to data from Supernova Legacy Survey (SNLS). The SNLS is a five year program to measure the expansion history of the universe using SN Ia. Since SN Ia is not perfect standard candle, selection and correction requirements applied in this sample. The selection requirements exclude known peculiar Supernovae, exclude SN Ia with peculiar spectroscopy, and require minimum redshift cut at $z = 0.010$, stretch (s), and color (c) factor. The final sample from [3] contain 123 low redshift data (≤ 0.1), 92 intermediate redshift data from SDSS II ($0.1 < z < 0.4$), 241 SNLS data ($0.3 < z < 1$), and 12 high redshift data from Hubble Space Telescope (HST) with $z > 1$.

2.2. Fitter

The importance of modelling Supernovae light curves is clear from the large number of methods that have been developed for this purpose. We fit all SN Ia light curve data with fitter SiFTO [2]. SiFTO is a new empirical method for modelling SN Ia light curves by manipulating a spectral template. This program will give three result parameters for each Supernova Ia. The results are peak magnitude at rest frame (m_B), shape that indicates stretch (s), color (c) that indicates maximum color at rest frame.

Confidence level for fitting cosmological parameter is illustrated with map contours. We use cosmoMC program to plot confidence level. CosmoMC decide confidence level contour by Monte Carlo method and measure likelihood for each cosmological parameters by SN Ia+CMB data.

3. Analysis

3.1. Light curve

Building SN Ia light curve is important step before constraining cosmological parameters. Shown below are examples of SiFTO fitting to SN Ia light curves (figure 1) for a low redshift (sn1997do) and a high redshift (Aphrodite). Table 1 shows the result of fitting parameter light curve with SiFTO. On the upper of x-axis there is effective day where 0.0 effective day corresponds to maximum flux, while negative and positive effective day correspond to interval time before and after maximum flux, respectively.

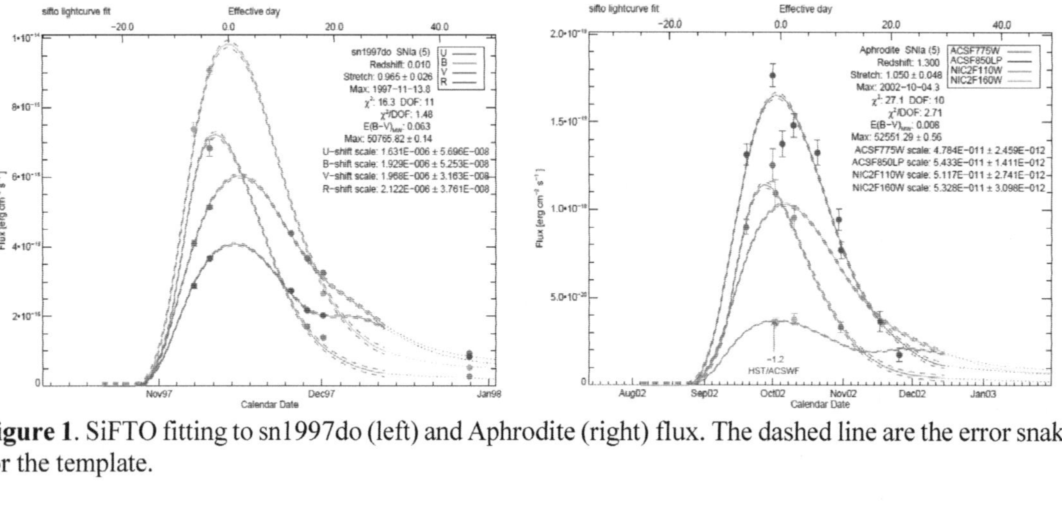

Figure 1. SiFTO fitting to sn1997do (left) and Aphrodite (right) flux. The dashed line are the error snake for the template.

Table 1. Light curve parameters for sn1997do and Aphrodite.

	Sn1997do	Aphrodite
Redshift (z)	0.01012 ± 0.00002	1.3000 ± 0.01000
Stretch (s)	0.947 ± 0.027	1.050 ± 0.048
Color (c)	-0.006 ± 0.027	-0.028 ± 0.056
Peak magnitude (m_B)	14.274 ± 0.039	25.673 ± 0.035

3.2. Hubble diagram

Hubble diagram is built from relation between redshift and distance modulus. Corrected magnitude, stretch (s) and color (c), use formula [3]:

$$m_{corr} = m_B + \alpha(s-1) - \beta c \quad (1)$$

$$\mu_{corr} = m_B + \alpha(s-1) - \beta c + 19.3 \quad (2)$$

where m_{corr} is corrected magnitude, m_B is peak magnitude at B band rest frame, α and β are coefficients for s and c, and μ_{corr} is magnitude if we changed the equation to distance modulus (M_B = -19.3 ± 0.3 [1]). We use value $\alpha = 1.43^{+0.12}_{-0.10}$ and $\beta = 3.26^{+0.12}_{-0.10}$, which is the best fitting results from [3]. Distance modulus can be transformed to luminosity distance:

$$\mu = 5 log d_L + 25 \quad (3)$$

$$d_L = cH_0^{-1}(1+z) \times \int_0^z dz\left[(1+z)^3(\Omega_m) + (1-\Omega_m)(1+z)^{3(1+w)}\right]^{-1/2} \quad (4)$$

where μ is distance modulus, d_L is luminosity distance, and w is equation of state. If w > -1/3, universe will experience accelerated expansion, but if w < -1/3 universe will experience decelerated expansion. Hubble diagram in luminosity distance is shown in figure 2 with flat model universe. Figure 2 shows SN Ia data fitted with flat model universe which is dominated by dark energy. Moreover, when SN Ia data is fitted with fixed $\Omega_m = 0.18$ and $\Omega_\Lambda = 0.82$, the result shows SN Ia data is fit well with $w = -1$, indicating accelerated expansion of the universe. The linear part of the Hubble diagram corresponding to low redshift SN Ia data ($z \leq 0.1$), is used for measuring Hubble constant (H_0). Luminosity distance at linear part can be approximated by $d_L \approx c(1+z)/H_0$, giving value of $H_0 = 69.78 \pm 2.10$.

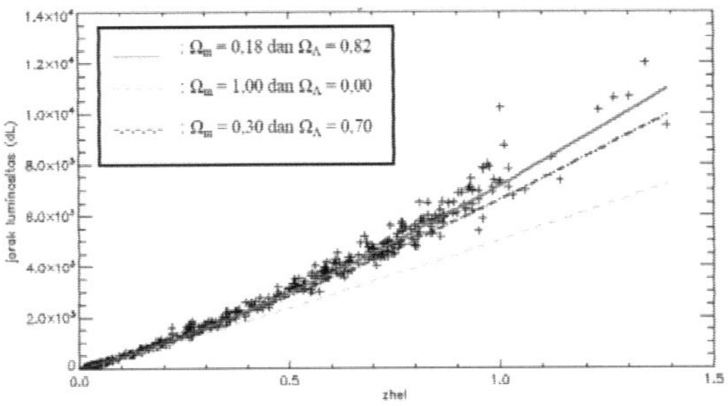

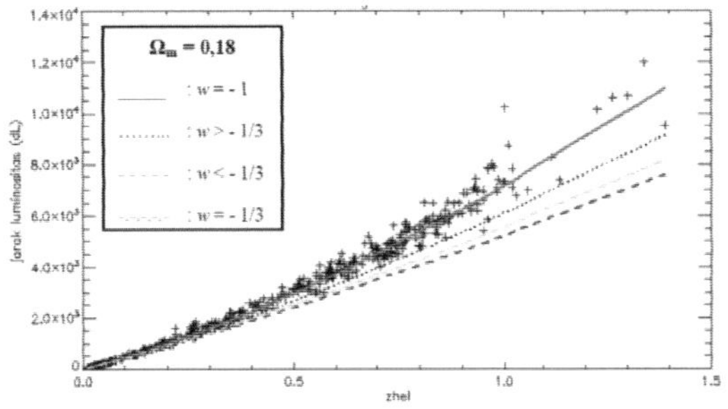

Figure 2. Hubble diagram in luminosity distance with three flat models (left) and with $\Omega_m = 0.18$ (right).

3.3. Confidence level

Contours of confidence level generated by using cosmoMC for SN Ia+CMB data from 9th year WMAP and Planck shown in figure 3. Best fitting of cosmological parameters from SN Ia+CMB data are shown

Journal of Physics: Conference Series **771** (2016) 012056 doi:10.1088/1742-6596/771/1/012056

in table 2. Dark energy density shows higher value than matter density, indicating accelerated expansion of the universe.

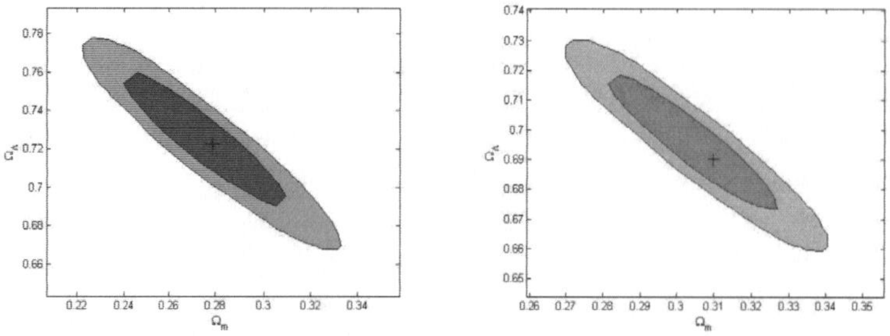

Figure 3. Contours of confidence level (68% and 95%) for SN Ia+CMB data from WMAP9 (left) and Planck (right).

Table 2. Best fits of cosmological parameters from Supernova Ia+CMB data.

	SN Ia + WMAP9	SN Ia + Planck
Ω_m	0.28	0.31
Ω_Λ	0.72	0.69
w	-1.00	
H_0	69.82	67.80
Age (Gyr)	13.75	13.78

4. Conclusion

We have presented determination of the cosmological parameters from data of the first three years of the SNLS and combining them with external SN data sets at higher and lower redshift. From SN Ia data only, we find $H_0 = 69.77$, $\Omega_m = 0.18$, $\Omega_\Lambda = 0.82$, and $w = -1$. From SN Ia+CMB data, we find $H_0 = 69.82$, $\Omega_m = 0.28$, $\Omega_\Lambda = 0.72$, $w = -1$ for SN Ia+WMAP9 and $H_0 = 67.80$, $\Omega_m = 0.31$, $\Omega_\Lambda = 0.69$ for SN Ia+Planck. Both results agree with SN Ia data only as analyzed in this work. All these results suggest an accelerated expansion of the universe, due to the dominance dark energy component.

References

[1] Astier P *et al* 2006 *Astronomy & Astrophysics* **447** 31-48

[2] Conley A *et al* 2008 *Astrophysical Journal* **681** 482-513

[3] Conley A *et al* 2011 *Astrophysical Journal* **192**:1-29

[4] Perlmutter S *et al* 1999 *Astrophysical Journal* **517** 565-586

[5] Riess A G *et al* 1998 *Astrophysical Journal* **116** 1009-1038

[6] Riess A G *et al* 2004 *Astrophysical Journal* **607** 665-687

Study of correlation between ultraluminous X-ray sources and their host galaxies

I G P M Priajana and H R T Wulandari

Bandung Institute of Technology, Bandung, Indonesia

E-mail: mahadipa@as.itb.ac.id

Abstract. Ultraluminous X-ray sources (ULXs) are defined as non-nuclear point-source objects with apparent X-ray luminosities, $L_X > 2\times10^{39}$ erg s^{-1}, in the 0.3-8 keV band. ULXs are often explained using two different scenarios, (1) ULXs as intermediate mass black hole (IMBH) with sub-Eddington accretion and (2) ULXs as stellar mass black hole with super-Eddington accretion. There are two methods that commonly used to study the characteristics of ULXs. One method is to study the X-ray spectra of ULXs, to determine the characteristics of their accretion flows from fitting their spectra using available spectral models. The other method is to investigate how population of ULXs correlate with their environment, in this case their host galaxies. Our goal is to find correlation between ULXs and the properties of its host galaxies, for example with Star Formation Rate (SFR), mass and morphology. From this study we found a positive correlation between ULXs number and SFR. From X-ray luminosity function, we found upper limit of black holes mass that power ULXs is about 100 $M_\odot$.

1. Introduction

Ultraluminous X-ray Sources (in short, ULXs) are point source objects that lie in non-nuclear part of a galaxy that typically emits X-ray luminosities higher than Eddington limit of black holes with mass 20 $M_\odot$ (Lx (0.3-10 keV) $\geq 2\times10^{39}$ erg s^{-1}) [1, 2]. The true characteristics of ULXs are still unknown, but the probability that ULXs are black hole is high [1]. We may know their characteristics from study about their X-ray spectra, but in this paper we shall focus our study to find their characteristics by correlating ULXs population with their host galaxies. In Section 2, we describe about catalog and data; Section 3, we present the results of our study and discuss the results; finally, we summarize our study in Section 4.

2. ULXs catalog and dataset

For this study, we used two ULXs candidate catalog, namely Liu (2011) [2] and Walton et al. (2011) [7] as our main data. [2] used data from Chandra mission, whereas [7] used data from XMM-Newton mission. We combined these two catalogs by selecting sources that meet our criteria. We selected sources with $L_x \geq 2\times10^{39}$ erg s^{-1} that reside in galaxies within 40 Mpc. For overlapping sources, we need to merge them to avoid double counting when doing correlation. From this selection process, we obtain a total of 599 ULXs in 234 host galaxies. We calculated mass of galaxies using mass-to-light ratio (M/L) and galactic color relation [3]. For Star Formation Rate (SFR), we estimated by using relation from [4]. We classified galaxies that have SFR higher than 10 $M_\odot$/yr as starburst galaxy. The information about galaxy morphology were obtained from Third Reference Catalog of Bright Galaxies (RC3) [10].

Journal of Physics: Conference Series **771** (2016) 012057 doi:10.1088/1742-6596/771/1/012057

CasJob (http://skyserver.sdss.org/casjobs/) was used to get apparent magnitude data of the host galaxies. We derive color of the galaxies using these magnitudes. To determine the type of the galaxies (normal or peculiar), we cross-correlated host galaxies data from our catalog with Arp catalog of peculiar galaxies [11]. For instance, if we found that some of the galaxies are listed in Arp catalog, then we defined them as peculiar galaxies. Isophotal contour was also used to confirm the peculiarity of galaxies. If their isophotal contour are asymmetric, they are confirmed as peculiar galaxies.

3. Result and discussion

Our goal is to determine the characteristic of ULXs population by finding correlation with properties of their host galaxies. For this analysis, we divided our galaxy sample from combined catalog into four sub-sample, by their morphology (spiral and elliptical), interaction (normal and peculiar), color (red and blue) and starburst galaxies. The detail of Sub-sample members is stated in table 1.

Table 1. List of sub-sample. Color of the table represent the sub-sample group.

Sub-Sample	Galaxy Count	ULXs Count
Spiral Galaxy	147	353
Elliptical Galaxy	42	125
Normal Galaxy	172	340
Peculiar Galaxy	62	219
Blue Galaxy	25	61
Red Galaxy	75	122
Starburst Galaxy	44	183

Next step, we compared the ULXs population for each sub-sample. Distribution of ULXs population relative with isophotal radius (R_{25}) of their host galaxies for each subsample can be seen in figure 1. From this figure, overall trend of ULXs population distribution is declining as they have more distance from center of their host galaxies. Several samples (Spiral, Normal, Red, and Blue sample) have flat ULXs distribution from $0.1R_{25}$ to $0.5R_{25}$ before they start to decline. We also notice there are increment of ULXs population as they are going to $0.5R_{25}$ in spiral sample. This range of R_{25} coincide with star forming area within disk of spiral galaxies. From literature check, we confirm ULXs population in $0.5R_{25}$ for some spiral galaxies (e.g. NGC 1313, NGC 5194) actually is located in star forming disk.

We notice that nearly 50% ULXs candidates in elliptical galaxies are located outside R_{25}. The ULXs candidates that reside outside R_{25} most likely are contaminant source, like background or foreground source. Walton [5] argue that about 60% of ULXs candidates outside R_{25} of the host galaxies in Liu catalog are contaminants. Fraction of contaminants in elliptical galaxies tend to be higher than in spiral galaxies. Contamination analysis performed by [6] show that large fraction of ULXs candidate in elliptical galaxies may be contaminant (e.g., ULXs in NGC 1399). For several sources that are confirmed as contaminant, they analyzed them spectroscopically and found that these sources are QSOs.

For peculiar and starburst galaxies sub-sample, we found high number of ULXs population near the galactic center (0.1R25). Galactic inclination might be take part in this case, especially for galaxy that has high inclination, we could misinterpret the ULXs position to be in the center of galaxies instead of in the disk or bulge. To get the actual distribution of ULXs in their host galaxies, using inclination class data from MCG catalog, we only choose galaxy that have face-on orientation and then plot the distribution like Figure 1 for normal and peculiar galaxies. We got peculiar galaxies still have higher numbers of ULXs candidate near the center than normal galaxies with difference is about 10%. ULXs are highly correlated with star formation rate of galaxies [7]. The higher SFR of the galaxy, the more ULXs we expect to find. Excess number of ULXs candidate near center of peculiar and starburst galaxies may be due to circumnuclear starburst area that tend to occur in galaxies that undergo interaction, but further investigations are needed to confirm this hypothesis. From table 1 we can calculate the average number of ULXs candidate that reside in each sub-sample galaxies. For spiral galaxies, we calculate the average number of ULXs candidate is about 2.4 ULXs per galaxy. In elliptical galaxies, we get slightly higher number than in spiral galaxies, about 2.9 ULXs per galaxy. The same result is also obtained by Colbert and Ptak [8]. But we must aware that the rate of contaminant in elliptical galaxies is higher than spirals. If we exclude ULXs candidate outside R25, the average number of ULXs candidate in spiral and elliptical galaxies respectively are 2.13 and 2.21 ULXs per galaxy. As for peculiar galaxies, they have higher ULXs per galaxies than normal galaxies (3.5 ULXs per galaxy for peculiar, 1.9 for normal). This strengthen hypothesis that ULXs tend to occur at galaxies that undergo interaction or merging [9]. Due to interaction, gas in galaxies will experience compression because of the shockwave. This event can trigger star forming activity. The excess of ULXs in interacting galaxies may be correlated with increment of SFR due to interaction. For complement, the highest ULXs per galaxy can be found in Starburst sample (4.15 ULXs per galaxy), which can be used to support idea that ULXs are correlated with SFR. Figure 2 shows how number ULXs/galaxy correlate with far infrared (FIR) luminosity for spiral and starburst galaxies (SFR positively correlated with LFIR).

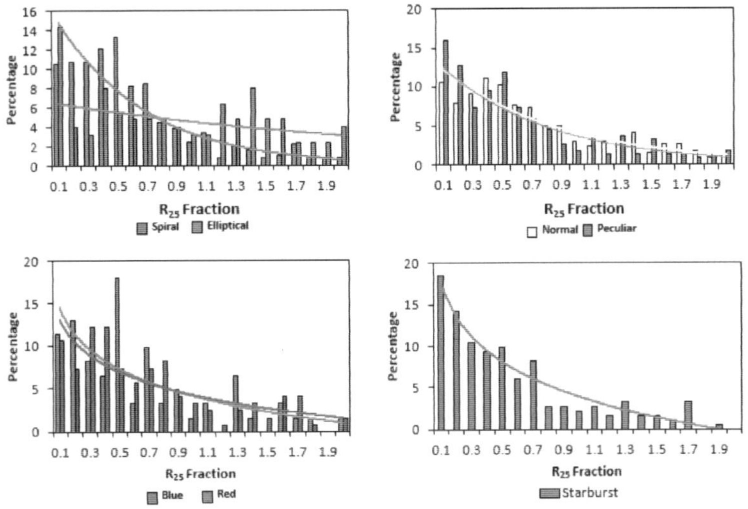

Figure 1. Distribution of ULXs in host galaxies relative of R_{25}

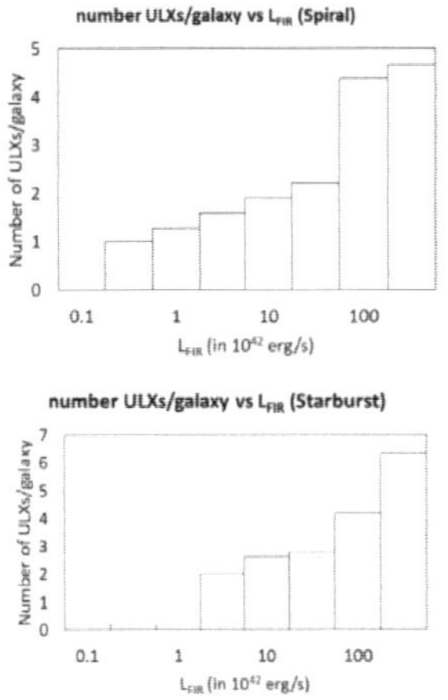

Figure 2. Number ULXs/gala xy vs L_{FIR}

We also constructed X-ray luminosity function (XLF) of ULXs for each sub-sample. We fit the luminosity function using power-law (CL_X^a) and power-law with cut-off model ($CL_X^a \exp(L_x/L_c)$). Parameter C is normalization coefficient, L_x is X-ray luminosity of the source, a is power-law index and L_c is cut-off luminosity. The results are presented in figure 3. The second model give better fit for ULXs XLF. We found that in general luminosity function of ULXs has cut-off when luminosity goes around 2×10^{40} erg s^{-1}.This indicates there are upper limit for black holes that power ULXs, about 100 $M_{\odot}$.

Next we tried to correlate specific luminosity of ULXs with stellar mass of galaxies to find how mass of galaxies correlate with ULXs populations. We defined specific luminosity of ULXs as total luminosity of ULXs in specific stellar mass bin divided by total mass of galaxies in the same stellar mass bin. We obtained a tendency that specific luminosity of ULXs goes lower with increasing stellar mass bin. There are two explanations for this tendencies [5]. First, lower mass galaxies have higher SFR specific than the massive one. We know that ULXs have positive correlation with SFR. This can be the reason why lower mass galaxies have higher specific luminosity of ULXs. Second, lower mass galaxies also tend to have lower metallicities than massive galaxies. Metallicity have a role to determine whether a star became low or high mass black hole. In lower metallicity region, it is easier to form more massive black holes [5] and ULXs prefer more massive black holes than less massive one.

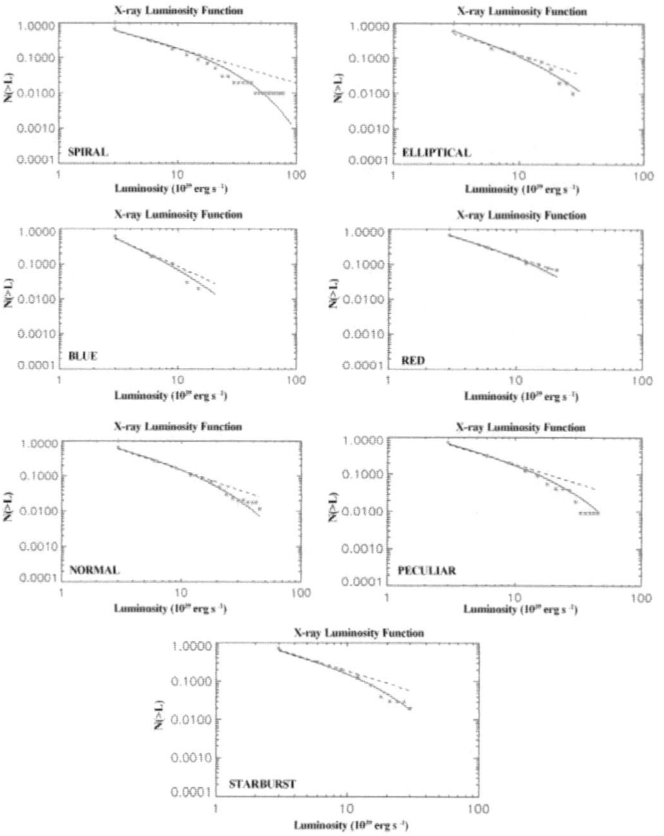

Figure 3. XLF for each sub-samples. Green asterix indicate the data, dash line indicate fitting using power-law model and solid line indicate fitting using power-law with exponential cut-off model.

4. Summary

From this study, we summarize several results: (1) ULXs tend to be found at spiral galaxies, (2) nearly 50% ULXs in elliptical galaxies are suspected as contaminant, (3) there are higher number of ULXs in peculiar and starburst galaxies, show that there are positive correlation between interaction and SFR with number of ULXs, (4) Excess of ULXs near center of peculiar and starburst galaxies may be due to circumnuclear starburst, (5) ULXs luminosity function give upper limit for black hole mass about 100 $M_{\odot}$, (6) distribution of specific luminosity of ULXs are declining with the increase of galaxies stellar mass.

References

[1] Feng H and Soria R 2011 *New Astronomy Reviews* **55** 166-183
[2] Liu J 2011 *Astrophysical Journal* **192** 10
[3] Bell E F 2003 *Astrophysical Journal Supplement Series* **194** 289
[4] Kennicut R C 1998 *Annual Review of Astronomy and Astrophysics* **36** 189-232
[5] Walton *et al* 2011 *Monthly Notices of the Royal Astronomy Society* **416** 1844
[6] Corredoira and Gutiérrez 2006 *Astronomy & Astrophysics* **454** 77-84
[7] Swartz *et al* 2011 *Astrophysical Journal* **741** 49
[8] Colbert and Ptak 2002 *Astrophysical Journal Supplement Series* **143** 25
[9] Smith *et al* 2012 *Astronomical Journal* **143** 6
[10] de Vaucouleurs *et al* 1991 *Third Reference Catalogue of Bright Galaxies (RC3)* (New York: Springer-Verlag)
[11] Arp H 1966 *Atlas of Peculiar Galaxies* (Caltech: Pasadena)